Continuous Groups for Physicists

The theory of groups and group representations is an important part of mathematics with applications in other areas of mathematics as well as in physics. It is basic to the study of symmetries of physical systems. Its mathematical concepts are equally significant in understanding complex physical systems. It offers the necessary tools to describe, for instance, crystal structures, elementary particles with spin, both Galilean symmetric and special relativistic quantum mechanics, the fundamental properties of canonical commutation relations and spinor representations of orthogonal groups extensively used in quantum field theory.

Continuous Groups for Physicists introduces the ideas of continuous groups and their applications to graduate students and researchers in theoretical physics. The book begins with an introduction to groups and group representations in the context of finite groups. This is followed by a chapter on the special algebraic features of the symmetric groups. The authors then present the theory of Lie groups, Lie algebras and in particular the classical families of compact simple Lie groups and their representations. Several interesting topics not often found in standard physics texts are then presented: the spinor representations of the real orthogonal groups, the real symplectic groups in even dimensions, induced representations, the Schwinger representation concept, the Wigner theorem on symmetry operations in quantum mechanics, and the Euclidean, Galilei, Lorentz and Poincaré groups associated with spacetime. The general methods and notions of quantum mechanics are used as background throughout.

Narasimhaiengar Mukunda is Adjunct Professor at the Indian Institute of Science Education and Research, Bhopal, India. He has been associated with the Tata Institute of Fundamental Research, Mumbai, and the Indian Institute of Science, Bangalore. He has co-authored six books in the area of theoretical physics. His areas of interest are classical and quantum mechanics, theoretical optics and mathematical physics.

Subhash Chaturvedi is Visiting Professor at the Indian Institute of Science Education and Research, Bhopal, India. Earlier, he was associated with the University of Hyderabad for almost three decades. He has co-authored a book on stochastic quantisation. His research interests are quantum mechanics, quantum information theory, statistical mechanics and stochastic processes.

Continuous Groups for Physicists

Narasimhaiengar Mukunda
Subhash Chaturvedi

CAMBRIDGE
UNIVERSITY PRESS

CAMBRIDGE
UNIVERSITY PRESS

University Printing House, Cambridge CB2 8BS, United Kingdom

One Liberty Plaza, 20th Floor, New York, NY 10006, USA

477 Williamstown Road, Port Melbourne, VIC 3207, Australia

314 to 321, 3rd Floor, Plot No.3, Splendor Forum, Jasola District Centre, New Delhi 110025, India

103 Penang Road, #05–06/07, Visioncrest Commercial, Singapore 238467

Cambridge University Press is part of the University of Cambridge.

It furthers the University's mission by disseminating knowledge in the pursuit of education, learning and research at the highest international levels of excellence.

www.cambridge.org
Information on this title: www.cambridge.org/9781009187053

First published 2022

Printed in India by Nutech Print Services, New Delhi 110020

A catalogue record for this publication is available from the British Library

ISBN 978-1-009-18705-3 Hardback

Cambridge University Press has no responsibility for the persistence or accuracy of URLs for external or third-party internet websites referred to in this publication, and does not guarantee that any content on such websites is, or will remain, accurate or appropriate.

In Memoriam
Sophus Lie, Elie Cartan, Hermann Weyl,
Eugene Wigner and Valentine Bargmann – who showed us the way

'Beauty is truth, truth beauty
That is all ye know on earth,
And all ye need to know'

– John Keats (1795–1821)

Contents

Preface

It has rightly been said that the mathematical theory of groups and group representations is a magnificent gift of nineteenth century mathematics to twentieth century physics. While this is particularly true within the framework of quantum mechanics, with the passage of time its relevance within classical physics has also become well understood and greatly appreciated. Today the importance of group theoretical ideas and methods for physics can hardly be overemphasised; and over the past century or so, a veritable profusion of books devoted to this theme, many of them gems of the literature, have appeared.

The present monograph is primarily based on lectures given by one of us (NM) at the Institute of Mathematical Sciences in Chennai, India, in the Fall of 2007. The lectures were prepared and presented at the invitation of Rajiah Simon, to whom both authors are indebted for his support and encouragement.

The course was titled 'Continuous Groups for Physicists' and consisted of about 45 extended lectures over a two month period. Its aim was to introduce the basic ideas of continuous groups and some of their applications to an audience of post graduate and doctoral students in theoretical physics. After an introduction to the basic ideas of groups and group representations (mainly in the context of finite groups and compact Lie groups), the course presented a selection of useful, interesting and quite sophisticated specific topics not often included in standard courses in physics curricula. The methods and concepts of quantum mechanics served as a backdrop for all the lectures.

The real rotation groups in two and three dimensions are followed by an account of the structures of Lie groups and Lie algebras, and then a description of the compact simple Lie groups. Their irreducible representations are described in some detail. Some of the 'non standard' topics that follow are: spinor representations of real orthogonal groups in both even and odd dimensions; the notion of the 'Schwinger' representation of a group with examples, induced representations, and systems of generalised coherent states; the properties and uses of the real symplectic groups, which are defined only in real even dimensions, and their metaplectic covering group, in a quantum mechanical setting; and the Wigner Theorem on the representation of symmetry operations in quantum mechanics. For the sake of completeness an account of the representation theory of the permutation groups, with its many algebraic features, has been added essentially at the beginning.

While this monograph is not intended to be a text book in the traditional sense, it is hoped that readers will find it useful in that a succinct account of the basics is followed by short accounts of the special topics mentioned above.

References have been listed at the end of each chapter. These include some classics of the literature, a few texts which we have found to be particularly well written, and in some cases a few journal publications. Some of the references appear at the conclusion of more than one chapter. Two types of references have been provided – those useful for the chapter as a whole, and those relevant at specific points in the chapter. Only the latter are indicated in the text, by author name and year of publication.

Some problems are given at the end of each chapter. To help with the more challenging ones, references to books or original papers are included.

Abbreviations

BI	Bargmann Invariant
CCR	Canonical Commutation Relation
CSLA	Compact Simple Lie Algebra
GCS	Generalised Coherent States
Irrep.	Irreducible Representation
ODE	Ordinary Differential Equation
PDE	Partial Differential Equation
UIR	Unitary Irreducible Representation
UR	Unitary Representation

Basic Group Theory and Representation Theory

In this chapter, we present the basic theory of finite groups and their representations as preparation for the discussion of continuous groups that starts from Chapter 3. It is assumed that readers know the basics of set theory, vector spaces, transformations, linear operators, matrix representations, inner products and such. These will be called upon as and when needed.

1.1 Definition of a Group

A group G is a set of elements $a, b, c, \cdots, g, g', \cdots, e, \cdots$ along with a composition (or 'multiplication') law obeying four conditions:

(i) Closure: $a, b \in G \to ab =$ unique product element $\in G$.

(ii) Associativity: for any $a, b, c \in G$,

$$a(bc) = (ab)c = abc \in G.$$

(iii) Identity: there is a unique element $e \in G$ such that

$$ae = ea = a, \qquad \text{for any } a \in G.$$

(iv) Inverses: for each $a \in G$, there is a unique $a^{-1} \in G$, the inverse of a, such that

$$a^{-1}a = aa^{-1} = e. \tag{1.1}$$

The composition rule or law can be called a binary law as the product is defined for any *pair* of elements. The conditions in (1.1) could be stated in more economical forms, for instance introducing only a left identity and left inverses, and then showing that the more general properties in (1.1) do hold.

One can immediately think of various qualitatively different possibilities. The number of (distinct) elements in G may be finite. Then this number, denoted by $|G|$, is called the order of G. Some other possibilities are that the number of elements may be a discrete infinity, or else a continuous infinity with G being a manifold of some dimension.

1.2 Some Examples

(i) The symmetric group, the group of permutations on n objects, is finite, of order $n!$, and is denoted by S_n. We mention only a few pertinent properties now, and go into some more detail in Chapter 2. Each $p \in S_n$ can be written in several convenient ways:

$$p = \begin{pmatrix} 1 & 2 & \cdots & n \\ p(1) & p(2) & \cdots & p(n) \end{pmatrix} = \begin{pmatrix} j \\ p(j) \end{pmatrix}_{j=1\cdots n}$$

$$= (1\ p(1)\ p(p(1)) \cdots)\ (j\ p(j)\ p(p(j)) \cdots) \cdots (k\ p(k)\ p(p(k)) \cdots). \tag{1.2}$$

In the first form the columns can be rearranged at will, while in the second form the factors and their entries are distinct. The composition law can be developed as follows:

$$qp = \begin{pmatrix} k \\ q(k) \end{pmatrix} \begin{pmatrix} j \\ p(j) \end{pmatrix} = \begin{pmatrix} p(j) \\ q(p(j)) \end{pmatrix} \begin{pmatrix} j \\ p(j) \end{pmatrix} = \begin{pmatrix} j \\ q(p(j)) \end{pmatrix},$$

i.e.,

$$(qp)(j) = q(p(j)), \quad j = 1, 2, \cdots, n. \tag{1.3}$$

We have followed here the rule of 'reading from right to left'. The identity and inverses are:

$$e = \begin{pmatrix} 1\ 2\ \cdots\ n \\ 1\ 2\ \cdots\ n \end{pmatrix} = \begin{pmatrix} j \\ j \end{pmatrix}_{j=1\cdots n} = (1)(2)\cdots(n);$$

$$p^{-1} = \begin{pmatrix} p(j) \\ j \end{pmatrix}_{j=1\cdots n}, \quad \text{i.e., } p^{-1}(p(j)) = j. \tag{1.4}$$

All the conditions in (1.1) can and should be checked.

(ii) All integers with respect to addition. Here the group 'multiplication' law is arithmetic addition. The identity is 0, and inverses are negatives.

(iii) All positive real (or rational) numbers with respect to multiplication. Now the identity is 1, and inverses are reciprocals.

(iv) All vectors in any (real or complex) linear vector space with respect to vector addition. Similar to (ii) above, the identity is the zero vector $\mathbf{0}$, and inverses are negatives.

(v) $SO(2)$ and $O(2)$, $SO(3)$ and $O(3)$: these are the proper and full groups of rotations in a plane or in three dimensions, respectively. These are continuous groups with infinitely many elements. In the $O(2)$ and $O(3)$ cases, the group is made up of two disjoint components, each of which is continuous and connected in an obvious intuitive sense. We study these in Chapter 3.

We can also consider discrete subsets of these, leaving invariant some given regular plane figure or solid. These are the point groups.

(vi) The rotation groups $SO(3), O(3)$ generalise to any dimension n and to the complex case. Thus we arrive at the groups $SO(n)$ and $O(n)$, $SU(n)$ and $U(n)$ for various integers n, which we will study in some detail later. Yet others are the complex orthogonal groups $SO(n, \mathbb{C})$.

(vii) Groups related to spacetime. Here we have the Euclidean, Galilean, Lorentz, and Poincaré groups. All of these are continuous groups with infinitely many elements. They have more than one connected component if discrete operations like space and/or time reflections are included. We will study some of these groups in Chapter 10.

A group G is said to be abelian if for any pair of elements a and b, $ab = ba$. Otherwise it is nonabelian. In the examples listed above, (ii), (iii) and (iv) are abelian.

Among the symmetric groups, S_2 is abelian while S_n for $n \geq 3$ is nonabelian. $SO(2)$ is abelian, and $O(2), SO(3)$ and $O(3)$ are all nonabelian.

The existence and properties of a unique identity element and unique inverse a^{-1} for each $a \in G$ lead to the useful cancellation rules:

$$ab = ac \Leftrightarrow b = c; \quad ba = ca \Leftrightarrow b = c. \tag{1.5}$$

For a finite group G the composition law or entire structure can be displayed in a *multiplication table* with $|G|$ 'rows' and 'columns' labelled by the group elements:

$$
\begin{array}{c|ccccc}
  & e & \cdots & b & \cdots \\
\hline
e & e & \cdots & b & \cdots \\
\vdots & \vdots & & \vdots & \\
a & a & \cdots & ab & \cdots \\
\vdots & \vdots & & \vdots &
\end{array}
\tag{1.6}
$$

At the intersection of row a and column b stands the product ab. These entries must be consistent with the group laws (1.1). So by (1.5) each row (column) contains every element of G exactly once.

1.3 Operations within a Group

For the present we continue to have in mind a finite group, though many concepts we will go on to introduce are more generally meaningful. For any elements $a, b \in G$, conjugation of a by b leads to another element $a' \in G$:

$$a' = bab^{-1}, \quad a = b^{-1}a'b. \tag{1.7}$$

We then say a and a' are conjugate to one another. For each $a \in G$, its *equivalence class or conjugacy class* $C(a)$ consists of all a' conjugate to a:

$$C(a) = \text{equivalence class of } a = \{bab^{-1}|\, b \in G, \text{ omit repetitions}\} \subset G. \tag{1.8}$$

Different classes are generally of different 'sizes'. For instance, $C(e) = \{e\}$ consists of one element alone. It is easy and important to check:

(*i*) $C(a) = C(bab^{-1})$, any b, so $C(a)$ is determined by any one of its elements;

(*ii*) $C(a) \cap C(b) = C(a)$ if $b \in C(a)$, null otherwise, so two classes cannot overlap partially;

(*iii*) G is the union of *disjoint* equivalence classes. $\tag{1.9}$

In S_n, for example, all elements in one class have common cycle structure and vice versa. Thus if we write any $p \in S_n$ in the form

$$p = \underbrace{(i_1 \ p(i_1) \ p(p(i_1)) \cdots)}_{m_1} \ \underbrace{(i_2 \ p(i_2) \cdots)}_{m_2} \cdots ,$$

$$n = m_1 + m_2 \cdots , \tag{1.10}$$

where i_2 is not one of the earlier m_1 entries, i_3 is not one of the earlier $m_1 + m_2$ entries, $\cdots$, then $m_1, m_2, \cdots$ is some partition of n. The cycle structure of p is denoted by $(m_1, m_2, \cdots)$, and cycle structures determine equivalence classes and conversely. All $p' \in S_n$ conjugate to p in (1.10) have the same cycle structure as p, and conversely.

In $SO(3)$, as we will recall later, each rotation is by some right handed angle about some axis. Then each class consists of all rotations by a given angle about all possible axes.

Subgroups

A subset $H \subset G$ is a *subgroup* if its elements obey all the conditions to be a group, given the composition law in G. $H = \{e\}$ or G are trivial cases. An elegant and compact criterion for a subset to be a subgroup is this:

$$H \subset G \text{ is a subgroup} \Leftrightarrow \text{for all } h_1, h_2 \in H, \ h_1^{-1} h_2 \in H. \tag{1.11}$$

As examples of subgroups in familiar groups we have: all even integers in the additive group of all integers; all even permutations in S_n making up the alternating subgroup A_n; S_{n-1} within S_n, i.e., permutations p with $p(n) = n$; all rotations in $SO(3)$ about a given axis.

In a finite group, each element leads via its 'powers' to a subgroup called its *cycle*:

$$a \in G \rightarrow \{e, a, a^2, \cdots, a^{q-1}\} = \text{cycle of } a = \text{subgroup in } G,$$

$$q = \text{least positive integer such that } a^q = e, q = \text{order of } a. \tag{1.12}$$

We will see that q divides $|G|$; it is clear that $a^{-1} = a^{q-1}$, etc.

Given a subgroup $H \subset G$ and any $g \in G$, conjugation leads to another *conjugate subgroup*:

$$H_g = gHg^{-1} = \{ghg^{-1} | \ h \in H \text{ varying}, \ g \text{ fixed}\} \subset G. \tag{1.13}$$

Of course, $g \in H$ implies $H_g = H$; and if $g \notin H$, H_g could be different from H.

Cosets

Given a subgroup $H \subset G$, there are two generally distinct and useful ways to break G up into subsets, based on two kinds of cosets:

$$aH = \{ah|\ a \text{ fixed, all } h \in H\} = \text{ right coset containing } a \in G;$$

$$Ha = \{ha|\ a \text{ fixed, all } h \in H\} = \text{ left coset containing } a \in G. \qquad (1.14)$$

As we saw with equivalence classes, here too it is easy to see the following: each coset is determined by any one of its elements; two cosets (both right or both left) either coincide fully or are disjoint; G is a union of (right or left) cosets. However, unlike classes, each coset is of the same 'size', namely as 'big' as H. It follows that $|G|/|H|$ is an integer, the number of disjoint right (or left) cosets. This is *Lagrange's theorem*. The claim made after (1.12) is now understood.

For a general subgroup $H \subset G$, right and left cosets are different, leading to different ways of expressing G as a union of disjoint subsets. However, if $H_g = H$ for all $g \in G$, i.e., H is self conjugate, then every right coset is also a left coset and vice versa. Then we say H is a *normal* or an *invariant* subgroup:

$$H \text{ is an invariant subgroup } \Leftrightarrow H_g = gHg^{-1} = H \text{ for every } g \in G$$

$$\Leftrightarrow aH = Ha \text{ for every } a \in G$$

$$\Leftrightarrow \text{ the two kinds of cosets coincide.} \qquad (1.15)$$

Then these (common) cosets can themselves be regarded as elements of a group, the *quotient group* or *factor group* G/H: the identity is the coset containing the identity element, $eH = H$, ie, H itself; the composition of two cosets is given by $aH \cdot bH = ab \cdot H$; and for inverses we take $(aH)^{-1} = a^{-1}H$. It is easy to check that all the group laws are obeyed. The orders obey $|G/H| = |G|/|H|$.

For any group G, there are two natural invariant subgroups, the *centre* and the *commutator* subgroup. The former consists of all those elements which 'commute' with all elements,

$$Z = \text{centre of } G = \{a \in G|\ ab = ba \text{ for all } b \in G\}, \qquad (1.16)$$

so obviously it is abelian. The latter is more intricate and in fact is a way of 'measuring' the extent to which G is nonabelian. If G is abelian, then always $ab = ba$. In general, we define the *commutator* of any two elements a and b as the element

$$q(a, b) = aba^{-1}b^{-1} = \text{measure of noncommutativity of } a \text{ and } b,$$

$$q(a, b) = e \Leftrightarrow ab = ba. \tag{1.17}$$

It is easy to see that under inversion and conjugation,

$$q(a, b)^{-1} = q(b, a), \quad cq(a, b)c^{-1} = q(cac^{-1}, cbc^{-1}). \tag{1.18}$$

The commutator subgroup $Q \subset G$ is now defined as consisting of products of any numbers of commutator factors:

$$Q = \{q(a_1, b_1)q(a_2, b_2)\cdots q(a_m, b_m)| \text{ any } m, a\text{'s, } b\text{'s}\}. \tag{1.19}$$

It is easy to see from (1.18) that Q is an invariant subgroup of G. Further, since $ab = q(a, b)ba = baq(a^{-1}, b^{-1})$, we see that G/Q is abelian. Thus in the quotient all noncommutativity in G has been removed.

There is a converse to this statement: if H is an invariant subgroup in G such that G/H is abelian, then H contains Q.

For the symmetric group S_n, Q is the alternating group A_n of all even permutations, and S_n/A_n is the two element abelian group.

We mention that the concept of the commutator subgroup is basic to the definitions of so-called solvable and nilpotent groups, but we do not go into them here.

Returning to the general concept of invariant subgroups, we have two important definitions:

(a) G is simple $\Leftrightarrow$ G has no (proper) invariant subgroup;

(b) G is semisimple $\Leftrightarrow$ G has no invariant abelian subgroup. $\tag{1.20}$

This leads to a one-way relationship: G is simple $\Rightarrow$ G is semisimple.

Among the symmetric groups S_n we always have the alternating subgroup A_n which is invariant, so all S_n are nonsimple. For $n = 3$ or $n \geq 5$, A_n is the only invariant subgroup in S_n. In the case of S_4, apart from A_4 there is one other invariant subgroup K of 4 elements, and the quotient $S_4/K \sim S_3$. In contrast, the rotation group $SO(3)$ is simple.

The last operation within a given group we consider is that of an *automorphism*. Given G, an automorphism of G is a map $\tau : G \to G$ which is one-to-one, onto, invertible and preserves products:

$$a \in G \rightarrow \tau(a) \in G: \tau(a)\tau(b) = \tau(ab), \ \tau^{-1} \text{ exists}, \tau(G) = G. \qquad (1.21)$$

It is easy to see that $\tau(a)^{-1} = \tau(a^{-1})$, $\tau(e) = e$. Conjugation by a fixed $g \in G$ is an automorphism:

$$g \in G, \text{ fixed: } \tau_g(a) = gag^{-1}, \quad a \in G. \qquad (1.22)$$

The conditions in (1.21) are immediately verified. These are called *inner* automorphisms, all others are *outer*. Clearly if G is abelian, every nontrivial automorphism is outer. We will later come across physically important examples of automorphisms.

1.4 Operations with and Relations between Groups

We consider four of these:

(A) *Homomorphism*   Given two groups G' and G, a homomorphism is a map $\Phi :$ $G' \rightarrow G$ such that images of products are products of images:

$$\Phi(a'b') = \Phi(a')\Phi(b'), \text{ all } a', b' \in G'. \qquad (1.23)$$

It is easy to see that $\Phi(e') = e \in G$, $\Phi(a'^{-1}) = \Phi(a')^{-1}$, $\Phi(G') \subset G$ is a subgroup in G. We may generally assume $\Phi(G') = G$ or else limit ourselves to $\Phi(G')$ in G. The kernel of a homomorphism is

$$K = \{g' \in G' | \ \Phi(g') = e\} = \text{invariant subgroup in } G', \qquad (1.24)$$

so we can form the quotient G'/K.

(B) *Isomorphism*   This is a particular case of homomorphism when Φ is one to one, onto and invertible, so $\Phi(G') = G$. Thus as groups G' and G are 'identical' or essentially the same. Returning to a general homomorphism, case (A), we see that, assuming $\Phi(G') = G$,

$$G'/K \text{ is isomorphic to } G. \qquad (1.25)$$

In an isomorphism G' and G play symmetric roles, but in a homomorphism this is not so.

(C) *Direct product*   Given two groups G_1 and G_2, their direct (or Cartesian) product $G_1 \times G_2$ is another group. Its elements are ordered pairs (a_1, a_2) with $a_1 \in G_1$, $a_2 \in G_2$. The group law is trivial, each entry 'minding its own business':

$$(a_1, a_2)(b_1, b_2) = (a_1 b_1, a_2 b_2);$$

$$\text{identity} = (e_1, e_2);$$

$$(a_1, a_2)^{-1} = (a_1^{-1}, a_2^{-1}). \tag{1.26}$$

The orders multiply: $|G_1 \times G_2| = |G_1||G_2|$. In a natural sense, G_1 and G_2 are invariant subgroups in $G_1 \times G_2$ and are recoverable as factor groups with respect to the appropriate kernels.

(D) *Semidirect product*   This is a more intricate way of combining G_1 and G_2, a direct product with a 'twist', in which G_1 and G_2 are not treated symmetrically. For each $a_1 \in G_1$, we need an automorphism τ_{a_1} of G_2 obeying certain conditions:

$$\tau_{a_1'}(\tau_{a_1}(a_2)) = \tau_{a_1' a_1}(a_2),$$

$$\tau_{a_1^{-1}} = \tau_{a_1}^{-1}. \tag{1.27}$$

Then the group law for ordered pairs is:

$$G_1 \rtimes G_2 : (a_1, a_2)(b_1, b_2) = (a_1 b_1, a_2 \tau_{a_1}(b_2)). \tag{1.28}$$

It is instructive to verify the laws of group structure here; it is not as trivial as with the direct product. In a natural manner we do find that G_1 is a subgroup, G_2 is an invariant one, and the quotient $G_1 \rtimes G_2 / G_2 \simeq G_1$.

We will see several physically important examples of semidirect products, especially in Chapter 10.

1.5 Realisations and Representations of Groups

A realisation of a group G arises in the following way. We have a set X, and for each $g \in G$, a map $\phi_g : X \to X$ obeying the group and other laws:

    (i) each ϕ_g one-to-one, onto, invertible;

    (ii) $\phi_e = \text{Id}_X$, identity map;

    (iii) $\phi_{g'} \circ \phi_g = \phi_{g'g}$, all $g', g \in G$;

so (iv) $\phi_{g^{-1}} = (\phi_g)^{-1}. \tag{1.29}$

(For composition of maps, associativity is automatic; in a group in the abstract, it is explicitly postulated). Then we say we have a realisation of G by maps on X.

In the context of realisation of a group G on a set X as defined above, the following notions naturally arise:

Orbit of $x \in X : \vartheta(x) = \{x' \in X \mid x' = \phi_g(x),$ some $g \in G\}$

Equivalence relation on $X : x \sim y \Leftrightarrow \phi_g(x) = y$ for some $g \in G$

Stability (Isotropy) group $H(x)$ of $x \in X : H(x) = \{g \in G \mid \phi_g(x) = x\}$

Fixed points X_g of $g \in G : X_g = \{x \in X \mid \phi_g(x) = x$ for a fixed $g \in G\}$

In classical Hamiltonian mechanics, groups usually act as canonical transformations on phase space. In quantum mechanics, the state space is a linear vector space, a Hilbert space, and we have the Superposition Principle. In this context, linear representations of groups become relevant. We will hereafter study only these.

1.6 Group Representations

These are particular cases of realisations, with added features. Given a group G and a (real or complex) linear vector space V, we have a representation D of G on V if the following hold:

(i) For each $g \in G$, $D(g) = $ invertible linear operator on V;

(ii) $D(e) = \mathbb{1} = $ identity or unit operator on V;

(iii) $D(g')D(g) = D(g'g)$, all $g',g \in G$;

so (iv) $D(g^{-1}) = D(g)^{-1}$. 　　　　　　　　　　　　　　　(1.30)

The $D(g)$ are also called linear transformations on V. The representation is faithful if

$$g' \neq g \Rightarrow D(g') \neq D(g),$$ 　　　　　　　　　　　(1.31)

otherwise it is nonfaithful. The dimension of the representation D is that of V; it may be finite or infinite.

A group G generally has many representations, on various V, of various dimensions.

Given a realisation of a group G on a set X, not naturally endowed with a vector space structure, one can elevate it to a representation of G by defining a vector space $\mathbb{F}[X]$ over a field $\mathbb{F}$ consisting of all formal linear combinations $\sum_i c_i x_i$ of elements

x_i of X with the coefficients c_i drawn from the field $\mathbb{F}$. A particularly interesting and important case arises when the set X is chosen to be G itself, which would be analysed in greater detail later in this chapter. Other instances where this device can be profitably employed are discussed in Chapter 2 in the context of the representations of the group S_n.

1.7 Equivalent Representations

Let D on V, D' on V' be two representations of G. (It may happen that $V' = V$ but $D' \neq D$.) We say they are *equivalent* if there is a linear invertible map $S : V \to V'$ such that

$$D'(g) = SD(g)S^{-1}, \quad \text{all } g \in G. \tag{1.32}$$

If no such S exists, they are *inequivalent*. We are often interested in representations determined up to equivalence.

For clarity we may assume that both V and V' are real vector spaces, or both complex ones. Cases where one is real and the other complex can be handled suitably.

1.8 Unitary/Orthogonal Cases – UR's

Let the complex (or real) space V of a representation of G carry a hermitian (or symmetric) inner product of vectors, written as (x, y) or $\langle x|y \rangle$ for $x, y \in V$. The representation $D(\,\cdot\,)$ of G on V is said to be unitary (or real orthogonal) if

$$(D(g)x, D(g)y) = (x, y), \text{ all } g \in G, \ x, y \in V. \tag{1.33}$$

Unitary representations are usually denoted as UR's. For the most part we will work with unitary group representations. Given a representation $D(\,\cdot\,)$ of G on V, it may happen that there is no choice of inner product on V such that (1.33) holds. Then $D(\,\cdot\,)$ is essentially nonunitary (or essentially nonorthogonal). For all finite groups as well as so-called continuous compact groups, every representation is equivalent to a unitary one. The Lorentz group $SO(3, 1)$ is an important case where this does not happen.

1.9 Matrices of a Representation

Assume for simplicity that we are dealing with finite dimensional representations. Given $D(\,\cdot\,)$ of G on V, choose a basis $\{e_j\}$ in V, and for each $g \in G$ write

$$D(g)e_j = D^k_{\ j}(g)e_k, \text{ sum on } k. \qquad (1.34)$$

The $n \times n$ matrices $(D^j_{\ k}(g))$, where $n =$ dimension of V, j is a row index and k a column index, are the matrices of the representation in this basis. We simply write them too as $D(g)$. They obviously obey (1.30) in matrix form:

(i) $D^j_{\ k}(g_1 g_2) = D^j_{\ l}(g_1)D^l_{\ k}(g_2)$;

(ii) $D^j_{\ k}(e) = \delta^j_{\ k}$;

(iii) $D^j_{\ k}(g^{-1}) =$ elements of inverse of matrix $(D^j_{\ k}(g))$. $\qquad (1.35)$

Any change of basis $\{e_j\} \rightarrow \{e'_j\}$ in V according to

$$e'_j = (S^{-1})^k_{\ j}e_k, \qquad (1.36)$$

where S is a (real or complex as the nature of V) nonsingular $n \times n$ matrix leads to the relation (compare (1.32))

$$\text{matrix } D'(g) = S \left(\text{matrix } D(g)\right) S^{-1}, \qquad (1.37)$$

so the representation matrices experience a similarity transformation.

If the representation $D(\,\cdot\,)$ on V is unitary or real orthogonal, and we use an orthonormal basis in V, then the representation matrices are themselves unitary or real orthogonal in the familiar senses.

1.10 Some Operations with Group Representations

We consider a few of these here. Picture a representation via its matrices $D(g)$, and do not assume unitarity or real orthogonality. We define three operations starting with $D(g)$:

Contragredient representation: $g \rightarrow (D(g)^T)^{-1} = D(g^{-1})^T$;

Adjoint representation: $g \rightarrow (D(g)^\dagger)^{-1} = D(g^{-1})^\dagger$;

Complex conjugate representation: $g \rightarrow D(g)^*$. $\qquad (1.38)$

In each case we clearly see that from the initial representation we have created a new one. The adjoint is the complex conjugate of the contragredient.

In case $D(g)$ is a UR, it is self adjoint, and the contragredient equals the complex conjugate. If $D(g)$ is real orthogonal, it is self contragredient.

Direct Sums and Products

Let D_1 on V_1 and D_2 on V_2 be two representations of G. Assume V_1 and V_2 are both complex or both real, but in general of different dimensions. Then the *direct sum* representation $D = D_1 \oplus D_2$ is a representation on $V = V_1 \oplus V_2$ defined in the natural way:

$$x_1 \in V_1, \ x_2 \in V_2 \to x = x_1 + x_2 \in V,$$

$$D(g)x = D_1(g)x_1 + D_2(g)x_2 \in V, \ \text{ all } g \in G. \qquad (1.39)$$

The representation property is immediate, and the dimension is the sum of the individual ones.

The *direct product* representation $D = D_1 \times D_2$ acts on the tensor product space $V_1 \times V_2$: for product vectors,

$$x \in V_1, \ y \in V_2, \ g \in G : \ D(g)(x \times y) = D_1(g)x \times D_2(g)y. \qquad (1.40)$$

This is then extended by linearity to general vectors in $V_1 \times V_2$.

1.11 Character of a Representation

This is an important concept which we will often come back to. It is most simply defined using the matrices of a representation in any basis. The character $\chi^{(D)}$ of a representation D of G on V, a function on G, is the trace of the representation matrices:

$$\chi^{(D)}(g) = D^j{}_j(g) = \text{Tr } D(g). \qquad (1.41)$$

By Eq. (1.37) this is actually basis independent. It is in general a complex valued function on the group, and constant over each conjugacy class:

$$\chi^{(D)}(g'gg'^{-1}) = \chi^{(D)}(g). \qquad (1.42)$$

So we say $\chi^{(D)}(g)$ is a class function. For a UR, $\chi(g^{-1}) = \chi(g)^*$. For sums and products of representations, the character behaves very simply:

$$\chi^{(D_1 \oplus D_2)}(g) = \chi^{(D_1)}(g) + \chi^{(D_2)}(g),$$
$$\chi^{(D_1 \times D_2)}(g) = \chi^{(D_1)}(g)\chi^{(D_2)}(g). \tag{1.43}$$

We will soon see that the character of a representation determines the representation completely up to equivalence.

1.12 Invariant Subspaces, Reducibility, Irreducibility – UIR's

Given a representation D of G on V, it is *reducible* if there is a nontrivial subspace $V_1 \subset V$ *invariant* under G action:

$$x \in V_1, \ g \in G \Rightarrow D(g)x \in V_1. \tag{1.44}$$

If no such V_1 exists, $D(\cdot)$ is an *irreducible* representation, or an irrep. If it is also unitary, we say it is an UIR – unitary irreducible representation.

The convenient abbreviations 'irreps', 'UR' and 'UIR' will be used throughout. In general we will keep in mind the complex case.

In the reducible case we can go further and ask: can we supplement V_1 with another (disjoint) subspace $V_2 \subset V$ which is also invariant and such that $V = V_1 \oplus V_2$? If we can, then $D(\cdot)$ is *decomposable*, otherwise it is *indecomposable*. So the arrangement of these ideas can be indicated in this way:

Irrep No nontrivial invariant subspace V_1

↗

Rep. D of G on V

↘

Reducible $V_1 \subset V$, nontrivial, invariant under D

↙ ↘

Decomposable   *Indecomposable*

$V = V_1 \oplus V_2$, V_2 also invariant   No such V_2

In matrix form, in suitable bases, these cases mean:

$$D(g) \text{ reducible } \Leftrightarrow D(g) \text{ can be brought to the form } \begin{pmatrix} D_1(g) & \vdots & B(g) \\ \cdots & \vdots & \cdots \\ 0 & \vdots & D_2(g) \end{pmatrix};$$

$D(g)$ reducible, decomposable $\Leftrightarrow$ can choose basis so that $B(g) = 0$;

$D(g)$ reducible, indecomposable $\Leftrightarrow$ always $B(g) \neq 0$. (1.45)

In the decomposable case we can 'look inside' V_1 and V_2 and try to repeat the process. For UR's, reducibility always implies decomposability, as we can simply take $V_2 = V_1^{\perp}$, the orthogonal complement of V_1. So by repeating this analysis we finally arrive at UIR's and can say: every UR is the direct sum of UIR's. This holds for all finite and all continuous compact groups. Thus in these cases, the basic building blocks for representation theory are the UIR's, which for these types of groups are also finite dimensional. Anticipating a little, we have: for a finite group, the number of inequivalent UIR's is finite; for a continuous compact group they are a denumerable infinity.

1.13 Schur's Lemma: Proof and Applications

This is a key result, a very powerful tool to deal with irreps, equivalences, etc. Let D on V, D' on V' be two given irreps of G, both assumed complex for definiteness. Suppose there exists, or we are able to somehow construct, a linear operator $T : V \to V'$ such that it 'intertwines' D and D' in the sense:

$$T D(g) = D'(g)T, \text{all } g \in G. \tag{1.46}$$

Then we can prove that either

(i) D and D' are inequivalent, and $T = 0$;

or (ii) T is nonsingular, so D and D' are equivalent. (1.47)

(We can leave aside the case where D and D' are equivalent, yet $T = 0$.)
We must appreciate that what *cannot happen* is $T \neq 0$, but T is singular, i.e., non invertible.

Proof.   Define subspaces in V and in V' as follows:

$$\mathcal{N}(T) = \text{null space of } T = \text{subspace of } V \text{ on which } T \text{ vanishes;}$$

$$\mathcal{R}(T) = \text{range of } T = \text{subspace of } V' = \text{image of } V \text{ under } T; \qquad (1.48)$$

i.e.,

$$T\mathcal{N}(T) = 0, \quad \mathcal{N}(T) \subset V; \quad \mathcal{R}(T) = T(V) \subset V'. \qquad (1.49)$$

(Again we can leave aside the case $T = 0, \mathcal{N}(T) = V$.) Then from Eq. (1.46) we easily find: $\mathcal{N}(T)$ is invariant under $D(G)$, $\mathcal{R}(T)$ under $D'(g)$. Therefore the irreducibility of both D and D' implies: $\mathcal{N}(T) = 0$, so T is one-to-one; $\mathcal{R}(T) = V'$, so T is onto. (Again we can discard $\mathcal{R}(T) = 0$ as then $T = 0$.) Then if $T \neq 0$, T^{-1} exists; and the two representations, both irreducible, are equivalent.

For a first application of the Lemma, we consider a single irrep, and take $V' = V$, $D' = D$. Then suppose for some T, a map $V \to V$, we find:

$$TD(g) = D(g)T, \quad \text{all } g \in G. \qquad (1.50)$$

Then form $T - \lambda \cdot \mathbb{1}$ where λ is (any) eigenvalue of T. Since $T - \lambda \cdot \mathbb{1}$ is singular and also obeys (1.50), it must vanish, so we conclude:

$$D(g) \text{ an irrep of } G, \; TD(g) = D(g)T, \text{ all } g \Rightarrow T = \lambda \cdot \mathbb{1}, \text{ some } \lambda. \qquad (1.51)$$

The converse can also be easily shown: if only multiples of $\mathbb{1}$ commute with $D(g)$ for all g, $D(\cdot)$ must be an irrep.

The next application of the Lemma reveals the inner structure of a reducible representation, and the sense in which the reduction is unique.   □

'Uniqueness' of Complete reduction

Let a representation $D(\cdot)$ of G on V of finite dimension be fully reducible into irreps, so $D(\cdot)$ is a direct sum of these. Imagine two reductions of V into irreducible invariant subspaces with corresponding irreps, namely,

$$V = \sum_{\rho=1,2,\cdots} \oplus \, \mathcal{L}_\rho, \quad \mathcal{L}_\rho \text{ carrying irrep } D^{(\rho)},$$

$$= \sum_{\sigma=1,2,\cdots} \oplus \, \mathcal{M}_\sigma, \quad \mathcal{M}_\sigma \text{ carrying irrep } D^{(\sigma)}. \qquad (1.52)$$

We want to examine the extent to which these two decompositions can differ, and what they must hold in common. Any vector $x_\rho \in \mathcal{L}_\rho$ can be expanded uniquely in terms of components in $\mathcal{M}_\sigma$:

$$x_\rho \in \mathcal{L}_\rho: x_\rho = \sum_\sigma y_\sigma, \quad y_\sigma \in \mathcal{M}_\sigma. \tag{1.53}$$

For each pair $\sigma\rho$, define the linear map $T_{\sigma\rho} : \mathcal{L}_\rho \to \mathcal{M}_\sigma$ by

$$T_{\sigma\rho} x_\rho = y_\sigma \tag{1.54}$$

as determined by (1.53). (Here there is no sum on ρ.) Apply $D(g)$ to both sides to get:

$$D(g)x_\rho \equiv D^{(\rho)}(g)x_\rho = D(g)\sum_\sigma y_\sigma$$

$$= \sum_\sigma D(g)y_\sigma = \sum_\sigma D^{(\sigma)}(g)y_\sigma,$$

$$\text{i.e., } D^{(\sigma)}(g)y_\sigma = T_{\sigma\rho} D^{(\rho)}(g)x_\rho,$$

$$\text{i.e., } D^{(\sigma)}(g)T_{\sigma\rho}x_\rho = T_{\sigma\rho} D^{(\rho)}(g)x_\rho, \text{ any } x_\rho \in \mathcal{L}_\rho. \tag{1.55}$$

Therefore for each pair $\sigma\rho$ we have an intertwining relation

$$T_{\sigma\rho} D^{(\rho)}(g) = D^{(\sigma)}(g) T_{\sigma\rho}, \tag{1.56}$$

as in Eq. (1.46). We can apply Schur's Lemma to conclude:

(i) $T_{\sigma\rho} = 0$ if $D^{(\rho)}, D^{(\sigma)}$ are not equivalent;

(ii) $T_{\sigma\rho} \neq 0$ only if they are equivalent. $\tag{1.57}$

We can now draw out the consequences. For each $\mathcal{L}_\rho$ carrying the irrep $D^{(\rho)}$, pick out all those $\mathcal{M}_\sigma$ such that $D^{(\sigma)}$ is equivalent to $D^{(\rho)}$. Then $x_\rho \in \mathcal{L}_\rho \Rightarrow y_\sigma \neq 0$ only in these $\mathcal{M}_\sigma$; $x_\rho \neq 0 \Rightarrow$ at least one such $y_\sigma \neq 0$, and at least one $T_{\sigma\rho} \neq 0$. Then by the Lemma, this $T_{\sigma\rho}$ is nonsingular.

So for each irrep $D^{(\rho)}$ in the first reduction, there is at least one equivalent irrep $D^{(\sigma)}$ in the second reduction, *and conversely*. Now we can examine the question of multiplicities, the number of times a given irrep. repeats itself in a complete reduction. Several subspaces $\mathcal{L}_\rho$ (similarly $\mathcal{M}_\sigma$) may carry equivalent irreps. Let us label irreps (up to equivalence) by $\alpha, \beta, \cdots$, and multiple appearances by $j, k, \cdots$; so we write

$$\mathcal{L}_\rho \equiv \mathcal{L}_{\alpha,j}, \quad \mathcal{M}_\sigma \equiv \mathcal{M}_{\beta,k}. \tag{1.58}$$

Now 'combine' subspaces according to 'representation type':

$$V = \sum_\rho \oplus \mathcal{L}_\rho = \sum_\alpha \oplus \left(\sum_j \oplus \mathcal{L}_{\alpha,j} \right) = \sum_\alpha \oplus \mathcal{L}^{(\alpha)},$$

$$\mathcal{L}^{(\alpha)} = \sum_j \oplus \mathcal{L}_{\alpha,j};$$

$$V = \sum_\sigma \oplus \mathcal{M}_\sigma = \sum_\beta \oplus \left(\sum_k \oplus \mathcal{M}_{\beta,k} \right) = \sum_\beta \oplus \mathcal{M}^{(\beta)},$$

$$\mathcal{M}^{(\beta)} = \sum_k \oplus \mathcal{M}_{\beta,k}. \tag{1.59}$$

Any $x \in \mathcal{L}_{\alpha,j}$ has components only in $\mathcal{M}_{\alpha,k}$ and vice versa. So

$$T_{\alpha j, \beta k} = 0 \;\; \text{if} \;\; \alpha \neq \beta;$$

$$x \in \mathcal{L}^{(\alpha)} \Leftrightarrow x \in \mathcal{M}^{(\alpha)}, \quad \mathcal{L}^{(\alpha)} = \mathcal{M}^{(\alpha)}. \tag{1.60}$$

We see that the irreps present in D on V and their multiplicities are unique, so are the subspaces $\mathcal{L}^{(\alpha)}$. Only the further break up of each $\mathcal{L}^{(\alpha)}$ into $\mathcal{L}_{\alpha,j}$ is arbitrary.

The Orthogonality Relations

This is our third and last application of Schur's Lemma. It leads to important properties of the matrix elements of the representation matrices in the various irreducible representations.

Consider a complex-valued function $f(g)$ on G, viewed as an element in a complex vector space of dimension $|G|$, which we assume is finite. (More generally we could consider $f(\cdot)$ belonging to some linear space of some dimension.) For any given $f(\cdot)$ we define its mean value by an averaging process over G:

$$\mathcal{M}(f) \equiv \mathcal{M}_g(f(g)) = \frac{1}{|G|} \sum_{g \in G} f(g). \tag{1.61}$$

The main properties which are evident upon inspection are linearity, nonnegativity, normalisation and 'translation' invariances:

$$\mathcal{M}(f_1 + f_2) = \mathcal{M}(f_1) + \mathcal{M}(f_2), \ \mathcal{M}(\lambda f) = \lambda \mathcal{M}(f);$$

$$f \geq 0 \Rightarrow \mathcal{M}(f) \geq 0; \ f = 1 \Rightarrow \mathcal{M}(f) = 1,$$

$$\mathcal{M}_g(f(g)) = \mathcal{M}_g(f(g_0 g)) \text{ or } f(g g_0) \text{ or } f(g^{-1})), \ g_0 \text{ fixed.} \qquad (1.62)$$

Now take two irreps $D(g)$ on V, $D'(g)$ on V'. Let $S : V \to V'$ be any linear map, and consider

$$T_S = \mathcal{M}_g(D'(g^{-1})SD(g)) = \text{linear map } V \to V'. \qquad (1.63)$$

(So here the quantity being averaged is not a complex valued function but something belonging to a linear space, namely, the space of all linear maps $V \to V'$.) Using Eq. (1.62) we see that it obeys:

$$D'(g'^{-1})T_S = \mathcal{M}_g\left(D'(g'^{-1})D'(g^{-1})SD(g)\right) = T_S D(g'^{-1}),$$

$$\text{i.e., } D'(g)T_S = T_S D(g), \text{any } g \in G. \qquad (1.64)$$

Thus T_S intertwines the two irreps, and we can use Schur's Lemma:

$$D' \text{ and } D \text{ inequivalent } \Rightarrow T_S = 0, \text{ any } S;$$

$$D' \text{ and } D \text{ equivalent, the same } \Rightarrow T_S = \lambda(S)\mathbb{1}, \text{some } \lambda(S). \qquad (1.65)$$

So for the case of inequivalent irreps, in terms of the matrices in any basis, as S is arbitrary, we get:

$$\mathcal{M}_g(D'_{j'k'}(g^{-1})D_{jk}(g)) = 0, \text{ all } j \ k \ j' \ k'. \qquad (1.66)$$

In case these matrices are unitary, this reads:

$$\mathcal{M}_g(D'_{j'k'}(g)^* D_{jk}(g)) = 0, \text{ all } j \ k \ j' \ k'. \qquad (1.67)$$

For equivalent irreps, we assume $D' = D$ as in the second line of Eq. (1.65), and get:

$$\mathcal{M}_g(D_{jk}(g^{-1})S_{kl}D_{lm}(g)) = \lambda(S)\delta_{jm}, \text{ sum on all } k, l. \qquad (1.68)$$

Setting $j = m$ and summing on j gives $\lambda(S) = \text{Tr } S/\text{dim } V$, so (1.68) becomes

$$\mathcal{M}_g(D_{jk}(g^{-1})D_{lm}(g)) = \delta_{jm}\delta_{kl}/\text{dim } V. \qquad (1.69)$$

And in the unitary case we have

$$\mathcal{M}_g(D_{jk}(g)^* D_{lm}(g)) = \delta_{jl}\delta_{km}/\dim V. \tag{1.70}$$

Incidentally, the last two equations imply that in any irrep, in any chosen basis, any *fixed* matrix element cannot vanish for all g.

We can regard the space of complex functions $f(g)$ on G as a Hilbert space $L^2(G)$ of dimension $|G|$, and we use $\mathcal{M}$ to define the inner product:

$$(f_1, f_2) = \mathcal{M}_g(f_1(g)^* f_2(g)). \tag{1.71}$$

At the same time, let us label the various inequivalent UIR's by $\alpha, \beta, \cdots$ as in Eq. (1.58), with dimensions $N_\alpha, N_\beta, \cdots$. Then Eqs. (1.67, 1.70) lead to the orthogonality relations

$$\left(D_{jk}^{(\alpha)}(\,\cdot\,),\ D_{lm}^{(\beta)}(\,\cdot\,) \right) = \delta_{\alpha\beta}\delta_{jl}\delta_{km}/N_\alpha. \tag{1.72}$$

Each fixed matrix element in each UIR thus gives one nonzero vector in $L^2(G)$, mutually orthogonal if the UIR's are inequivalent or the row or column indices differ. This leads to the inequality

$$\sum_\alpha N_\alpha^2 \le |G|. \tag{1.73}$$

From Eq. (1.72) we get important relations for the characters $\chi^{(\alpha)}(g)$ of the various UIR's: summing over $j = k$ and $l = m$,

$$\left(\chi^{(\alpha)}(\,\cdot\,),\ \chi^{(\beta)}(\,\cdot\,) \right) = \delta_{\alpha\beta}. \tag{1.74}$$

Now the χ's are class functions, constant over each class. In general such functions on G form a linear subspace in $L^2(G)$, and within this Eq. (1.74) means the $\chi^{(\alpha)}(g)$ are an orthonormal set. This gives another inequality to accompany (1.73):

Number of inequivalent irreps or UIR′s $\le$ number of equivalence classes in G
$$\tag{1.75}$$

Let the character of a general UR $D(\,\cdot\,)$ be written as $\chi(g)$, and upon reduction let $D(\,\cdot\,)$ contain $D^{(\alpha)}(\,\cdot\,)$ with multiplicity ν_α. Then we easily find from Eq. (1.74) and the definition of $\chi(g)$:

$$\chi(g) = \sum_\alpha v_\alpha \chi^{(\alpha)}(g),$$

$$v_\alpha = \text{multiplicity of occurrence of } D^{(\alpha)} \text{ in } D = (\chi^{(\alpha)}, \chi),$$

$$(\chi, \chi) = \sum_\alpha v_\alpha^2 \geq 1. \tag{1.76}$$

Thus the UR $D(\cdot)$ with character $\chi(g)$ is irreducible or reducible according as the squared norm (χ, χ) of χ is or exceeds unity.

It was mentioned earlier that for any finite group (as well as for any continuous compact group) every representation may be assumed without loss of generality to be unitary. We leave it as an exercise to prove this (in the finite group case) using the idea of averaging over G as introduced in Eq. (1.61).

The Regular Representations of G

As the last but one item in this chapter, we explicitly construct two special UR's of G, each of which contains all UIR's, and which show that both inequalities $(1.73, 1.75)$ are equalities. These are the two regular representations, both acting on $L^2(G)$. For finite G, the space $L^2(G)$ is finite dimensional, with inner product (1.71). For now we consider this case. Later for continuous G we will use an integral version of this inner product, possessing invariances as in the last line of (1.62).

We write $L(g), R(g)$ or L_g, R_g for the operators on $L^2(G)$ corresponding to the left and right regular representations of G. On a general complex function $f(g)$ on G, their actions are:

$$(L_{g'}f)(g) = f(g'^{-1}g), \quad (R_{g'}f)(g) = f(gg'). \tag{1.77}$$

It is easy to check that these are unitary operators, and they give two mutually commuting UR's of G:

$$L_{g_1}L_{g_2} = L_{g_1 g_2}, \quad R_{g_1}R_{g_2} = R_{g_1 g_2}, \quad L_{g_1}R_{g_2} = R_{g_2}L_{g_1}. \tag{1.78}$$

If we introduce Dirac notation with kets and bras according to

$$f \in L^2(G) : f(g) = \langle g|f\rangle, \quad \langle g'|g\rangle = \delta_{g',g}, \tag{1.79}$$

then the actions of $L_{g'}, R_{g'}$ on these basis kets are:

$$L_{g'}|g\rangle = |g'g\rangle, \quad R_{g'}|g\rangle = |gg'^{-1}\rangle. \tag{1.80}$$

As these are UR's, they are fully reducible, and their characters determine the contents. From Eq. (1.80) we obtain:

$$\chi^{(\text{left reg})}(g) = \text{Tr } L_g = \chi^{(\text{right reg})}(g) = \text{Tr } R_g = |G|\delta_{g,e}. \tag{1.81}$$

Then we ask: how often does each UIR $D^{(\alpha)}(\,\cdot\,)$ occur in each of these UR's? From Eq. (1.76) we get the answer:

$$\nu_\alpha = \left(\chi^{(\alpha)}, \chi^{(\text{left or right reg})}\right) = \frac{1}{|G|}\sum_g \chi^{(\alpha)}(g)^* \chi^{(\cdots)}(g) = \chi^{(\alpha)}(e)^* = N_\alpha. \tag{1.82}$$

Thus every UIR α is present as often as its dimension N_α. This allows us to strengthen (1.73) to an equality,

$$\sum_\alpha N_\alpha^2 = |G|. \tag{1.83}$$

It is important to observe that even though at this point the only UR's that have been constructed are the regular ones, and the various UIR's are as yet 'unknown', we have the information that each of the latter is definitely contained in (each of) the former.

From Eqs. (1.72, 1.83) we have that the 'normalised' matrix elements $\{N_\alpha^{1/2} D_{jk}^{(\alpha)}(g)\}$ form an orthonormal basis for $L^2(G)$. Any $f(g)$ thus has a unique expansion:

$$f(g) = \sum_\alpha \sum_{jk} N_\alpha^{1/2} D_{jk}^{(\alpha)}(g) f_{jk}^{(\alpha)},$$

$$f_{jk}^{(\alpha)} = N_\alpha^{1/2}\left(D_{jk}^{(\alpha)}(\,\cdot\,), f(\,\cdot\,)\right),$$

$$\|f\|^2 = \frac{1}{|G|}\sum_g |f(g)|^2 = \sum_\alpha \sum_{jk} |f_{jk}^{(\alpha)}|^2. \tag{1.84}$$

In this basis, $L_{g'}$-action alters only the first index j, while $R_{g'}$-action alters only k: This is consistent with their mutually commuting. For the former action, k is a multiplicity index; for the latter, j plays this role.

Since the collection $\{N_\alpha^{1/2} D_{jk}^{(\alpha)}(g)\}$ is both orthonormal and complete, we see that any class function can be expanded in terms of the characters of the UIR's:

$$f(gg'g^{-1}) = f(g'), \text{all } g \text{ and } g' \Rightarrow$$

$$f(g) = \sum_\alpha f_\alpha \chi^{(\alpha)}(g), \ f_\alpha = (\chi^{(\alpha)}(\,\cdot\,), f(\,\cdot\,)). \tag{1.85}$$

As a result, (1.75) too can be strengthened to an equality:

Number of inequivalent irreps or UIR's = number of equivalence classes in G

$$(1.86)$$

Complex Conjugation, Direct Products, of UIR's

Given G, we know that in principle all the UIR's $D^{(\alpha)}(\,\cdot\,)$ can be 'extracted' from the regular representations. We know how many there are, and have some information, (1.83), on their dimensions.

Now suppose all the $D^{(\alpha)}(\cdot)$ have been constructed, each up to unitary equivalence. From (1.76) in the irreducible case, we see that $D^{(\alpha)^*}(\,\cdot\,)$, the complex conjugate of $D^{(\alpha)}(\,\cdot\,)$, is also a UIR. (This is one of the three operations listed in Eq. (1.38).) By judicious use of Schur's Lemma, we can easily find out the qualitatively different cases that can arise. We describe them briefly, omitting the details of derivations.

To begin with, and this is obvious,

$$\chi^{(\alpha)}(g)^* \neq \chi^{(\alpha)}(g) \Leftrightarrow D^{(\alpha)}(\,\cdot\,)^*, \; D^{(\alpha)}(\,\cdot\,) \text{ are inequivalent UIR's;}$$

$$\chi^{(\alpha)}(g) \text{ real } \Leftrightarrow D^{\alpha}(\,\cdot\,)^*, \; D^{(\alpha)}(\,\cdot\,) \text{ are equivalent UIR's.} \qquad (1.87)$$

In the latter case one can ask whether the matrices $D^{(\alpha)}(g)$ can all be made real by a suitable choice of orthonormal basis in the representation space. Here one finds the results:

$$\chi^{(\alpha)}(g) \text{ real } \Leftrightarrow D^{(\alpha)}(g)^* = CD^{(\alpha)}(g)C^{-1},$$

$$C^{\dagger}C = \mathbb{1}, C^T = \lambda C, \; \lambda = \pm 1;$$

$$\lambda = +1 \Leftrightarrow D^{(\alpha)}(g)^* = D^{(\alpha)}(g) \text{ in suitable basis – real case;}$$

$$\lambda = -1 \Leftrightarrow D^{(\alpha)}(g)^* \neq D^{(\alpha)}(g) \text{ in any basis – pseudoreal case.} \qquad (1.88)$$

(In the last case, of course, we mean that there are some g for which $D^{(\alpha)}(g)$ is complex.) Thus the three possibilities are – complex, pseudoreal, and real. There is a nice criterion in terms of $\chi^{(\alpha)}(\,\cdot\,)$ to directly distinguish between the last two, due to Wigner, but we omit the details.

As a final item in representation theory we consider direct products of UIR's, the Clebsch–Gordan problem. For any pair α, β we have a direct product UR (1.40)

$$D(g) = D^{(\alpha)}(g) \times D^{(\beta)}(g), \text{ dimension } N_\alpha N_\beta,$$

$$\text{Character } \chi(g) = \chi^{(\alpha)}(g)\chi^{(\beta)}(g). \qquad (1.89)$$

In the complete reduction of this product into UIR's, each $D^{(\gamma)}$ occurs with some multiplicity:

$$D^{(\alpha)} \times D^{(\beta)} = \sum_{\gamma} \oplus \, \nu_{\alpha\beta,\gamma} D^{(\gamma)}, \quad \nu_{\alpha\beta,\gamma} = \text{integer} \geq 0, \qquad (1.90)$$

where by Eq. (1.76)

$$\nu_{\alpha\beta,\gamma} = (\chi^{(\gamma)}, \, \chi^{(\alpha)} \chi^{(\beta)}). \qquad (1.91)$$

The series on the right in (1.90) is called the Clebsch–Gordan series, and the change of basis in the space of the product representation $D^{(\alpha)} \times D^{(\beta)}$ to effect this block diagonalisation involves Clebsch–Gordan coefficients which are basis dependent.

For an abelian group G, every UIR is one dimensional, every element is a class by itself, and UIR's are usually called 'characters'. If G is nonabelian, at least some UIR's are multi dimensional.

1.14 Group Algebra

Given a finite group G and a field $\mathbb{F}$, we can construct out of them a new set, denoted by $\mathbb{F}[G]$, consisting of all formal linear combinations of the form

$$a = \sum_{i} a_i x_i, a_i \in F, x_i \in G. \qquad (1.92)$$

The set $\mathbb{F}[G]$ can be viewed in several different ways:

1. As a vector space over $\mathbb{F}$: elements of $\mathbb{F}[G]$ can be added and multiplied by elements of $\mathbb{F}$ in a natural way. This is the vector space which underlies the regular representation of G as stated earlier.

2. As a unital ring: elements of $\mathbb{F}[G]$ can be added and (using the composition law of G) multiplied together to yield elements of $\mathbb{F}[G]$, with the identity element $1.e_G$ of G serving as the multiplicative unit of the ring. If G is abelian (nonabelian) $\mathbb{F}[G]$ is a commutative (noncommutative) ring.
   With the identification of $c \in \mathbb{F}$ with $ce_G \in \mathbb{F}[G]$ one can consider $\mathbb{F}$ as sitting in $\mathbb{F}[G]$.

3. As an algebra over $\mathbb{F}$: in addition to being a ring, $\mathbb{F}[G]$ is also closed under multiplication by elements of $\mathbb{F}$.

1.15 Representations of G and Its Group Algebra $\mathbb{F}[G]$

Recall that by a linear representation of a group G on a vector space V over a field $\mathbb{F}$ we mean a collection of linear operators $\{D(g)\}$ on V respecting the composition law in G. Given a representation $\{D(g)\}$, we can define multiplication of elements v of the vector space by elements a of $\mathbb{F}[G]$ to obtain other elements v' of V as follows:

$$a.v = \left(\sum_i a_i x_i\right).v$$

$$\rightarrow \sum_i a_i D(x_i) v = \sum_i a_i v_i \in V. \tag{1.93}$$

Here v_i are the vectors in V to which v is mapped by $D(x_i)$. We are thus led to a new mathematical structure – a module M over the ring $\mathbb{F}[G]$. (The notion of a module is same as that of a vector space with the field replaced by a ring.)

With this at hand, all aspects of the representation theory of finite groups can be couched in the language of modules – a subspace V_1 of the vector space V invariant under all of $\{D(g)\}$ corresponds to a submodule M_1 of M, an irreducible representation of G corresponds to a simple or irreducible submodule of M and so on. A crucial advantage of this language for studying group representations lies in the fact that constructing submodules of M (and hence subrepresentations of G) can be shown to reduce to the task of finding idempotent (essentially idempotent) elements in the ring $\mathbb{F}[G]$ – elements a in $\mathbb{F}[G]$ such that $a^2 = a$ ($a^2 = ca, c \in \mathbb{F}$). Further, constructing irreducible submodules (and hence irreducible representations of G) can be shown to reduce to the task of finding idempotents in $\mathbb{F}[G]$ which are primitive, i.e., those idempotents in $\mathbb{F}[G]$ which can not be expressed as a sum of two 'orthogonal' idempotents. (Two idempotents a and b in $\mathbb{F}[G]$ are said to be orthogonal provided $ab = ba = 0$.)

We shall put this machinery to use in Chapter 2 in the context of the irreducible representations of the symmetric group. Note also that the considerations here apply to any field $\mathbb{F}$. However, later in this book we will exclusively be concerned with the case when $\mathbb{F}$ is the complex field $\mathbb{C}$.

Problems

P1.1   Given an associative group composition law, and left inverses and left identity only, i.e.,

   (i) unique e such that $ea = a$ for all a,
   (ii) for each a, unique a^{-1} such that $a^{-1}a = e$;

   derive all the results in Eq. $(1.1(iii),(iv))$.

P1.2   For a finite group of prime order, show that there are no nontrivial subgroups.

P1.3   If H is an invariant subgroup of G such that G/H is abelian, show that the commutator subgroup Q of G is a subgroup of H as well.

P1.4   The group of translations in one real dimension, $x \to x + a$, is abelian. Show that

$$a \to \begin{pmatrix} 1 & a \\ 0 & 1 \end{pmatrix}$$

is a representation, and find out whether it is reducible or irreducible, decomposable or indecomposable.

P1.5   For a finite group G acting on a set X, show that

(i)   If $x \in X$ and $y \in X$ lie on the same orbit then their stability subgroups are conjugate subgroups in G, i.e., $H(y) = gH(x)g^{-1}$, some $g \in G$,

(ii) For any $x \in X$, there is a natural bijective map from $\vartheta(x)$ to the coset space $G/H(x)$ and hence $|\vartheta(x)| = |G/H(x)|$,

(iii) If the set X is decomposed into disjoint orbits using the equivalence relation noted in Section 1.5 then one has

$$\text{Number of orbits} = \frac{1}{|G|} \sum_{g \in G} |X_g| \quad \text{(Burnside's Lemma)}.$$

Bibliography

Boerner, H. (1963). *Representations of Groups with Special Consideration for the Needs of Modern Physics*. Amsterdam: North Holland.

Hamermesh, M. (1962). *Group Theory and Its Application to Physical Problems*. Reading, Massachusetts: Addison-Wesley.

Jansen, L. and Boon, M. (1967). *Theory of Finite Groups: Applications in Physics*. Amsterdam: North Holland.

Weyl, H. (1931). *The Theory of Groups and Quantum Mechanics*. New York: Dover.

Wigner, E. P. (1959). *Group Theory and Its Application to the Quantum Mechanics of Atomic Spectra*. New York: Academic Press.

2

The Symmetric Group

In Chapter 1 symmetric group S_n; the group of permutations of n objects, was introduced as an example of a finite group. An interesting feature of the symmetric group is that every finite group of order n is isomorphic to some subgroup of S_n. This group plays an important role in very many branches of physics and mathematics and has been investigated in great detail right from the early days of group theory. Here we confine ourselves to discussing some known facts and results concerning S_n mainly to illustrate the concepts introduced in Chapter 1. The proofs are quite intricate, and involve ingenious algebraic arguments which could occupy a whole volume. Even to see these results described can be illuminating.

2.1 Cycle Structure Notation

In Chapter 1 we learnt two useful notations for an element $p \in S_n$:

$$p = \begin{pmatrix} 1 & 2 & \cdots & n \\ p(1) & p(2) & \cdots & p(n) \end{pmatrix}; \qquad (2.1)$$

$$p = (1\ p(1)\ p(p(1)) \cdots)(j\ p(j)\ p(p(j)) \cdots) \cdots (k\ p(k)\ p(p(k)) \cdots). \qquad (2.2)$$

While the first form is convenient for stating the composition rule for S_n, the rule for computing inverses and such, as in (1.4), the second, referred to as the cycle structure notation, proves to be extremely useful in delivering a great deal of information about the symmetric group as we shall see. To go from the first form to the second, start from 1 and list the sequence of numbers it visits under repeated applications of p until one is back to the starting point, namely 1. This yields the cycle of 1. Next pick a number j from $1, 2, \cdots, n$ which has not already appeared in the first cycle and construct its cycle. Then pick a number which has not appeared in the first two cycles and so on. Repeat this process until all numbers are exhausted. Thus, for instance, the element

$$p = \begin{pmatrix} 1\ 2\ 3\ 4\ 5\ 6 \\ 6\ 4\ 3\ 5\ 2\ 1 \end{pmatrix} \tag{2.3}$$

of S_6 reads, in the second form, as

$$(16)(245)(3). \tag{2.4}$$

Conversely one can go from the second form to the first by viewing $(ijk \cdots l)$ as a set of replacements: $i \to j,\ j \to k, \cdots, l \to i$. Thus

$$(13)(64)(25) \in S_6 \tag{2.5}$$

reads in the first form as

$$p = \begin{pmatrix} 1\ 2\ 3\ 4\ 5\ 6 \\ 3\ 5\ 1\ 6\ 2\ 4 \end{pmatrix}. \tag{2.6}$$

Note that, in the cycle notation, by construction,

1. every number $1, 2, \cdots, n$ appears in some cycle and no two cycles have a common entry
2. the cycles may be arranged in any order, and
3. the entries in any cycle may be replaced by a cyclic permutation thereof.

Thus, $(16)(245)(3)$ could be variously written as $(245)(3)(16)$ or as $(452)(16)(3)$, etc. They all contain the same set of instructions and hence represent the same element of S_6.

Every cycle $(ijk \cdots l)$ is characterised by its length – the number of entries that appear in it. A cycle of length k is also referred to as a k-cycle.

Note also that the cycle $(ijk \cdots l)$ may naturally be identified with an element of the symmetric group which cyclically permutes the entries $i, j, k, \cdots l$ leaving the rest unchanged. A k-cycle under this correspondence goes to an element of S_n of order k.

Thus, in the example that we are considering,

$$(16) \rightarrow \begin{pmatrix} 1\ 2\ 3\ 4\ 5\ 6 \\ 6\ 2\ 3\ 4\ 5\ 1 \end{pmatrix}, (245) \rightarrow \begin{pmatrix} 1\ 2\ 3\ 4\ 5\ 6 \\ 1\ 4\ 3\ 5\ 2\ 6 \end{pmatrix},$$

$$(3) \rightarrow \begin{pmatrix} 1\ 2\ 3\ 4\ 5\ 6 \\ 1\ 2\ 3\ 4\ 5\ 6 \end{pmatrix} \qquad (2.7)$$

and one can easily verify that composing them as elements of S_n using (1.3) one recovers the element in (2.3). One may thus view the cycle structure notation as a minimal decomposition of elements of S_n into a commuting set of elements of S_n each of which corresponds to a cyclic permutation of a specific order.

2.2 Signature of a Permutation: Alternating Subgroup

We have so far learnt how to decompose an element of S_n into cycles. Every element of S_n can therefore be characterised by a sequence of nonnegative integers, $(m(1), m(2), \cdots)$ which displays its cycle structure, i.e., it gives the number $m(1)$ of 1-cycles, the number $m(2)$ of 2-cycles, etc., present in it. Of course we must have $\sum_i im(i) = n$. The simplest nontrivial cycles are the 2-cycles. As elements of S_n 2-cycles correspond to transpositions and square to the unit element. One may wonder if the higher cycles can be broken up into these simpler units, the 2-cycles. This is indeed true. Every k-cycle $(123\cdots k)$ can be expressed in terms of 2-cycles as

$$(123\cdots k) = (1k)\cdots(13)(12) \qquad (2.8)$$

involving $(k-1)$ 2-cycles. This decomposition though not unique is minimal in the sense that we can not reduce the number of 2-cycles appearing on the right hand side to less than $k-1$. (We may, of course, arbitrarily increase the number of transpositions appearing on the right by inserting squares of transpositions.) This fact permits us to assign a signature equal to $(-1)^{k-1}$ to each k-cycle, and hence to any element $p \in S_n$ with cycle structure $(m(1), m(2), \cdots)$ through:

$$\text{sign}(p) = (-1)^{\sum_{i=1}^{n} (i-1)m(i)} = (-1)^{n-\sum_{i=1}^{n} m(i)}, \qquad (2.9)$$

from which it is easy to see that an element $p \in S_n$ has the signature $+1$ (-1) if it contains an even (odd) number of even cycles. Further, we may divide the elements $p \in S_n$ into two categories, even or odd, according to their signatures being $+1$ or -1. Even (odd) elements of S_n, when decomposed into transpositions contain an even (odd) number of factors. The signature of an element of $p \in S_n$ is intrinsic to p and is independent of the decomposition into transpositions used for calculating it. $(\text{sign}(p)$ can be read off also from the effect of p on the product $\prod_{j<k=2}^{n}(x_j - x_k)$ involving n independent variables $x_1, x_2, \cdots, x_n$.) Further, for any $p, q \in S_n$,

$$\text{sign}(p) = \text{sign}(p^{-1}); \quad \text{sign}(pq) = \text{sign}(p)\text{sign}(q), \tag{2.10}$$

which in turn implies the elements of S_n with even signature form a subgroup of S_n. This subgroup is referred to as the alternating subgroup and is denoted by A_n. Further, since for any $q \in S_n$, $qpq^{-1} \in A_n$ whenever $p \in A_n$, A_n is an invariant subgroup of S_n. It contains $n!/2$ elements, i.e., half as many as in S_n. The corresponding factor group S_n/A_n is an abelian group of order 2. In fact, A_n coincides with the commutator subgroup of S_n as was noted earlier. What about the commutator subgroup of A_n? A_n certainly inherits the 3-cycles of S_n. It can be shown that for $n \geq 5$, all 3-cycles of A_n are also present in its commutator subgroup and the trend continues; the commutator subgroup chain never ends up consisting of the identity element alone. This circumstance is expressed by saying that for $n \geq 5$, S_n is not solvable.

Having learnt that every $p \in S_n$ can be expressed in terms of transpositions, one may wonder if one could do with a subset thereof. This is indeed so and such a special subset consists of the $n - 1$ adjacent transpositions σ_i:

$$\sigma_i = (i, i+1), \quad i = 1, 2, \cdots, n-1. \tag{2.11}$$

These obey the following algebraic relations:

$$\sigma_i^2 = 1, \quad i = 1, 2, \cdots, n-1, \tag{2.12}$$

$$\sigma_i\sigma_{i+1}\sigma_i = \sigma_{i+1}\sigma_i\sigma_{i+1}, \quad i = 1, 2, \cdots, n-2, \tag{2.13}$$

$$\sigma_i\sigma_j = \sigma_j\sigma_i, \quad \text{for } |i - j| \geq 2. \tag{2.14}$$

and furnish a *presentation* of S_n, i.e., the set of all *words* constructed out of the σ_i's when *reduced* using the algebraic relations above yields precisely the group S_n. Modifying (or 'deforming') the first of these relations while retaining the latter two, referred to as braid relations, leads to interesting mathematical structures which of late have found many useful applications in several areas in physics.

2.3 Conjugacy Classes

Consider an element $p \in S_n$ written in the cycle structure notation as:

$$p = (1\ p(1)\ p(p(1)) \cdots)\ (j\ p(j)\ p(p(j)) \cdots) \cdots (k\ p(k)\ p(p(k)) \cdots). \tag{2.15}$$

Viewing the cycles that appear in this decomposition as elements of S_n, we have for its conjugate qpq^{-1} by q

$$qpq^{-1} = q(1 \ p(1) \ p(p(1)) \cdots) \cdots (k \ p(k) \ p(p(k)) \cdots)q^{-1}$$
$$= q(1 \ p(1) \ p(p(1)) \cdots)q^{-1} \cdots q(k \ p(k) \ p(p(k)) \cdots)q^{-1}. \quad (2.16)$$

which expresses qpq^{-1} in terms of the conjugates of each of the cycles by q. Since each k-cycle corresponds to an element of order k of S_n and since the order of a group element does not change under conjugation we can immediately conclude that all the elements in the conjugacy class of $p \in S_n$ must have the same cycle structure as p. Conversely if two elements p and r of S_n have the same cycle structure then one can always construct an element q of S_n such that $qpq^{-1} = r$ and hence belong to the same conjugacy class. The construction is rather simple: lay out the cycles in p and r as

$$p = (a)(b)(cd)(ef)\,(ghi) \cdots$$
$$r = (\alpha)(\beta)(\gamma\delta)(\mu\nu)(\lambda\sigma\rho) \cdots \quad (2.17)$$

by putting together 1-cycles first followed by 2-cycles and so on. Then the element q such that $qpq^{-1} = r$ written in the first notation is

$$q = \begin{pmatrix} a & b & c & d & e & f & g & h & i \cdots \\ \alpha & \beta & \gamma & \delta & \mu & \nu & \lambda & \sigma & \rho \cdots \end{pmatrix}. \quad (2.18)$$

This can easily be verified using the composition law for S_n in (1.3).

From the considerations above we have the following result:

Conjugacy classes of S_n are completely characterised by a sequence of nonnegative integers, $(m(1), m(2), \cdots)$ with $\sum_i im(i) = n$ and are conventionally labelled as $1^{m(1)}2^{m(2)}3^{m(3)} \cdots$.

This result immediately enables us to conclude that

- The number of elements in the conjugacy class $1^{m(1)}2^{m(2)}3^{m(3)} \cdots$ is given by

$$\frac{n!}{\prod_{i=1}^{n} i^{m(i)}m(i)!}. \quad (2.19)$$

This can be easily seen by laying out a typical element of this class as follows:

$$\underbrace{(\cdot)(\cdot)\cdots(\cdot)}_{m(1)} \quad \underbrace{(\cdot\cdot)(\cdot\cdot)\cdots(\cdot\cdot)}_{m(2)} \quad \underbrace{(\cdot\cdot\cdot)(\cdot\cdot\cdot)\cdots(\cdot\cdot\cdot)}_{m(3)} \quad \cdots \cdots \quad . \quad (2.20)$$

All elements in this class are obtained by distributing $1, 2, \ldots, n$ among the various parentheses in all possible ways. This number is clearly $n!$. However, of these, those which are related by permutation of the $m(i)$ parentheses in the ith group for each i, or by cyclic permutations of the entries in each of the parentheses, correspond to the same element. This accounts for the factor in the denominator.

- The number of conjugacy classes of S_n and hence the number of inequivalent irreducible representations of S_n equals the number of partitions of n. To see this write $\sum_i im(i)$ as

$$\underbrace{1 + 1 + \cdots + 1}_{m(1)} + \underbrace{2 + 2 + \cdots + 2}_{m(2)} + \underbrace{3 + 3 + \cdots + 3}_{m(3)} + \quad \cdots \cdots \tag{2.21}$$

and rewrite it in the reverse order. The result is a partition of n, i.e., a set of positive integers $\lambda \equiv (\lambda_1, \lambda_2, \cdots)$, $\lambda_1 \geq \lambda_2 \geq \lambda_3 \cdots \geq 0$ adding up to n. In symbols a partition λ of n is expressed as $\lambda \vdash n$. Hence we conclude that the classes of S_n are in one to one correspondence with the partitions of n. Their number therefore equals the number of partitions of n. (There is no closed form expression for this number. One only has an asymptotic formula for large n due to Ramanujan.) One may label the classes and hence the irreducible representations of S_n either by λ, a partition of n or, as before, by $1^{m(1)} 2^{m(2)} 3^{m(3)} \cdots$ where $m(i)$ denotes the multiplicity of occurrence of i in the partition λ. Conventionally, one uses λ to label the irreducible representations of S_n and the cycle structure notation for the classes.

Note that $\sum_i m(i)$, gives the 'length' $\ell(\lambda)$ of the partition λ, i.e., the number of non zero parts λ_i.

- All elements in the same class, have the same signature. This is evident from (2.9). Further, the signature of the elements in the conjugacy class λ of S_n is $+1$ or -1 depending on whether $n - \ell(\lambda)$ is even or odd. It therefore follows that the conjugacy classes of A_n, the alternating subgroup, are labelled by those partitions λ of n for which $n - \ell(\lambda)$ is even. Equivalently, only those classes of S_n survive the transition to A_n which have an even number of even cycles.

2.4 Young Frames and Young Tableaux

We have seen that the conjugacy classes are in one to one correspondence with the partitions of n. It proves convenient and also very useful to represent partitions $\lambda = (\lambda_1, \lambda_2, \cdots, \lambda_\ell)$ through what are called *Young frames* : ℓ left adjusted rows with λ_i

boxes in the row i. Thus, for the partition $(3, 2)$ of 5 we have

$$\lambda = (3, 2) \rightarrow \quad \boxed{} \quad . \tag{2.22}$$

The Young frame associated with λ is also referred to as the *shape* of λ. The boxes in a Young frame can be assigned coordinates in the usual way: a box lying in row i and column j has coordinates (i, j).

By taking the transpose of the Young frame for the partition λ of n, i.e., by arranging the boxes in the ith column as boxes of the ith row, one obtains the Young frame corresponding to a partition λ' of n. The partitions λ and λ' are said to be *conjugates* of each other, i.e., they are such that their Young frames are transposes of each other. Thus, for the partition $(3, 2)$ of 5 we have

$$\lambda = (3, 2) \rightarrow \quad \xrightarrow{\text{Transpose}} \quad \rightarrow \lambda' = (2, 2, 1). \tag{2.23}$$

Of course, it may happen that $\lambda = \lambda'$ and such a λ is said to be self conjugate.

To each box (i, j) in a Young frame one can associate an integer $h(i, j)$, the *hook length*, equal to the number of boxes lying on the right-angled path with the box (i, j) as vertex as shown

$$\quad , \quad h(2, 2) = 5. \tag{2.24}$$

A Young frame with boxes filled with integers in a prescribed manner is referred to as a *Young tableau*. Three types of Young tableaux are of particular interest:

1. A Young tableau corresponding to a partition λ of n, or simply a λ-tableau, denoted by t^λ is obtained by distributing $1, 2, \cdots, n$ in the n boxes in the associated frame. Clearly such tableaux are $n!$ in number. Thus, for the partition $\lambda = (2, 1)$ of 3, the six λ-tableaux are

$$\boxed{\begin{array}{cc}1&2\\3\end{array}}, \boxed{\begin{array}{cc}2&1\\3\end{array}}, \boxed{\begin{array}{cc}1&3\\2\end{array}}, \boxed{\begin{array}{cc}3&1\\2\end{array}}, \boxed{\begin{array}{cc}2&3\\1\end{array}}, \boxed{\begin{array}{cc}3&2\\1\end{array}}. \tag{2.25}$$

A useful derived concept here is that of a *tabloid* $\mathbf{t}^\lambda$, equivalence classes of t^λ which have the same set of entries in the corresponding rows. For $\lambda = (2, 1)$, the three tabloids, indicated by a thicker frame, are

$$
\boxed{\begin{array}{cc} 1 & 2 \\ 3 \end{array}} = \left\{ \begin{array}{cc} 1 & 2 \\ 3 \end{array}, \begin{array}{cc} 2 & 1 \\ 3 \end{array} \right\},
$$

$$
\boxed{\begin{array}{cc} 1 & 3 \\ 2 \end{array}} = \left\{ \begin{array}{cc} 1 & 3 \\ 2 \end{array}, \begin{array}{cc} 3 & 1 \\ 2 \end{array} \right\}, \tag{2.26}
$$

$$
\boxed{\begin{array}{cc} 2 & 3 \\ 1 \end{array}} = \left\{ \begin{array}{cc} 2 & 3 \\ 1 \end{array}, \begin{array}{cc} 3 & 2 \\ 1 \end{array} \right\}.
$$

The number of $\mathbf{t}^\lambda$ is clearly

$$
\frac{n!}{\lambda_1! \lambda_2! \cdots}. \tag{2.27}
$$

2. A Young tableau corresponding to a partition λ of n is said to be *standard* if the boxes in the associated frame are assigned numbers $1, 2, \cdots n$ such that the numbers are increasing from left to right in rows and also increasing from top to bottom in columns. We use the symbol T^λ to denote such tableaux and $\mathbf{T}^\lambda$ for the corresponding tabloids. (Note that the tableaux contained in a tabloid corresponding to a chosen standard tableau will contain nonstandard tableaux as well.) Thus, for the partition $(2, 1, 1)$ of 4 one has three standard Young tableaux $T^{(2,1,1.)}$

$$
\begin{array}{cc} 1 & 2 \\ 3 \\ 4 \end{array}, \quad \begin{array}{cc} 1 & 3 \\ 2 \\ 4 \end{array}, \quad \begin{array}{cc} 1 & 4 \\ 2 \\ 3 \end{array}. \tag{2.28}
$$

For a given partition $\lambda = (\lambda_1, \lambda_2, \cdots, \lambda_\ell)$ of n the number f^λ of standard tableaux is given by

$$
f^\lambda = n! \, \det[1/\Gamma(\lambda_i - i + j + 1)]; \;\; 1 \leq i, j \leq \ell. \tag{2.29}
$$

and involves calculating the determinant of an $\ell \times \ell$ matrix $1/\Gamma(\lambda_i - i + j + 1)$, $i, j. = 1, 2, \cdots, \ell$. A computationally convenient way of finding the number of standard tableaux of shape λ is the celebrated Frame–Robinson–Thrall hook-length formula:

$$
f^\lambda = \frac{n!}{\prod_{(i,j) \in \lambda} h(i,j)}. \tag{2.30}
$$

For example, the number of standard tableaux of shape $\lambda = (2,2,1)$:

$$\lambda = (2,2,1) \rightarrow \quad\quad \tag{2.31}$$

is easily found to be $5!/4 \cdot 2 \cdot 3 \cdot 1 \cdot 1 = 5$.

3. A Young tableau corresponding to a partition λ of n is said to be *semistandard* if the boxes in the associated frame are assigned numbers $1, 2, \cdots d$ such that the numbers are non-decreasing from left to right in rows and increasing from top to bottom in columns. Clearly, the set of semistandard Young tableaux corresponding to a given partition λ of n will be nonempty provided $d \geq \ell(\lambda)$. We denote such tableaux by $\mathcal{T}_d^\lambda$. Thus, for the partition $(2,1)$ of 3 and $d = 3$ one has eight semistandard Young tableaux.

$$\begin{array}{cc} 1&1\\2 \end{array}, \begin{array}{cc} 1&1\\3 \end{array}, \begin{array}{cc} 1&2\\2 \end{array}, \begin{array}{cc} 1&2\\3 \end{array}, \begin{array}{cc} 1&3\\2 \end{array}, \begin{array}{cc} 1&3\\3 \end{array}, \begin{array}{cc} 2&2\\3 \end{array}, \begin{array}{cc} 2&3\\3 \end{array}. \tag{2.32}$$

The number f_d^λ of semistandard tableau $\mathcal{T}_d^\lambda$ corresponding to the partition λ of n is given by

$$f_d^\lambda = \frac{\prod_{(i,j)\in\lambda}(d+i-j)}{\prod_{(i,j)\in\lambda}h(i,j)}. \tag{2.33}$$

With each semistandard Young tableau $\mathcal{T}_d^\lambda$ one may associate a unique monomial, symbolically denoted by $x^{\mathcal{T}_d^\lambda}$, by replacing each integer i by a variable x_i. Thus the monomial associated with the semistandard Young tableau

$$\begin{array}{ccc} 1&1&2\\3&3&5 \end{array} \tag{2.34}$$

corresponding to the partition $\lambda = (3,3)$ of 6 reads $x_1^2 x_2 x_3^2 x_5$.

This association in turn permits one to assign, to each partition λ of n, certain symmetric homogeneous polynomials in $x_1 \cdots, x_d$ of degree n, the Schur functions $s_\lambda(x)$, $x \equiv (x_1, \cdots, x_d)$, as follows:

$$s_\lambda(x_1, \cdots, x_d) = \sum_{\mathcal{T}_d^\lambda} x^{\mathcal{T}_d^\lambda} \tag{2.35}$$

where the sum is over all the semistandard Young tableaux of shape λ. For example, for $\lambda = (2, 1)$ and $d = 3$, referring to (2.32), we have

$$s_{(2,1)}(x) = x_1^2 x_2 + x_1^2 x_3 + x_1 x_2^2 + x_1 x_2 x_3 + x_1 x_2 x_3 + x_1 x_3^2 + x_2^2 x_3 + x_2 x_3^2$$

$$= x_1^2 x_2 + x_1 x_2^2 + x_2^2 x_3 + x_2 x_3^2 + x_3^2 x_1 + x_3 x_1^2 + 2 x_1 x_2 x_3. \qquad (2.36)$$

Historically, Schur functions were originally defined as the ratio of two determinants:

$$s_\lambda(x) = \frac{\det[x_i^{\lambda_j + n - j}]}{\det[x_i^{n-j}]}, \ 1 \leq i, j \leq d. \qquad (2.37)$$

Their association with semistandard Young tableaux came later.

2.5 Young Subgroups of S_n

The group S_n is the set of permutations of n objects, say, $\{1, 2, \cdots, n\}$. If we divide this set into two as $\{1, 2, \cdots, \lambda_1\}\{\lambda_1 + 1, \cdots, n\}$ then clearly the set of elements of S_n which only permute $1, \cdots \lambda_1$ form a subgroup of S_n and so do those which permute only $\lambda_1 + 1, \cdots, n$. These two subgroups, suggestively denoted by $S_{\{1,2,\cdots,\lambda_1\}}$ and $S_{\{\lambda_1+1,\cdots,n\}}$, obviously commute with each other, as they operate on disjoint subsets of $1, 2, \cdots, n$ and clearly have no elements in common except the identity. It therefore follows that their direct product is also a subgroup of S_n. This idea can be extended to any partition λ of n leading to the notion of Young subgroups S_λ of S_n.

To each partition $\lambda = (\lambda_1, \lambda_2, \cdots, \lambda_\ell)$ of n we can associate a subgroup S_λ of S_n as follows:

$$S_\lambda = S_{\{1,2,\cdots,\lambda_1\}} \times S_{\{\lambda_1+1,\lambda_1+2,\cdots,\lambda_1+\lambda_2\}} \times \cdots. \qquad (2.38)$$

The order of the Young subgroup of S_n corresponding to $\lambda = (\lambda_1, \lambda_2, \cdots, \lambda_\ell)$, being a direct product of symmetric groups, $S_{\lambda_1}, \cdots, S_{\lambda_\ell}$ is clearly

$$\prod_{\alpha=1}^{\ell} \lambda_\alpha! \qquad (2.39)$$

and for the same reason, its classes can be specified by ℓ sequences of nonnegative integers

$$(m_\alpha(1), m_\alpha(2), \cdots), \sum_{i=1}^{\lambda_\alpha} i m_\alpha(i) = \lambda_\alpha, \ \alpha = 1, \cdots, \ell. \qquad (2.40)$$

one for each symmetric group that appears in the direct product. The number of elements in such a class is then

$$\prod_{\alpha=1}^{\ell} \left(\frac{\lambda_\alpha!}{1^{m_\alpha(1)} m_\alpha(1)! 2^{m_\alpha(2)} m_\alpha(2)! \cdots} \right). \tag{2.41}$$

Further, the number of elements of the Young subgroup S_λ of S_n present in the class $(m(1), m(2), \cdots)$ of S_n, is obtained by summing the expression above over all $m_\alpha(i)$ satisfying

$$\sum_{i=1}^{\lambda_\alpha} i m_\alpha(i) = \lambda_\alpha; \quad \alpha = 1, \cdots, \ell, \tag{2.42}$$

$$\sum_{\alpha=1}^{\ell} m_\alpha(i) = m(i), \ i = 1, \cdots, n. \tag{2.43}$$

The Young subgroup of S_n corresponding to a partition λ of n can be used to decompose S_n into

$$\frac{n!}{\lambda_1! \lambda_2! \cdots} \tag{2.44}$$

cosets which in turn can be used to obtain certain interesting reducible representations of S_n. We will return to this topic later in this chapter.

2.6 Young Symmetrisers

Consider a tableau t^λ corresponding to a partition λ of n. The group S_n has a natural action on t^λ to yield another tableau t'^λ obtained by replacing each entry in t^λ by $p(i)$. Thus, for example,

$$p = (123) \in S_3 : p \begin{array}{|c|c|} \hline 1 & 3 \\ \hline 2 \\ \cline{1-1} \end{array} = \begin{array}{|c|c|} \hline 2 & 1 \\ \hline 3 \\ \cline{1-1} \end{array}. \tag{2.45}$$

Corresponding to each tableau t^λ, two subgroups of S_n may naturally be identified: the row subgroup and the column subgroup.

$$R_{t^\lambda} = S_{r_1} \times S_{r_2} \times \cdots, \quad C_{t^\lambda} = S_{c_1} \times S_{c_2} \times \cdots \tag{2.46}$$

where S_{r_i} (S_{c_i}) denote the symmetric group associated with the entries in the ith row (column) of t^λ. Thus the row subgroup R_{t^λ} consists of elements of S_n which preserve each row of t^λ. Likewise the column subgroup C_{t^λ} consists of elements of S_n which preserve each column of t^λ.

Note that R_{t^λ} and C_{t^λ} have no elements in common except the identity element of S_n. As a result each element in $\Upsilon_{t^\lambda} \equiv C_{t^\lambda} R_{t^\lambda}$ can be expressed uniquely as qp, $q \in C_{t^\lambda}, p \in R_{t^\lambda}$.

Corresponding to each of the subgroups we define elements in the group algebra of S_n

$$r_{t^\lambda} = \sum_{p \in R_{t^\lambda}} p; \quad c_{t^\lambda} = \sum_{q \in C_{t^\lambda}} \text{sign}(q) q. \tag{2.47}$$

The Young symmetriser y_{t^λ} associated with the tableau t^λ is then defined as follows:

$$y_{t^\lambda} = c_{t^\lambda} r_{t^\lambda} = \sum_{q \in C_{t^\lambda},\, p \in R_{t^\lambda}} \text{sign}(q)\, qp. \tag{2.48}$$

Two properties of y_{t^λ} follow immediately from its definition:

$$q y_{t^\lambda} = \text{sign}(q)\, y_{t^\lambda}, \quad q \in C_{t^\lambda} \tag{2.49}$$

$$y_{t^\lambda} p = \quad y_{t^\lambda} \qquad p \in R_{t^\lambda} \tag{2.50}$$

Consider now the set of all tableau corresponding to the same partition λ of n. If t_0^λ and t_1^λ are two tableau related to each other by the action of an element g in S_n,

$$t_1^\lambda = g t_0^\lambda \tag{2.51}$$

then it can be shown that

- The row and column subgroups associated with t_1^λ are related to those of t_0^λ as follows:

$$R_{t_1^\lambda} = g R_{t_0^\lambda} g^{-1}, C_{t_1^\lambda} = g C_{t_0^\lambda} g^{-1}. \tag{2.52}$$

As a result

$$r_{t_1^\lambda} = g r_{t_0^\lambda} g^{-1}, c_{t_1^\lambda} = g c_{t_0^\lambda} g^{-1} \tag{2.53}$$

and hence

$$y_{t_1^\lambda} = g y_{t_0^\lambda} g^{-1}. \tag{2.54}$$

- The element g relating the two tableau is not in $C_{t_0^\lambda} R_{t_0^\lambda}$ if and only if there are two elements in the same row of t_1^λ which are also in the same column of t_0^λ. Thus $g \notin C_{t_0^\lambda} R_{t_0^\lambda}$ if and only if there is a transposition $\sigma \in R_{t^\lambda}$ and a transposition $\tau \in C_{t^\lambda}$ such that

$$\sigma g \tau = g \tag{2.55}$$

holds.

This in turn leads to the following useful statement: For a given tableau t^λ, for any element g in S_n not contained in $C_{t^\lambda} R_{t^\lambda}$ there is a transposition $\sigma \in R_{t^\lambda}$ and a transposition $\tau \in C_{t^\lambda}$ such that $\sigma g \tau = g$.

2.6.1 Primitive idempotence of y_{t^λ}

To establish the result that y_{t^λ} is an (essentially) idempotent as well as a primitive element of the algebra $\mathbb{C}[S_n]$ one makes use of the following result:

If uxu is a scalar multiple of u for every $x \in \mathbb{C}[S_n]$ then u is an essential idempotent of the algebra $\mathbb{C}[S_n]$.

Clearly since every $x \in \mathbb{C}[S_n]$ is a complex linear combination of elements in S_n, it suffices to show that

$$z \equiv y_{t_\lambda} g \, y_{t^\lambda} = c y_{t^\lambda}. \tag{2.56}$$

z may be expressed as a linear combination of the elements g of S_n thus:

$$z = \sum_g z(g) \, g \tag{2.57}$$

and from the way z is defined it is evident that the coefficients $z(g)$ appearing therein must be integers. Now

$$z = \sum_g z(g) \, g \tag{2.58}$$

$$= \sum_{g \in C_{t^\lambda} R_{t^\lambda}} z(g) \, g + \sum_{g \notin C_{t^\lambda} R_{t^\lambda}} z(g) \, g \tag{2.59}$$

$$= \sum_{q \in C_{t^\lambda}, p \in R_{t^\lambda}} z(qp) \, qp + \sum_{g \notin C_{t^\lambda} R_{t^\lambda}} z(g) \, g. \tag{2.60}$$

From (2.49) and (2.50) it follows that

$$qzp = \text{sign}(q)\, z, \quad \text{i.e.,} \tag{2.61}$$

$$q^{-1}zp^{-1} = \text{sign}(q)\, z. \tag{2.62}$$

Hence

$$z(qp) = \text{Coeff of 1 in } q^{-1}zp^{-1} \tag{2.63}$$

$$= \text{sign}(q)\, z(1) \tag{2.64}$$

which on using (2.60) yields

$$z = z(1) y_{t\lambda} + \sum_{g \notin C_{t\lambda} R_{t\lambda}} z(g)\, g. \tag{2.65}$$

Using the fact stated above that for any element in S_n not contained in $C_{t\lambda} R_{t\lambda}$ there is a transposition $\sigma \in R_{t\lambda}$ and a transposition $\tau \in C_{t\lambda}$ such that $\sigma g \tau = g$ we have from (2.61) and (2.65)

$$\tau z \sigma = -z = -z(1) y_{t\lambda} - \sum_{g \notin C_{t\lambda} R_{t\lambda}} z(g)\, g, \tag{2.66}$$

$$\tau z \sigma = -z(1) y_{t\lambda} + \sum_{g \notin C_{t\lambda} R_{t\lambda}} z(g)\, g, \tag{2.67}$$

which imply that

$$\sum_{g \notin C_{t\lambda} R_{t\lambda}} z(g)\, g = 0 \tag{2.68}$$

and hence

$$z = z(1) y_{t\lambda}, \quad \text{i.e.,} \tag{2.69}$$

$$y_{t\lambda}\, g\, y_{t\lambda} = z(1) y_{t\lambda}, \quad z(1) = \text{integer}. \tag{2.70}$$

It can further be shown that $z(1) \neq 0$.

Putting $g = e$ in the above equation, we obtain $y_{t\lambda}^2 = z(1) y_{t\lambda}$ showing that $y_{t\lambda}$ is essentially idempotent. Further, as $y_{t\lambda}\, x\, y_{t\lambda}, x \in \mathbb{C}[S_n]$ is a scalar multiple of $y_{t\lambda}$, it establishes the primitivity of $y_{t\lambda}$.

In the light of what we said in Section 1.15 of Chapter 1, the primitive idempotence of $y_{t\lambda}$ means that $\mathbb{C}[S_n y_{t\lambda}]$ is an irreducible (simple) submodule of $\mathbb{C}[S_n]$ and thus gives rise to an irreducible representation of S_n. Recall further that the construction of $y_{t\lambda}$

was based on an arbitrary but fixed Young tableau t^λ corresponding to the partition λ of n. However, owing to (2.54) it is clear that a different choice of the Young tableau would have given rise to an equivalent primitive idempotent and hence to an equivalent irreducible representation of S_n. We may therefore conclude that

- Given a partition λ of n and a tableau t^λ thereof, the associated Young symmetriser y_{t^λ} permits one to construct an irreducible representation S^λ of S_n. The choice of t^λ does not matter as they all give rise to essentially the same, i.e., equivalent irreducible representations.

2.6.2 Orthogonality properties of Young symmetrisers

In this subsection we list some of the well established orthogonality properties of Young symmetrisers:

- The integer $z(1)$ appearing in (2.70) can be shown to be related to n and the irreducible representation generated by y_{t^λ} as

$$z(1) = \frac{n!}{\dim(S^\lambda)}. \tag{2.71}$$

  Some times it proves convenient to 'normalise' y_{t^λ} by defining $e_{t^\lambda} = \frac{1}{z(1)} y_{t^\lambda}$ so that one has $e_{t^\lambda}^2 = e_{t^\lambda}$.

- $y_{t^\lambda} y_{t'^\mu} = 0$ if and only if $\lambda \neq \mu$ or when $\lambda = \mu$ some row of t^λ has two or more entries in common with some column of t'^λ.
  In view of this, two Young symmetrisers y_{t^λ}, $y_{t'^\mu}$ are orthogonal, i.e., $y_{t^\lambda} y_{t'^\mu} = y_{t'^\mu} y_{t^\lambda} = 0$ if and only if $\lambda \neq \mu$ or when $\lambda = \mu$ some row of t^λ has two or more entries in common with some column of t'^λ and conversely some row of t'^λ has two or more entries in common with some column of t^λ.

- One may naively expect the Young symmetrisers corresponding to different standard tableaux T^λ to be orthogonal. This is not so as was noted by Young himself. Thus for $\lambda = \mu = (3,2)$ and T^λ and T'^λ as shown

$$T^\lambda = \begin{array}{|c|c|c|} \hline 1 & 2 & 3 \\ \hline 4 & 5 \\ \cline{1-2} \end{array} \qquad T'^\lambda = \begin{array}{|c|c|c|} \hline 1 & 3 & 5 \\ \hline 2 & 4 \\ \cline{1-2} \end{array} \tag{2.72}$$

one can explicitly check that the corresponding Young symmetrisers are not orthogonal. While $y_{T^\lambda} y_{T'^\lambda} = 0$, $y_{T'^\lambda} y_{T^\lambda} \neq 0$.

2.7 Irreducible Representations of S_n

We learnt earlier that for each partition λ of n we can construct an irreducible representation S^λ of S_n through the associated Young symmetriser y_{t^λ}. We further learnt that the Young symmetrisers corresponding to distinct partitions λ and μ of n are orthogonal which in turn implies that the corresponding irreducible representations S^λ and S^μ are inequivalent. Thus for each λ one has a distinct irreducible representation S^λ. Since the number of S^λ equals the number of classes, this correspondence between partitions λ of n and the irreducible representations of S_n is exhaustive. Thus

- The number of irreducible representations S^λ of S_n is equal to the number of partitions λ of n

- As to the dimension of S^λ, one has the following result:

$$\dim(S^\lambda) = f^\lambda, \text{the number of standard Young tableaux of shape } \lambda \qquad (2.73)$$

and hence from Eqs. (2.29, 2.30)

$$\dim(S^\lambda) = \frac{n!}{\prod_{(i,j)\in\lambda} h(i,j)} \qquad (2.74)$$

$$= n! \det[1/\Gamma(\lambda_i - i + j + 1)]. \qquad (2.75)$$

Thus S_4 has 5 irreducible representations

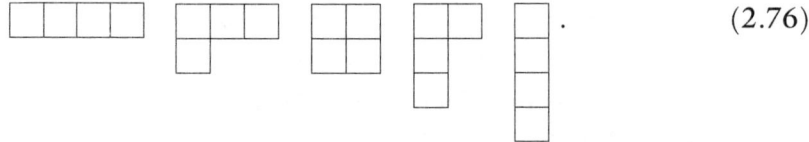

$$(2.76)$$

Their dimensions, obtained using, for instance, the hook-length formula are 1, 3, 2, 3, 1, respectively.

For any n, S_n always has two (and only two) one dimensional representations corresponding to the two extremes of the partitions of n: $\lambda = (n)$ and $\lambda = (1^n)$. The first corresponds to the trivial representation in which every element is represented by 1. The second is called the sign representation where every element is represented by its signature.

Turning now to the characters χ_μ^λ of the irreducible representations S^λ in the class μ, there are two methods for computing them, one graphical and the other algebraic. The first is formulated as a set of rules – Murnaghan–Nakayama rule – defined in terms of the Young frames corresponding to λ and μ. In the second method, the irreducible characters arise as coefficients of expansions relating two specific bases in the vector space of symmetric homogeneous polynomials of degree n. We will now discuss them both.

2.7.1 Murnaghan–Nakayama rule for irreducible characters of S_n

To state the Murnaghan–Nakayama rule in a compact manner we need to define the notion of a *rim hook*. A rim hook is a continuous strip of boxes along the boundary of a Young frame (only to the right or below) such that its removal leads to a legitimate Young frame. The number of boxes in the rim hook is called its length. Its vertical extent, i.e., the number of rows it encompasses, denoted by v permits us to associate a sign $(-1)^{v-1}$ with each rim hook. Thus the rim hook below, indicated by $\bullet$, has a length 6 and its sign is -1.

$$(2.77)$$

Armed with the definition of a rim hook and its sign, the Murnaghan–Nakayama rule for computing the character χ_μ^λ of the irreducible representation S^λ, $\lambda = (\lambda_1, \lambda_2, \cdots)$ in the class $\mu = (\mu_1, \mu_2, \cdots)$ consists in carrying out the following steps:

1. Draw the Young frame for λ and ascribe it a sign $+1$.

2. Remove from it a rim hook of length μ_1 in all possible ways.

3. If there is no way this can be done, then $\chi_\mu^\lambda = 0$.

4. If this is not the case, one would be left with one or more 'smaller' Young frames. Update their signs by multiplying the initial sign, namely $+1$, by the sign of the rim hook removed.

5. Next remove a rim hook of length μ_2 in all possible ways from each of the Young frame or frames in the previous step and update their signs as before.

6. Keep doing it until all the parts of μ are exhausted. χ_μ^λ is then given by the sum of the signs at the final stage.

44   Continuous Groups for Physicists

As an illustration of this rule, we compute below $\chi_{2,1^3}^{2,1^3}$. (Here, to keep track of the rim hooks being removed at each stage we have assigned the symbols $(\bullet, \circ, \times, *)$ to the four parts $(2,1,1,1)$ of μ, respectively.)

- **0th stage**

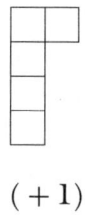

$$(+1)$$

- **1st stage**

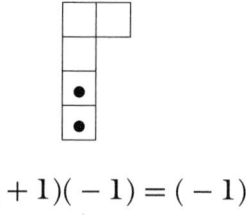

$$(+1)(-1) = (-1)$$

- **2nd stage**

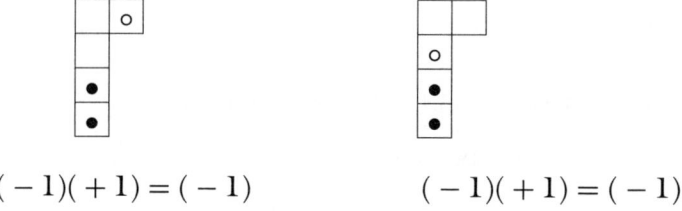

$$(-1)(+1) = (-1) \qquad (-1)(+1) = (-1)$$

- **3rd stage**

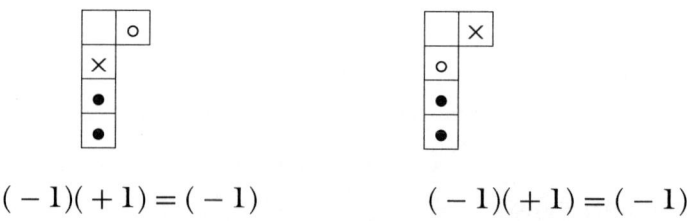

$$(-1)(+1) = (-1) \qquad (-1)(+1) = (-1)$$

- **4th stage**

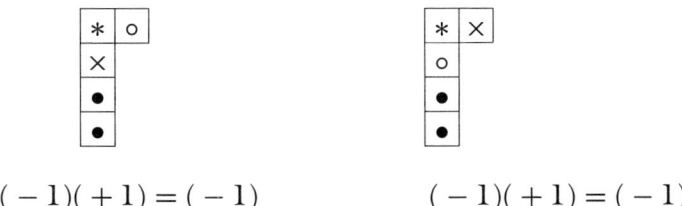

$$(-1)(+1) = (-1) \qquad\qquad (-1)(+1) = (-1)$$

- **Final stage**

$$\chi^{2,1^3}_{2,1^3} = -1 - 1 = -2$$

Using this rule one can explicitly construct the character table for, say, S_5:

	1	10	15	20	20	30	24
	(1^4)	$(2,1^3)$	$(2^2,1)$	$(3,1^2)$	$(3,2)$	$(4,1)$	(5)
$\chi^{(5)}$	1	1	1	1	1	1	1
$\chi^{(4,1)}$	4	2	0	1	-1	0	-1
$\chi^{(3,2)}$	5	1	1	-1	1	-1	0
$\chi^{(3,1^2)}$	6	0	-2	0	0	0	1
$\chi^{(2^2,1)}$	5	-1	1	-1	-1	1	0
$\chi^{(2,1^3)}$	4	-2	0	1	1	0	-1
$\chi^{(1^4)}$	1	-1	1	1	-1	-1	1

2.7.2 Symmetric functions and the irreducible characters of S_n

Earlier in this chapter we learnt how Schur functions $s_\lambda(x_1, \cdots, x_d)$ arise from the set of semistandard Young tableaux of shape λ. It turns out that the set $\{s_\lambda(x)\}$, where $x \equiv (x_1, \cdots, x_d)$ and λ's are the partitions of n, provide a basis for the vector space of all symmetric homogeneous polynomials of degree n in d variables $x_1 \cdots, x_d$. Four other well studied 'classical' choices for a basis in this vector space are (see, for instance, Macdonald, (1995)):

- monomial symmetric functions $m_\lambda(x)$

$$m_\lambda(x) = \sum_{\substack{\text{distinct permutations} \\ \text{of } \lambda_1, \cdots, \lambda_d}} x_1^{\lambda_1} \cdots x_d^{\lambda_d}, \qquad (2.78)$$

- symmetric functions $e_\lambda(x)$ defined through elementary symmetric functions $e_r(x)$:

$$e_\lambda(x) = e_{\lambda_1}(x) \cdot e_{\lambda_2}(x) \cdots e_{\lambda_d}(x), \tag{2.79}$$

$$e_r(x) = m_{1^r}(x) = \sum_{i_1 < i_2 < \cdots < i_r} x_{i_1} x_{i_2} \cdots x_{i_r}; \quad 1 \le i_1, \cdots, i_r \le d, \tag{2.80}$$

- symmetric functions $h_\lambda(x)$ defined through completely symmetric functions $h_r(x)$:

$$h_\lambda(x) = h_{\lambda_1}(x) \cdot h_{\lambda_2}(x) \cdots h_{\lambda_d}(x), \tag{2.81}$$

$$h_r(x) = \sum_{\substack{\text{all partitions} \\ \lambda \text{ of } r}} m_\lambda(x)$$

$$= \sum_{i_1 \le i_2 \le \cdots \le i_r} x_{i_1} x_{i_2} \cdots x_{i_r}; \quad 1 \le i_1, \cdots, i_r \le d, \tag{2.82}$$

- symmetric functions $p_\lambda(x)$ based on power symmetric functions $p_r(x)$:

$$p_\lambda(x) = p_{\lambda_1}(x) \cdot p_{\lambda_2}(x) \cdots p_{\lambda_d}(x), \tag{2.83}$$

$$p_r(x) = m_r(x) = \sum_{i=1}^{d} x_i^r. \tag{2.84}$$

Since each of them provides a basis for the same vector space, each one of them can be expressed as a linear combination of any other. A particularly interesting relation is that between $p_\lambda(x)$ and $s_\lambda(x)$:

$$p_\mu(x) = \sum_\lambda \chi_\mu^\lambda s_\lambda(x) \tag{2.85}$$

where the coefficient $\chi_\mu^\lambda(x)$ is just the character of the irreducible representation S^λ in the class μ. The inverse of the above relation reads

$$s_\lambda(x) = \sum_\mu z_\mu^{-1} \chi_\mu^\lambda p_\mu(x); \quad z_\mu = \prod_{i=1} i^{m(i)} m(i)! \tag{2.86}$$

where $m(i)$ denote the multiplicity of occurrence of i in the partition μ.

Another interesting relation is that between the Schur functions and the monomial symmetric functions:

$$s_\lambda(x) = \sum_\mu K_{\lambda\mu} m_\mu(x). \tag{2.87}$$

where $K_{\lambda\mu}$ are the elements of an upper triangular matrix K called the Kostka matrix. The Kostka numbers $K_{\lambda\mu}$ can be computed using the following algorithm:

- Draw the Young frame corresponding to λ.

- Count the semistandard Young tableaux that arise by filling this frame with μ_1 1's, μ_2 2's and so on.

Thus, for example, $K_{(3^2)(2^2,1^2)}$ equals 2:

$$
\begin{array}{|c|c|c|}
\hline 1 & 1 & 2 \\
\hline 2 & 3 & 4 \\
\hline
\end{array},
\qquad
\begin{array}{|c|c|c|}
\hline 1 & 1 & 3 \\
\hline 2 & 2 & 4 \\
\hline
\end{array}.
\tag{2.88}
$$

For the special cases, when $\mu = (1^n)$, semistandard filling reduces to standard filling and hence

$$
K_{\lambda(1^n)} = \dim S^\lambda .
\tag{2.89}
$$

2.8 Some Useful Explicit Constructions of Representations of S_n

As was discussed in Section 1.6, given a set X on which G acts we can construct a representation of G by its action on the vector space $\mathbb{C}[X]$ consisting of formal complex linear combinations of elements of X. Here, by making use of the natural action of S_n on Young tableaux, we list some of the useful representations of S_n that can be constructed by taking the set X as consisting of Young tableaux themselves or of related structures such as tabloids and paratabloids. For each such choice of X we also display the irreducible representation content of the reducible representations of S_n thus obtained.

2.8.1 $X =$ set of all Young tableaux t^λ

In this case, since the Young tableaux t^λ for any partition λ are in one-to-one correspondence with the elements of S_n, the representation that one obtains is just the regular representation containing every irreducible representation as many times as its dimension.

2.8.2 $X =$ set of all tabloids $\mathfrak{t}^\lambda$ associated with t^λ

Recall that for a fixed partition λ of n, a tabloid $\mathbf{t}^\lambda$ corresponding to a Young tableau t^λ consists of equivalence classes of $\{t^\lambda\}$ which have the same set of entries in the corresponding rows. They can thus be easily seen to be in one-to-one correspondence with the cosets of S_n with respect to the Young subgroup. The representation M^λ thus obtained is clearly of dimension $n!/(\lambda_1!\lambda_2!\cdots)$ and, unlike the previous case, depends on the choice of λ. The irreducible representation content of M^λ turns out to be as given below:

$$M^\lambda = S^\lambda \bigoplus_{\mu > \lambda} K_{\mu\lambda} S^\mu. \tag{2.90}$$

Stated in words, it contains the irreducible representation S^λ once and the irreducible representations S^μ, with $\mu > \lambda$ (in the lexicographic order sense) with multiplicity $K_{\lambda\mu}$ where $K_{\lambda\mu}$ are the Kostka numbers encountered earlier.

Some simple particular cases are listed below:

- $\lambda = (n)$. Here there is only one tabloid

$$\boxed{1\,|\,2\,|\,3\,|\,\cdot\,|\,\cdot\,|\,\cdot\,|\,n} \tag{2.91}$$

and the representation $M^{(n)}$ is just the one dimensional symmetric representation $S^{(n)}$ of S_n.

- In the other extreme case $\lambda = (1^n)$, (1^n)-tableau being the same as (1^n)-tabloid, we obtain the regular representation:

$$M^{(1^n)} = \text{regular representation of } S_n. \tag{2.92}$$

- Further, when $\lambda = (n-1,1)$, $(n-1,1)$-tabloids, n in number, yield the defining representation of S_n.

$$M^{(n-1,1)} = \text{defining representation of } S_n. \tag{2.93}$$

This is because the tabloids in this case have the structure

$\tag{2.94}$

and the action of S_n on such tabloids is completely specified by the action on the entry in the second row.

2.8.3 X = set of all paratabloids $\mathbf{e}_{t^\lambda}$ one associated with each t^λ

Consider a Young tableau t^λ and let C_{t^λ} be its column subgroup. Then the paratabloid $\mathbf{e}_{t^\lambda}$ associated with t^λ is defined as

$$\mathbf{e}_{t^\lambda} = \sum_{q \in C_{t^\lambda}} \text{sign}(q) \, \{qt^\lambda\} \tag{2.95}$$

where $\{qt^\lambda\}$ denotes the tabloid corresponding to the tableau obtained by the action of $q \in C_{t^\lambda}$ on t^λ. The paratabloids are then certain linear combinations of tabloids with coefficients ± 1.

Example Take $\lambda = (2, 1)$ and consider the tableau t^λ

$$t^\lambda = \begin{array}{|c|c|} \hline 1 & 2 \\ \hline \end{array} \!\!\begin{array}{|c|} \hline 3 \\ \hline \end{array}. \tag{2.96}$$

Its column group C_{t^λ} is $\{e, (13)\}$. The corresponding paratabloid is then

$$\mathbf{e}_{t^\lambda} = \{t^\lambda\} - \{(13)t^\lambda\}$$

$$= \begin{array}{|c|c|} \hline 1 & 2 \\ \hline 3 \\ \hline \end{array} - \begin{array}{|c|c|} \hline 3 & 1 \\ \hline 2 \\ \hline \end{array}. \tag{2.97}$$

The action of an element $p \in S_n$ on $\mathbf{e}_{t^\lambda}$ is given by

$$p\mathbf{e}_{t^\lambda} = \mathbf{e}_{pt^\lambda} \tag{2.98}$$

as can easily be seen: The action of an element $p \in S_n$ on $\mathbf{e}_{t^\lambda}$ is given by

$$p\mathbf{e}_{t^\lambda} = \sum_{q \in C_{t^\lambda}} \text{sign}(q) \, \{pqt^\lambda\}$$

$$= \sum_{q \in C_{t^\lambda}} \text{sign}(pqp^{-1}) \, \{pqp^{-1}pt^\lambda\}$$

$$= \sum_{q \in C_{pt^\lambda}} \text{sign}(q) \, \{qpt^\lambda\}$$

$$= \mathbf{e}_{pt^\lambda}. \tag{2.99}$$

where in the third line we have used the fact that $C_{pt^\lambda} = pC_{t^\lambda}p^{-1}$

The linear span of the paratabloids can be shown to carry an irreducible representation S^λ of S_n. It can further be shown that the set $\{\mathbf{e}_{T^\lambda}\}$ where T^λ denotes the set of all standard tableaux corresponding to the partition λ form a basis in this space. This of course agrees with the fact stated earlier that the dimension of S^λ equals the number of standard tableaux corresponding to the partition λ.

Problems

P2.1   Show that the number of elements of S_n which square to the identity element is given by

$$\sum_{\beta=0}^{[n/2]} \frac{n!}{2^\beta \beta! (n-2\beta)!}$$

P2.2   For S_5, verify that the sum of the dimensions of the irreducible representations equals the number of involutions in S_5, i.e., the number of elements which square to the identity. (This result which in fact holds for S_n for all n such as for instance, Hamermesh (1964) for an accessible proof, has also been used in Kodiyalam and Verma (2004) to construct a representation of S_n in which all irreducible representations of S_n occur once each. Also see Chaturvedi et al. (2008).)

P2.3   Decompose the representation

of S_5 into its irreducible representations.

[4]   Consider the S_3 subgroup of S_4 by restricting to those elements of S_4 in which, say 4, stays put. How does the representation

of S_4 when restricted to S_3 decompose into irreducible representations of S_3?

Bibliography

Boerner, H. (1963). *Representations of Groups, with Special Consideration for the Needs of Modern Physics*. Amsterdam: North Holland.

Chaturvedi, S. (1996). Canonical Partition Functions for Parastatistical Systems of Any Order. *Phys. Rev.* E 54: 1378.

Chaturvedi, S., Marmo, G., Mukunda, N. and Simon, R. (2008). Schwinger Representation for the Symmetric Group: Two Explicit Constructions for the Carrier Space, *Phys. Lett.* A 372: 3763.

Coleman, A. J. (1968). The Symmetric Group Made Easy. *Advances in Quantum Chemistry* 4: 83.

Hamermesh, M. (1962). *Group Theory and Its Application to Physical Problems*. Reading, Massachusetts: Addison-Wesley.

Kodiyalam, V. and Verma, D. N. (2004). A Natural Representation Model for Symmetric Groups, *arXiv:math*/0402216.

**Macdonald, I. G. (1995). *Symmetric Functions and Hall Polynomials.* 2nd ed. Oxford: Clarendon Press.

Sagan, B. E. (2013). *The Symmetric Group: Representations, Combinatorial Algorithms, and Symmetric Functions.* New York: Springer.

Sengupta, A. N. (2012). *Representing Finite Groups: A Semisimple Introduction.* New York: Springer.

Stoilova, N. I. and Van der Jeugt, J. (2020). Partition Functions and Thermodynamic Properties of Paraboson and Parafermion systems. *Phys. Lett.* A 384: 126421.

Wybourne, B. G. (2000). Symmetric Functions and the Symmetric Group. *Lectures, Institute of Physics.* Nicolaus Copernicus University, Torun, Poland. https://fizyka. umk.pl/~bgw/symfn1.pdf.

**Symmetric functions play an important role in the context of canonical and grand canonical partition functions for systems of identical and indistinguishable particles.

3

Rotations in 2 and 3 Dimensions, SU(2)

The general theory of groups and group representations for finite groups was outlined in Chapter 1. This theory was utilised in Chapter 2 to describe some features of symmetric permutation groups. Here, we will move on to continuous groups, beginning with the results from the finite case and generalising them in a heuristic manner. For continuous groups we go beyond algebraic methods and use calculus very effectively.

As our first examples we look at rotations in two and three dimensions which are important for physics. When needed we will call upon results from the quantum theory of angular momentum. We will look at definitions, structures, topologies and representations mainly in a descriptive manner.

3.1 The Group $SO(2)$

This is the group of proper rotations in a plane, with respect to a chosen origin, with the defining matrix representation

$$SO(2) = \{A = 2 \times 2 \text{ real matrix} | A^T A = \mathbb{1}, \det A = 1\}$$

$$= \left\{ A(\alpha) = \begin{pmatrix} \cos \alpha & -\sin \alpha \\ \sin \alpha & \cos \alpha \end{pmatrix}, \ \alpha \in [0, 2\pi) \right\}. \tag{3.1}$$

This is a one parameter or one dimensional continuous abelian group. The group space or manifold is the circle $\mathbb{S}^1$. The composition law, identity and inverses are:

$$A(\alpha)A(\beta) = A(\alpha + \beta), \ A(0) = \mathbb{1} = \text{identity}, \ A(\alpha)^{-1} = A(-\alpha), \tag{3.2}$$

all arguments taken modulo 2π. As this is an abelian group, each element is a class by itself, the centre is the whole group, and the commutator subgroup is trivial.

For every positive integer N, the set of elements $\{A(\frac{2\pi n}{N}), n = 0, 1, \cdots, N - 1\}$ is a discrete subgroup of order N.

The topology of $SO(2)$ is nontrivial. While it is certainly connected – one can move continuously from any element to any other without leaving the group manifold – it has infinite connectivity. There is one class of continuous closed loops starting and ending at the identity for each 'winding number' $n = 0, \pm 1, \pm 2, \cdots$.

The matrices $A(\alpha)$ in Eq. (3.1) are simple exponentials, because the composition law is just addition of angles (modulo 2π):

$$A(\alpha) = e^{-i\alpha\sigma_2}, \ \sigma_2 = \begin{pmatrix} 0 & -i \\ i & 0 \end{pmatrix} = \text{second Pauli matrix.} \tag{3.3}$$

Due to the abelian nature, all UIR's are one dimensional. There is one UIR, $D^{(m)}(A(\alpha))$, for each $m = 0, \pm 1, \pm 2, \cdots$, a countable infinity of inequivalent UIR's:

$$D^{(m)}(A(\alpha)) = e^{im\alpha}. \tag{3.4}$$

There are the characters of $SO(2)$. The defining UR in Eq. (3.1) is the direct sum, in the complex sense, of $D^{(+1)}$ and $D^{(-1)}$.

If $f(\alpha)$ is a complex valued function on $SO(2)$, its squared norm involves an invariant integral over the group:

$$f(\alpha) \in L^2(SO(2)) \ : \ \|f\|^2 = \frac{1}{2\pi} \int_0^{2\pi} d\alpha |f(\alpha)|^2 < \infty. \tag{3.5}$$

As $SO(2)$ is abelian, there is only one regular representation acting in the obvious way on this space:

$$(L_{A(\alpha)}f)(\beta) = (R_{A(\alpha)^{-1}}f)(\beta) = f(\beta - \alpha). \tag{3.6}$$

The orthogonality and completeness of the $D^{(m)}$'s in the space $L^2(SO(2))$ amount to the familiar Fourier series expansion for any function on $\mathbb{S}^1$:

$$\frac{1}{2\pi}\int_0^{2\pi} d\alpha\, D^{(m')}(A(\alpha))^* D^{(m)}(A(\alpha)) = \delta_{m',m};$$

$$f(\alpha) = \sum_{m=0,\pm 1,\cdots} f_m e^{im\alpha},$$

$$f_m = \frac{1}{2\pi}\int_0^{2\pi} d\alpha\, e^{-im\alpha} f(\alpha),$$

$$\|f\|^2 = \sum_{m=0,\pm 1,\cdots} |f_m|^2. \tag{3.7}$$

So these are the continuous case analogues of Eqs. (1.72, 1.84) in a simple case.

3.2 The Group $O(2)$

Now we include improper rotations, so we have the full orthogonal group in the plane. The defining matrix representation is again two dimensional:

$$O(2) = \{A = 2 \times 2 \text{ real matrix} \,| A^T A = \mathbb{1},\ \det A = \pm 1\}$$

$$= \{A(\alpha) \in SO(2)\} \cup$$

$$\left\{ B(\alpha) = A(\alpha)P = \begin{pmatrix} \cos\alpha & \sin\alpha \\ \sin\alpha & -\cos\alpha \end{pmatrix}, P = B(0) = \begin{pmatrix} 1 & 0 \\ 0 & -1 \end{pmatrix}, \alpha \in [0, 2\pi) \right\} \tag{3.8}$$

We see that the group space is $\mathbb{S}^1 \cup \mathbb{S}^1$, the union of two disjoint individually connected components. The composition law and inverses are given by the algebraic relations

$$A(\alpha)A(\beta) = A(\alpha + \beta),\ A(\alpha)B(\beta) = B(\alpha + \beta),$$

$$B(\alpha)A(\beta) = B(\alpha - \beta),\ B(\alpha)B(\beta) = A(\alpha - \beta),$$

$$A(\alpha)^{-1} = PA(\alpha)P = A(-\alpha),$$

$$B(\alpha)^{-1} = B(\alpha). \tag{3.9}$$

where we have used $P^2 = A(0) = \mathbb{1}$. We see that in contrast to $SO(2)$, $O(2)$ is nonabelian. Using these relations we find results contrasting greatly with $SO(2)$; we display the two sets of results in a table:

	Centre	Equivalence classes	Commutator subgroup
SO(2)	SO(2)	each element $A(\alpha)$	trivial, $A(0) = \mathbb{1}$
O(2)	$A(0) = \mathbb{1}$	$A(0); \{A(\alpha), A(-\alpha)\},$ $0 < \alpha < 2\pi;$ $\{B(\alpha), 0 \leq \alpha < 2\pi\}$	SO(2)

$$(3.10)$$

Thus $SO(2)$ is an invariant subgroup in $O(2)$; and the quotient is the abelian two element group, essentially $\{\mathbb{1}, P\}$.

The pattern of UIR's is as follows: For $m = 0$, we have the trivial one dimensional UIR

$$D^{(0)}(A(\alpha) \text{ or } B(\alpha)) = 1. \qquad (3.11)$$

Next for each $m = 1, 2, \cdots$, we have a two dimensional UIR $D^{(m)}(\cdot)$:

$$D^{(m)}(A(\alpha)) = \begin{pmatrix} e^{im\alpha} & 0 \\ 0 & e^{-im\alpha} \end{pmatrix}, \quad D^{(m)}(B(\alpha)) = \begin{pmatrix} 0 & e^{im\alpha} \\ e^{-im\alpha} & 0 \end{pmatrix}$$

$$= D^{(m)}(A(\alpha))D^{(m)}(P),$$

$$D^{(m)}(P) = \begin{pmatrix} 0 & 1 \\ 1 & 0 \end{pmatrix}. \qquad (3.12)$$

The defining representation (3.8) is equivalent to $D^{(1)}$. We leave the analysis of the regular representation as an exercise.

3.3 The Group $SO(3)$

This is the group of proper rotations in three dimensional real Euclidean space, with respect to a chosen origin, of obvious importance in physics. Many new features naturally show up compared to $SO(2)$. The defining matrix representation is:

$$SO(3) = \{A = 3 \times 3 \text{ real matrix} \,|\, A^T A = \mathbb{1}, \ \det A = 1\}. \qquad (3.13)$$

One can convince oneself that this is nonabelian, the centre is trivial; and as there are no invariant subgroups, $SO(3)$ is simple. Group composition is given by matrix multiplication, the identity is the unit 3×3 matrix, and A^{-1} is A^T.

3.3.1 Parametrisations and topology of the $SO(3)$ manifold

By counting the number of independent constraints on the nine real matrix elements of A, we see that $SO(3)$ is a continuous nonabelian three dimensional or three parameter group. We describe three useful ways of choosing parameters or coordinates to label group elements, look at the structure of the group manifold, and several other properties.

1. **Axis-angle parameters**

   Each proper rotation is a right handed rotation by some angle α about some axis $\hat{\boldsymbol{n}}$ in three dimensional space. From simple geometry, we find that the corresponding matrix is

   $$A_{jk}(\hat{\boldsymbol{n}}, \alpha) = \cos\alpha\, \delta_{jk} + (1 - \cos\alpha)\, n_j n_k - \sin\alpha\, \epsilon_{jkl} n_l, \quad j, k = 1, 2, 3,$$

   $$A(\hat{\boldsymbol{n}}, \alpha)^{-1} = A(\hat{\boldsymbol{n}}, 2\pi - \alpha) = A(-\hat{\boldsymbol{n}}, \alpha). \tag{3.14}$$

   We can combine $\hat{\boldsymbol{n}}$ and α to form a real 3-vector $\boldsymbol{\alpha} = \hat{\boldsymbol{n}}\alpha$ and write

   $$A_{jk}(\boldsymbol{\alpha}) = \cos\alpha\, \delta_{jk} + \frac{(1 - \cos\alpha)}{\alpha^2}\alpha_j\alpha_k - \frac{\sin\alpha}{\alpha}\epsilon_{jkl}\alpha_l, \quad \alpha = |\boldsymbol{\alpha}|;$$

   $$A(\boldsymbol{\alpha})^{-1} = A(-\boldsymbol{\alpha}). \tag{3.15}$$

   The range $\hat{\boldsymbol{n}} \in \mathbb{S}^2$ is clear. For α we note that since from (3.14)

   $$A(\hat{\boldsymbol{n}}, \alpha) = A(-\hat{\boldsymbol{n}}, 2\pi - \alpha), \tag{3.16}$$

   it is adequate to limit α to the range $0 \leq \alpha \leq \pi$. However, for $\alpha = \pi$ we have from (3.14) or (3.16)

   $$A_{jk}(\hat{\boldsymbol{n}}, \pi) = A_{jk}(-\hat{\boldsymbol{n}}, \pi) = 2n_j n_k - \delta_{jk}, \tag{3.17}$$

   so we must understand that the elements labeled $(\hat{\boldsymbol{n}}, \pi)$ and $(-\hat{\boldsymbol{n}}, \pi)$ are identical! Thus the $SO(3)$ group space or manifold consists of the interior and the surface of a sphere of radius π in a fictitious three dimensional parameter space, except that diametrically opposite points *on the surface* are identified to be 'the same'. It is this rule which makes it rather difficult to visualise the $SO(3)$ manifold. We come back to this later.

2. Homogeneous Euler parameters

In terms of $\hat{\boldsymbol{n}}, \alpha$ define four real quantities a_0, a_j by

$$a_0 = \cos \alpha/2, \ a_j = n_j \sin \alpha/2 : \ a_0^2 + \boldsymbol{a} \cdot \boldsymbol{a} = 1. \tag{3.18}$$

So $a_0, \boldsymbol{a}$ define a point $a \in \mathbb{S}^3$, the surface of the unit sphere in the Euclidean space $\mathbb{R}^4$. Then from (3.14) we have

$$A_{jk}(\hat{\boldsymbol{n}}, \alpha) = A_{jk}(a) = (a_0^2 - \boldsymbol{a}^2)\delta_{jk} + 2a_j a_k - 2a_0 \epsilon_{jkl} a_l, \ a \in \mathbb{S}^3;$$

$$A(a_0, \boldsymbol{a})^{-1} = A(a_0, -\boldsymbol{a}). \tag{3.19}$$

Since this is quadratic in a, $A(-a) = A(a)$, now the group manifold is seen to be $\mathbb{S}^3$ with antipodal points identified.

3. Euler angles

This parametrisation is convenient for classical rigid body dynamics and also quantum angular momentum theory. First we note that there is a 'half $\mathbb{S}^2$' worth of $SO(2)$ subgroups in $SO(3)$, one subgroup for each 'director' (or headless vector) in $\mathbb{R}^3$:

$$SO(2, \pm\hat{\boldsymbol{n}}) = \{A(\hat{\boldsymbol{n}}, \alpha) | \ \hat{\boldsymbol{n}} \text{ fixed}, 0 \le \alpha < 2\pi\} \subset SO(3). \tag{3.20}$$

Taking $\hat{\boldsymbol{n}} = \hat{\boldsymbol{e}}_j, \ j = 1, 2, 3$ in turn we get the matrices of $SO(2)$ rotations in the three coordinate planes:

$$A(\hat{\boldsymbol{e}}_1, \alpha) = \begin{pmatrix} 1 & 0 & 0 \\ 0 & \cos \alpha & -\sin \alpha \\ 0 & \sin \alpha & \cos \alpha \end{pmatrix}$$

$$A(\hat{\boldsymbol{e}}_2, \alpha) = \begin{pmatrix} \cos \alpha & 0 & \sin \alpha \\ 0 & 1 & 0 \\ -\sin \alpha & 0 & \cos \alpha \end{pmatrix}$$

$$A(\hat{\boldsymbol{e}}_3, \alpha) = \begin{pmatrix} \cos \alpha & -\sin \alpha & 0 \\ \sin \alpha & \cos \alpha & 0 \\ 0 & 0 & 1 \end{pmatrix}. \tag{3.21}$$

Analogous to Eq. (3.3) in the $SO(2)$ case, each of these can be written as an exponential:

$$A(\hat{\boldsymbol{e}}_j, \alpha) = e^{-i\alpha S_j}, \ (S_j)_{kl} = -i\epsilon_{jkl}. \tag{3.22}$$

The three 'generator matrices' S_j are 3×3 pure imaginary antisymmetric, hence hermitian. The general rotation $A(\hat{\boldsymbol{n}}, \alpha)$ is also an exponential:

$$A(\hat{\boldsymbol{n}}, \alpha) = A(\boldsymbol{\alpha}) = e^{-i\boldsymbol{\alpha}\cdot S}. \qquad (3.23)$$

Now it can be seen by geometrical arguments that every $A \in SO(3)$ is expressible as a product of three factors involving two of the subgroup elements (3.21):

$$A = A(\theta, \psi, \phi) = A(\hat{\boldsymbol{e}}_3, \theta) A(\hat{\boldsymbol{e}}_2, \psi) A(\hat{\boldsymbol{e}}_3, \phi),$$

$$0 \le \theta, \ \phi \le 2\pi, \ 0 \le \psi \le \pi. \qquad (3.24)$$

The ranges are chosen so that all of $SO(3)$ is covered; however, for some elements (amounting to a 'set of measure zero') these angles are nonunique: when $\psi = 0$, only $\theta + \phi$ is relevant.

The $SO(3)$ composition law can in principle be found in any parametrisation but is somewhat complicated. In infinitesimal form or to leading order in axis angle parameters we can find it as follows. From (3.15) in the $\boldsymbol{\alpha}$ description we have up to quadratic terms:

$$A_{jk}(\boldsymbol{\alpha}) \simeq \delta_{jk} - \epsilon_{jkl}\alpha_l + \frac{1}{2}(\alpha_j \alpha_k - \boldsymbol{\alpha}^2 \delta_{jk}) + \cdots. \qquad (3.25)$$

Write $A(\boldsymbol{\alpha})A(\boldsymbol{\beta}) = A(\boldsymbol{\alpha} + \boldsymbol{\beta} + \boldsymbol{\gamma})$ where $\boldsymbol{\gamma}$ starts with cross terms α times β, and compare the two sides retaining terms up to linear in $\boldsymbol{\alpha}$, in $\boldsymbol{\beta}$ and cross terms only. Taking the jl matrix element on both sides and using (3.25) gives:

$$(\delta_{jk} - \epsilon_{jkr}\alpha_r)(\delta_{kl} - \epsilon_{kls}\beta_s) \simeq \delta_{jl} - \epsilon_{jlr}(\alpha_r + \beta_r + \gamma_r)$$

$$+ \frac{1}{2}(\alpha_j \beta_l + \alpha_l \beta_j - 2\boldsymbol{\alpha} \cdot \boldsymbol{\beta} \delta_{jl}) \cdots,$$

$$\text{i.e.,} \ \ \epsilon_{jlr}\gamma_r = \frac{1}{2}(\alpha_j \beta_l - \alpha_l \beta_j) \ \text{ or } \ \boldsymbol{\gamma} = \frac{1}{2}\boldsymbol{\alpha} \wedge \boldsymbol{\beta}. \qquad (3.26)$$

So to leading nontrivial order the composition law is

$$A(\boldsymbol{\alpha})A(\boldsymbol{\beta}) = A(\boldsymbol{\alpha} + \boldsymbol{\beta} + \frac{1}{2}\boldsymbol{\alpha} \wedge \boldsymbol{\beta} \cdots), \qquad (3.27)$$

and the nonabelian nature is evident.

We use the axis-angle description for two more purposes at this point. The first is to study the connectivity properties of $SO(3)$. It is easy to see that it is connected: one can go from any element to any other along a continuous path within $SO(3)$. But if we consider continuous closed loops starting and ending at the identity, which is the origin $\alpha = 0$, then with respect to continuous deformations there are two classes of loops: those which never reach the surface $\alpha = \pi$ or which use the identification of antipodal points an even number of times; and those which use it an odd number of times. Thus there are two equivalence classes of closed loops, or $SO(3)$ is doubly connected. It has a homotopy group $\pi_1(SO(3)) = \mathbb{Z}_2$. A representative of the trivial class is the constant or null loop. Another representative of this class is indicated in the figure: go radially outwards from e to A on the surface $\alpha = \pi$, jump to A' which is 'the same' as A, travel A' to B' on $\alpha = \pi$, jump to B which is 'the same' as B', then travel B to A on $\alpha = \pi$, and finally return radially to e. One can convince oneself that this closed loop can be continuously shrunk to the null loop. For a representative from the other nontrivial class we use the same figure: go from e to A radially, jump to A', return to e radially inwards. This loop is not shrinkable to the null loop.

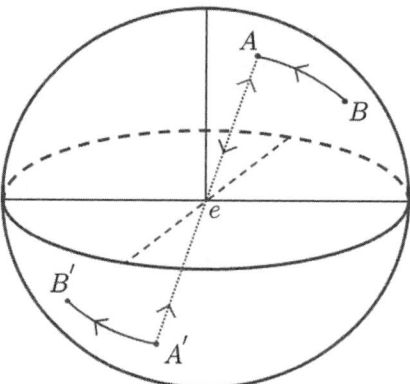

Figure 3.1   Closed loops in $SO(3)$. Two jumps: $e \to A = A' \to B' = B \to A \to e$, One jump: $e \to A = A' \to e$

The next use of axis-angle coordinates is to describe conjugation and equivalence classes. Conjugate $A(\boldsymbol{\alpha})$ by some $B \in SO(3)$, use Eq. (3.15) for $A(\boldsymbol{\alpha})$, and find:

$$BA(\boldsymbol{\alpha})B^{-1} = A(\boldsymbol{\alpha}'), \quad \boldsymbol{\alpha}' = B\boldsymbol{\alpha}, \text{ i.e., } \alpha'_j = B_{jk}\alpha_k. \tag{3.28}$$

Thus under conjugation $\boldsymbol{\alpha}$ experiences a rotation and α is preserved. This means α labels equivalence classes, they can be exhibited as follows:

$\alpha = 0 \; : \; C(0) = \{\mathbb{1}\}$, single element;

$0 < \alpha < \pi \; : \; C(\alpha) = \{A(\hat{\boldsymbol{n}}, \alpha) | \alpha \text{ fixed}, \ \hat{\boldsymbol{n}} \in \mathbb{S}^2 \text{ varying}\}$;

$\alpha = \pi \; : \; C(\pi) = \{A(\hat{\boldsymbol{n}}, \pi) | \hat{\boldsymbol{n}} \in \mathbb{S}^2, \text{ each element counted twice}\}. \tag{3.29}$

As representative elements we can take for example rotations around $\hat{\boldsymbol{e}}_3$:

$$0 \leq \alpha \leq \pi \ : \ \mathcal{C}(\alpha) \rightarrow A(\hat{\boldsymbol{e}}_3, \alpha) = \begin{pmatrix} \cos\alpha & -\sin\alpha & 0 \\ \sin\alpha & \cos\alpha & 0 \\ 0 & 0 & 1 \end{pmatrix}. \qquad (3.30)$$

A 'class function' is a function $f(\alpha)$ for $0 \leq \alpha \leq \pi$.

The nature of the coset space $SO(3)/SO(2)$ is nicely seen from the Euler angle parametrisation (3.24), when we regard the last factor as an element of the subgroup $SO(2, \hat{\boldsymbol{e}}_3)$. Then the first two factors amount to a coset representative, and their product 'runs over' $\mathbb{S}^2$, with polar angle ψ and azimuth θ. Thus we see that

$$SO(3)/SO(2) \simeq \mathbb{S}^2. \qquad (3.31)$$

A neater argument is this. Any $A \in SO(3)$ carries $\hat{\boldsymbol{e}}_3$ to some $\hat{\boldsymbol{n}} \in \mathbb{S}^2$:

$$A\hat{\boldsymbol{e}}_3 = \hat{\boldsymbol{n}} \in \mathbb{S}^2, \ A(\hat{\boldsymbol{e}}_3, \alpha)\hat{\boldsymbol{e}}_3 = \hat{\boldsymbol{e}}_3. \qquad (3.32)$$

Then all the elements in the coset containing A, namely $\{AA(\hat{\boldsymbol{e}}_3, \alpha), 0 \leq \alpha \leq 2\pi\}$ have the same effect on $\hat{\boldsymbol{e}}_3$, they all lead to $\hat{\boldsymbol{n}}$. Thus $\hat{\boldsymbol{n}}$ is a coset label, and Eq. (3.31) follows. We should carefully note, however, that $SO(3) \not\simeq SO(2) \times \mathbb{S}^2$.

3.3.2 Invariant integration over $SO(3)$

Let us denote a (real or complex) function over $SO(3)$ by $f(A)$. Its average over $SO(3)$, in the sense of Eq. (1.61), with invariance properties and normalisation, can be defined. Formally we write and require:

$$\int_{SO(3)} dA\, f(A) = \int_{SO(3)} dA(f(A_0 A) \text{ or } f(AA_0) \text{ or } f(A^{-1})), \ A_0 \text{ fixed.} \qquad (3.33)$$

Then along with normalisation it has the explicit forms

$$\int_{SO(3)} dA\, f(A) = \int_{0 \leq \alpha \leq \pi} d^3\alpha \frac{(1 - \cos\alpha)}{4\pi^2 \alpha^2} f(\boldsymbol{\alpha})$$

$$= \frac{1}{\pi^2} \int_{\mathbb{S}^3} d^4 a\, \delta(a_0^2 + \boldsymbol{a}^2 - 1) f(a), \quad f(a) = f(-a),$$

$$= \frac{1}{8\pi^2} \int_0^{2\pi} d\theta \int_0^{2\pi} d\phi \int_0^\pi \sin\psi\, d\psi\, f(\theta, \psi, \phi). \qquad (3.34)$$

The Hilbert space $L^2(SO(3))$ is then defined by

$$L^2(SO(3)) = \left\{ f(A) \mid \|f\|^2 = \int_{SO(3)} dA |f(A)|^2 < \infty \right\}. \qquad (3.35)$$

This is the carrier space for the left and right regular representations which of course naturally commute.

Going back to Eq. (3.34), if $f(A)$ is independent of the third Euler angle ϕ, it is constant over each $SO(2)$ coset. Then its average over $SO(3)$ reduces to an average over the space of cosets $\mathbb{S}^2$ identified in Eq. (3.31):

$$f(AA(\hat{e}_3, \phi)) = f(A) \text{ for all } \phi \Rightarrow$$

$$\int_{SO(3)} dA f(A) = \frac{1}{4\pi^2} \int_{\mathbb{S}^2} d\Omega(\psi, \theta) f(\psi, \theta), \qquad (3.36)$$

where $d\Omega(\psi, \theta) = \sin \psi \, d\psi \, d\theta$ is the usual element of solid angle (in unfamiliar variables).

3.3.3 Some important properties of $SO(3)$ representations

We will recall these properties, omitting derivations and relying on what is normally covered in quantum angular momentum theory.

i) Every representation is equivalent to, and so may be assumed to be, a UR; therefore we need only find all UIR's up to unitary equivalence.

ii) Every UIR is finite dimensional.

iii) There is one UIR in each odd dimension $(2l + 1)$ for $l = 0, 1, 2, \cdots$, written as $D^{(l)}(A)$. The defining representation (3.13) corresponds to $l = 1$.

iv) Every UIR can be brought to real form in a suitable orthonormal basis in the representation space; however, especially in quantum mechanics the UIR's are used in complex form.

v) Each UIR $D^{(l)}(A)$, $l \geq 1$, is faithful.

In a general UR we write the operators as $D(\boldsymbol{\alpha})$ or $D(\theta, \psi, \phi)$ depending on the parameters used. In axis-angle parameters, for instance, these operators obey

$$D(\boldsymbol{\alpha})D(\boldsymbol{\beta}) = D(\boldsymbol{\alpha} + \boldsymbol{\beta} + \frac{1}{2}\boldsymbol{\alpha} \wedge \boldsymbol{\beta} + \cdots). \qquad (3.37)$$

The hermitian generators of the UR are determined by $D(\boldsymbol{\alpha})$ near $\boldsymbol{\alpha} = 0$:

$$J_j = i\frac{\partial}{\partial \alpha_j}D(\boldsymbol{\alpha})\Big|_{\boldsymbol{\alpha}=0}, \quad D(\boldsymbol{\alpha}) = \mathbb{1} - i\alpha_j J_j + O(\alpha^2). \tag{3.38}$$

Thus from Eq. (3.23), in the defining representation in real form the generators are the S_j. From the composition law expressed in (3.37) for infinitesimal elements it can be shown that the generators J_j obey the commutation relations

$$[J_j, J_k] = i\epsilon_{jkl}J_l. \tag{3.39}$$

These have the same form in all UR's, and the ϵ_{jkl} are the 'structure constants' of $SO(3)$. By 'integration' of Eq. (3.38) the $D(\boldsymbol{\alpha})$ for finite group elements are single exponentials:

$$D(\boldsymbol{\alpha}) = e^{-i\boldsymbol{\alpha}\cdot\boldsymbol{J}}. \tag{3.40}$$

Let us express the composition law, say in the defining representation, as

$$A(\boldsymbol{\alpha})A(\boldsymbol{\beta}) = A(\boldsymbol{f}(\boldsymbol{\alpha},\boldsymbol{\beta})). \tag{3.41}$$

So the right hand side of (3.27) gives the initial terms in a Taylor expansion of $\boldsymbol{f}(\boldsymbol{\alpha},\boldsymbol{\beta})$. Then general theory shows that Eqs. (3.39, 3.40) together guarantee that we have a UR

$$D(\boldsymbol{\alpha})D(\boldsymbol{\beta}) = D(\boldsymbol{f}(\boldsymbol{\alpha},\boldsymbol{\beta})), \tag{3.42}$$

except for a 'global' condition which J_j must obey, and which we give later.

Under conjugation by $D(A)$ the generators transform in a particular way. From Eq. (3.28) (replacing B by A) and (3.40) we have

$$D(A)e^{-i\boldsymbol{\alpha}\cdot\boldsymbol{J}}D(A)^{-1} = e^{-iA\boldsymbol{\alpha}\cdot\boldsymbol{J}} = e^{-i\boldsymbol{\alpha}\cdot A^T\boldsymbol{J}} \Rightarrow$$

$$D(A)J_j D(A)^{-1} = A_{kj}J_k. \tag{3.43}$$

This is called the 'adjoint transformation rule', common to all $SO(3)$ UR's; we say the generators always 'belong' to the defining 3×3 representation.

The global condition on the J_j needed to ensure (3.42) is that any rotation by amount 2π must be represented by the identity operator:

$$A(\hat{\boldsymbol{n}}, 2\pi) = \mathbb{1}_{3\times3}, \quad \text{so} \quad D(\hat{\boldsymbol{n}}, 2\pi) = e^{-2\pi i\hat{\boldsymbol{n}}\cdot\boldsymbol{J}} = \mathbb{1}. \tag{3.44}$$

So we must select those solutions to the commutation relations (3.39) which obey (3.44).

The axis-angle parametrisation is special because $D(\boldsymbol{\alpha})$ is a single exponential and the generators are defined in a simple way by Eq. (3.38). In the Euler angle description, which is very convenient in quantum mechanics, a general $SO(3)$ element is represented by a product of three noncommuting exponentials:

$$D(\theta, \psi, \phi) = e^{-i\theta J_3} e^{-i\psi J_2} e^{-i\phi J_3}. \tag{3.45}$$

3.3.4 The UIR's of $SO(3)$

We recall some features which are familiar from quantum angular momentum theory. To find irreducible solutions to Eq. (3.39), we exploit the fact that $J^2 = J_j J_j$ commutes with all J_k:

$$[J^2, J_k] = 0. \tag{3.46}$$

By Schur's Lemma, in a UIR J^2 is a multiple of the unit matrix. For each $l = 0, 1, \cdots, J^2$ has corresponding 'value' $l(l+1)\mathbb{1}$. An orthonormal basis in the space of the lth UIR is provided by the eigenvectors of J_3, labelled by its eigenvalues $m = l, l-1, \cdots, -l$. We write the basis vectors as $|l, m\rangle$ where l is fixed and m varies. The arbitrary phases in these vectors can be adjusted so that the J_k act on them as follows:

$$J_3|l, m\rangle = m|l, m\rangle, \quad J^2|l, m\rangle = l(l+1)|l, m\rangle, \quad \langle l, m'|l, m\rangle = \delta_{m'm};$$

$$(J_1 \pm iJ_2)|l, m\rangle = \sqrt{(l \mp m)(l \pm m + 1)} \, |l, m \pm 1\rangle. \tag{3.47}$$

J_1 and J_3 are real symmetric, J_2 is pure imaginary antisymmetric.

The meaning of the spectrum of values of m is this: under the restriction from $SO(3)$ to the subgroup $SO(2, \hat{\boldsymbol{e}}_3)$,

$$\text{UIR } D^{(l)}(\,\cdot\,) \text{ of } SO(3)\Big|_{SO(2,\hat{\boldsymbol{e}}_3)} = \sum_{m=-l}^{l}{}_{\oplus} D^{(m)}(\,\cdot\,) \text{ of } SO(2, \hat{\boldsymbol{e}}_3), \tag{3.48}$$

in the sense of Eq. (3.4). From Eq. (3.30) we see that the l-th irreducible character is a geometric series:

$$\chi^{(l)}(\alpha) = \sum_{m=-l}^{l} e^{-im\alpha} = \frac{\sin\left(l + \frac{1}{2}\right)\alpha}{\sin\frac{1}{2}\alpha}. \tag{3.49}$$

Calling upon Eq. (1.88), the reality of $\chi^{(l)}(\alpha)$ means that $D^{(l)}(\,\cdot\,)^*$ is equivalent to $D^{(l)}(\,\cdot\,)$; it is indeed possible to choose a suitable orthonormal basis (different from the

$|l, m\rangle)$ in which the representation matrices become real. Now from (3.34), among class functions $f(\alpha)$ we have the scalar product

$$(f_1, f_2) = \frac{2}{\pi} \int_0^\pi d\alpha \, \sin^2 \frac{\alpha}{2} \, f_1(\alpha)^* f_2(\alpha). \tag{3.50}$$

With this it is then indeed true that

$$(\chi^{(l')}, \chi^{(l)}) = \delta_{l', l};$$

$$f(\alpha) = \sum_{l=0}^\infty f_l \chi^{(l)}(\alpha), \quad (f, f) = \sum_{l=0}^\infty |f_l|^2. \tag{3.51}$$

3.3.5 The D-matrices, orthogonality and completeness

In axis-angle parameters the UIR matrices for finite $SO(3)$ elements,

$$D^{(l)}_{m'm}(\boldsymbol{\alpha}) = \langle l, m' | e^{-i\boldsymbol{\alpha}\cdot\boldsymbol{J}} | l, m \rangle \tag{3.52}$$

are rather complicated. It is here that the Euler angle description is simpler:

$$\langle l, m' | D^{(l)}(\theta, \psi, \phi) | l, m \rangle = \langle l, m' | e^{-i\theta J_3} e^{-i\psi J_2} e^{-i\phi J_3} | l, m \rangle$$

$$= D^{(l)}_{m'm}(\theta, \psi, \phi)$$

$$= e^{-im'\theta - im\phi} d^{(l)}_{m'm}(\psi),$$

$$d^{(l)}_{m'm}(\psi) = \langle l, m' | e^{-i\psi J_2} | l, m \rangle$$

$$= \sqrt{\frac{(l+m')!(l-m')!}{(l+m)!(l-m)!}} \sum_k (-1)^{l-m'-k} \binom{l+m}{l-m'-k} \binom{l-m}{k}$$

$$\times \left(\cos \frac{\psi}{2}\right)^{2k+m'+m} \left(\sin \frac{\psi}{2}\right)^{2l-2k-m'-m}. \tag{3.53}$$

The $d^{(l)}_{m'm}(\psi)$ are called the 'little d-functions'; recalling that J_2 is pure imaginary antisymmetric, we see that they are real and form an orthogonal matrix.

The collection of all the D-functions running over all UIR's has orthogonality and completeness properties. In the notation (3.34) for invariant integration we have

$$\int_{SO(3)} dA \, D_{m_1' m_2'}^{(l')}(A)^* D_{m_1 m_2}^{(l)}(A) = \frac{1}{(2l+1)} \delta_{l'l} \delta_{m_1' m_1} \delta_{m_2' m_2};$$

$$f(A) \in L^2(SO(3)) : f(A) = \sum_{l=0}^{\infty} \sum_{m_1, m_2 = -l}^{l} (2l+1)^{1/2} f_{m_1 m_2}^{(l)} D_{m_1 m_2}^{(l)}(A),$$

$$f_{m_1 m_2}^{(l)} = (2l+1)^{1/2} \int_{SO(3)} dA \, D_{m_1 m_2}^{(l)}(A)^* f(A);$$

$$\|f\|^2 = \int_{SO(3)} dA \, |f(A)|^2 = \sum_{l, m_1, m_2} |f_{m_1 m_2}^{(l)}|^2. \quad (3.54)$$

In case $f(A)$ is constant over each $SO(2, \hat{e}_3)$ right coset, $f(\theta, \psi, \phi)$ is ϕ-independent. Then from (3.53, 3.54) only the $m_2 = 0$ terms are needed in the expansion of f. But these D-functions are just the spherical harmonics over $\mathbb{S}^2$. From (3.36),

$$f(\theta, \psi, \phi) \in L^2(SO(3)), \text{ independent of } \phi \Leftrightarrow$$

$$f(\theta, \psi) \in L^2(SO(3)/SO(2)) = L^2(\mathbb{S}^2);$$

$$\|f\|^2 = \frac{1}{4\pi} \int_{\mathbb{S}^2} d\Omega(\theta, \psi) |f(\theta, \psi)|^2, \ 0 \leq \theta \leq 2\pi, \ 0 \leq \psi \leq \pi;$$

$$f(\theta, \psi) = \sum_{lm} f_m^{(l)} Y_{lm}(\psi, \theta), \ Y_{lm}(\psi, \theta) = (2l+1)^{1/2} D_{m0}^{(l)}(\theta, \psi, \cdot);$$

$$\frac{1}{4\pi} \int_{\mathbb{S}^2} d\Omega(\theta, \psi) Y_{l'm'}(\psi, \theta)^* Y_{lm}(\psi, \theta) = \delta_{l'l} \delta_{m'm};$$

$$f_m^{(l)} = \frac{1}{4\pi} \int_{\mathbb{S}^2} d\Omega(\theta, \psi) Y_{lm}(\psi, \theta)^* f(\theta, \psi), \ \|f\|^2 = \sum_{lm} |f_m^{(l)}|^2. \quad (3.55)$$

Note that the usual spherical polar coordinates θ, ϕ are here replaced by ψ, θ; and the usual normalisation by $\sqrt{4\pi}$ factors.

3.3.6 Cartesian tensors

We have studied the pattern and properties of $SO(3)$ UIR's via direct solutions to the commutation relations (3.39) for the generators. Another route, older than quantum angular momentum theory and important in both classical and quantum contexts, is via Cartesian tensors.

An element $A \in SO(3)$ describes the change from one right handed triad of unit vectors in Euclidean three dimensional space to another:

$$A \in SO(3) : \hat{e}'_j = A_{jk}\hat{e}_k, \ A_{jk} = \hat{e}'_j \cdot \hat{e}_k,$$

$$\hat{e}'_j \cdot \hat{e}'_k = \hat{e}_j \cdot \hat{e}_k = \delta_{jk}. \tag{3.56}$$

For a 3-vector $\boldsymbol{x}$ with respective components x_j and x'_j:

$$\boldsymbol{x} = x'_j\hat{e}'_j = x_k\hat{e}_k \Rightarrow x'_j = \boldsymbol{x} \cdot \hat{e}'_j = A_{jk}x_k. \tag{3.57}$$

We now generalise this transformation law to obtain higher order objects. A Cartesian tensor $T_{...}$ of rank l is described in each right handed rectangular coordinate system by a set of 3^l (real) numerical components $T_{j_1 \cdots j_l}$ such that under the change of axes (3.56) these components transform to $T'_{j_1 \cdots j_l}$ by the rule

$$T'_{j_1 j_2 \cdots j_l} = A_{j_1 k_1} A_{j_2 k_2} \cdots A_{j_l k_l} T_{k_1 k_2 \cdots k_l}. \tag{3.58}$$

Since $A^T A = \mathbb{1}$, these tensors admit an invariant trace operation. Choose any two indices, say j_r and j_s for $r < s$, set them equal and sum over them. The result is a Cartesian tensor of rank $(l - 2)$, dependent in general on the choices of r and s. Next since the same matrix A appears l times in (3.58), any symmetry properties $T_{...}$ may have under permutation of the indices are preserved in $T'_{...}$. In particular, complete symmetry of $T_{...}$ implies the same for $T'_{...}$.

Combining symmetry with tracelessness we get irreducible Cartesian tensors. Such a tensor follows the transformation rule (3.58) and in addition obeys:

$$T_{j_{p(1)} j_{p(2)} \cdots j_{p(l)}} = T_{j_1 j_2 \cdots j_l}, \text{ any } p \in S_l;$$

$$T_{jj j_3 \cdots j_l} = 0. \tag{3.59}$$

The successive reductions in the numbers of independent components is as follows:

General $T_{...}$with 3^l components $\rightarrow$

Symmetric $T_{...}$with $\dfrac{1}{2}(l + 1)(l + 2)$ components $\rightarrow$

Irreducible $T_{...}$with $(2l + 1)$ components. $\tag{3.60}$

Thus we are back to the UIR's $D^{(l)}$ and see easily why they can be made real.

3.4 Inclusion of Parity – The Group $O(3)$

The group $O(3)$ of all three dimensional orthogonal rotations is defined as in (3.13) but dropping the determinant condition:

$$O(3) = \{A = 3 \times 3 \text{ real matrix} \,|A^T A = \mathbb{1}\}$$
$$= SO(3) \cup \{PA \text{ or } AP \,| A \in SO(3), \ P = -\mathbb{1}\} \tag{3.61}$$

The first component has $\det A = 1$, the second has $\det AP = -1$. So $SO(3)$ and $PSO(3)$ are disjoint, each piece being connected. The passage $SO(3) \to O(3)$ is both geometrically and algebraically much simpler than the passage $SO(2) \to O(2)$ since P commutes with $SO(3)$ as a result of odd dimensionality. So a UIR of $O(3)$ is just some UIR of $SO(3)$ along with a choice $\epsilon = \pm 1$ for P since $P^2 = \mathbb{1}$:

$$UIR \ (l, \epsilon) \text{ of } O(3) : l = 0, 1, 2, \cdots , \ ; \ \epsilon = \pm 1 :$$
$$A \in SO(3) \to D^{(l,\epsilon)}(A) = D^{(l)}(A), \ D^{(l,\epsilon)}(PA) = \epsilon D^{(l)}(A). \tag{3.62}$$

Thus $(0, 1)$ corresponds to scalars; $(0, -1)$ to pseudoscalars; $(1, 1)$ to polar vectors; $(1, -1)$ to axial vectors; etc.

3.5 The Group $SU(2)$

This group is closely related to $SO(3)$ and is important in many contexts: classical polarisation optics, spin of elementary particles, quantum mechanics of two level atoms or two mode radiation fields, symmetries of basic interactions.

The defining representation is two dimensional,

$$SU(2) = \{u = 2 \times 2 \text{ complex matrix} | \ u^\dagger u = \mathbb{1}, \ \det u = 1\}, \tag{3.63}$$

so it is called the unitary unimodular group in two dimensions. In all work related to $SU(2)$ the Pauli matrices and their properties are very important:

$$\sigma_1 = \begin{pmatrix} 0 & 1 \\ 1 & 0 \end{pmatrix}, \sigma_2 = \begin{pmatrix} 0 & -i \\ i & 0 \end{pmatrix}, \sigma_3 = \begin{pmatrix} 1 & 0 \\ 0 & -1 \end{pmatrix};$$
$$\sigma_j^\dagger = \sigma_j, \operatorname{Tr} \sigma_j = 0;$$
$$\sigma_j \sigma_k = \delta_{jk} + i\epsilon_{jkl}\sigma_l, \operatorname{Tr} \sigma_j \sigma_k = 2\delta_{jk};$$
$$\text{3-vectors } \boldsymbol{a}, \boldsymbol{b} : \ \boldsymbol{a} \cdot \boldsymbol{\sigma} \, \boldsymbol{b} \cdot \boldsymbol{\sigma} = \boldsymbol{a} \cdot \boldsymbol{b} + i\boldsymbol{a} \wedge \boldsymbol{b} \cdot \boldsymbol{\sigma}. \tag{3.64}$$

There are four useful parametrisations of $SU(2)$, three being extensions of those for $SO(3)$:

a) **Parametrisation specific to $SU(2)$:**

$$u \in SU(2) \Leftrightarrow u = \begin{pmatrix} \alpha & \beta \\ -\beta^* & \alpha^* \end{pmatrix}, \; \alpha \text{ and } \beta \text{ complex,}$$

$$|\alpha|^2 + |\beta|^2 = 1. \tag{3.65}$$

So $SU(2)$ is a three parameter continuous group, its manifold structure being $\mathbb{S}^3$. Recall from Eq. (3.19) that as a manifold $SO(3) \sim \mathbb{S}^3/$ antipodal points identified.

b) **Axis-angle parameters**

Any $u \in SU(2)$ can be written in the form

$$u(\hat{\boldsymbol{n}}, \alpha) \;\; = \cos\frac{\alpha}{2} - i\hat{\boldsymbol{n}} \cdot \boldsymbol{\sigma} \sin\frac{\alpha}{2}, \; \hat{\boldsymbol{n}} \in \mathbb{S}^2, \; \alpha \in [0, 2\pi];$$

$$\boldsymbol{\alpha} = \alpha\hat{\boldsymbol{n}} \; : \; u(\boldsymbol{\alpha}) = \cos\frac{\alpha}{2} - i\frac{\boldsymbol{\alpha} \cdot \boldsymbol{\sigma}}{\alpha} \sin\frac{\alpha}{2}. \tag{3.66}$$

Note that the real angle α here is to be distinguished from the complex matrix element α in (3.65).

c) **Homogeneous Euler parameters**

For each $a = (a_0, \boldsymbol{a}) \in \mathbb{S}^3$ we have a unique $SU(2)$ element

$$u(a) = a_0 - i\boldsymbol{a} \cdot \boldsymbol{\sigma}, \;\; a_0 = \cos\frac{\alpha}{2}, \; \boldsymbol{a} = \hat{\boldsymbol{n}} \sin\frac{\alpha}{2}. \tag{3.67}$$

d) **Euler angles**

As in Eq. (3.20) in the $SO(3)$ case, we first identify the general $U(1)$ subgroup in $SU(2)$. Remembering from Eq. (3.66) that in contrast to Eq. (3.16) now we have

$$u(\hat{\boldsymbol{n}}, \; 4\pi - \alpha) = u(-\hat{\boldsymbol{n}}, \alpha), \tag{3.68}$$

we define one $U(1)$ subgroup for each fixed $\hat{\boldsymbol{n}}$ up to sign (director in $\mathbb{R}^3$):

$$U(1, \pm\hat{\boldsymbol{n}}) = \{u(\hat{\boldsymbol{n}}, \alpha) | \hat{\boldsymbol{n}} \text{ fixed}, \; 0 \le \alpha < 4\pi\} \subset SU(2). \tag{3.69}$$

Each of these is abelian and isomorphic to $SO(2)$. Taking $\hat{n} = \hat{e}_j$, $j = 1, 2, 3$, we get three basic $U(1)$ subgroups:

$$u(\hat{e}_1, \alpha) = \begin{pmatrix} \cos\frac{\alpha}{2} & -i\sin\frac{\alpha}{2} \\ -i\sin\frac{\alpha}{2} & \cos\frac{\alpha}{2} \end{pmatrix},$$

$$u(\hat{e}_2, \alpha) = \begin{pmatrix} \cos\frac{\alpha}{2} & -\sin\frac{\alpha}{2} \\ \sin\frac{\alpha}{2} & \cos\frac{\alpha}{2} \end{pmatrix},$$

$$u(\hat{e}_3, \alpha) = \begin{pmatrix} e^{-i\frac{\alpha}{2}} & 0 \\ 0 & e^{i\frac{\alpha}{2}} \end{pmatrix}. \tag{3.70}$$

Each of these, and a general $u \in SU(2)$, can be written in exponential form:

$$u(\hat{e}_j, \alpha) = e^{-\frac{i}{2}\alpha\sigma_j}, \quad u(\hat{n}, \alpha) = e^{-\frac{i}{2}\alpha\hat{n}\cdot\boldsymbol{\sigma}}. \tag{3.71}$$

The three 'generator matrices' are now $\frac{1}{2}\sigma_j$, each hermitian and traceless. It is now an easy exercise in matrix algebra to check that any $u \in SU(2)$ is expressible as the product of three simple factors:

$$u = \begin{pmatrix} \alpha & \beta \\ -\beta^* & \alpha^* \end{pmatrix} = u(\theta, \psi, \phi) = u(\hat{e}_3, \theta)u(\hat{e}_2, \psi)u(\hat{e}_3, \phi),$$

$$\alpha = e^{-i(\phi+\theta)/2}\cos\frac{\psi}{2}, \beta = -e^{i(\phi-\theta)/2}\sin\frac{\psi}{2},$$

$$0 \le \theta \le 4\pi, \ 0 \le \phi \le 2\pi, \ 0 \le \psi \le \pi. \tag{3.72}$$

The range for θ is double that for $SO(3)$, while for ψ and ϕ they are the same as for $SO(3)$. Equally well we could have taken

$$0 \le \theta \le 2\pi, \ 0 \le \psi \le \pi, \ 0 \le \phi \le 4\pi.$$

Then it is easy to see that the coset space $SU(2)/U(1, \hat{e}_3) = \mathbb{S}^2$, as in the $SO(3)$ case (3.31).

As with $SO(3)$, here too the axis-angle description is convenient for three purposes: the infinitesimal composition law; topological aspects; conjugation and equivalence classes. Up to quadratic terms,

$$u(\boldsymbol{\alpha}) \simeq 1 - \frac{i}{2}\boldsymbol{\alpha}\cdot\boldsymbol{\sigma} - \frac{\alpha^2}{8}. \tag{3.73}$$

Writing $u(\boldsymbol{\alpha})u(\boldsymbol{\beta}) = u(\boldsymbol{\alpha} + \boldsymbol{\beta} + \boldsymbol{\gamma})$ with $\boldsymbol{\gamma}$ starting off with terms of the form $\alpha\beta$, and retaining only terms linear in $\boldsymbol{\alpha}$, linear in $\boldsymbol{\beta}$, and cross terms $\alpha\beta$, we find

$$(1 - \frac{i}{2}\boldsymbol{\alpha} \cdot \boldsymbol{\sigma})(1 - \frac{i}{2}\boldsymbol{\beta} \cdot \boldsymbol{\sigma}) \simeq 1 - \frac{i}{2}(\boldsymbol{\alpha} + \boldsymbol{\beta} + \boldsymbol{\gamma}) \cdot \boldsymbol{\sigma} - \frac{1}{4}\boldsymbol{\alpha} \cdot \boldsymbol{\beta},$$

$$\text{i.e.,} \quad -\frac{1}{4}(\boldsymbol{\alpha} \cdot \boldsymbol{\beta} + i\boldsymbol{\alpha} \wedge \boldsymbol{\beta} \cdot \boldsymbol{\sigma}) \simeq -\frac{i}{2}\boldsymbol{\gamma} \cdot \boldsymbol{\sigma} - \frac{1}{4}\boldsymbol{\alpha} \cdot \boldsymbol{\beta},$$

$$\text{i.e.,} \quad \boldsymbol{\gamma} = \frac{1}{2}\boldsymbol{\alpha} \wedge \boldsymbol{\beta} \cdots ,$$

$$u(\boldsymbol{\alpha})u(\boldsymbol{\beta}) = u(\boldsymbol{\alpha} + \boldsymbol{\beta} + \frac{1}{2}\boldsymbol{\alpha} \wedge \boldsymbol{\beta} \cdots). \tag{3.74}$$

For elements near the identity, $SO(3)$ and $SU(2)$ have 'isomorphic' composition laws, their differences are global. The nonabelian nature of $SU(2)$ is also seen clearly.

We saw that topologically $SU(2) \sim \mathbb{S}^3$. Therefore $SU(2)$ is simply connected – every closed loop starting and ending at the identity can be continuously shrunk to the trivial or constant loop. In axis-angle parameters we see that

$$u(\hat{\boldsymbol{n}}, 2\pi) = -\mathbb{1}, \quad \text{all } \hat{\boldsymbol{n}} \in \mathbb{S}^2. \tag{3.75}$$

Therefore the space of $SU(2)$ can also be pictured as the interior and surface of a sphere of radius 2π in the fictitious three dimensional $\boldsymbol{\alpha}$-space, except that *all* points on the surface are identified by (3.75) to be *the same*. In this picture too one can draw loops and see that they are all collapsible to the trivial loop.

3.5.1 Relation to $SO(3)$, conjugation, classes

The action of conjugation and description of classes become easy once the $SU(2)$-$SO(3)$ relationship is seen, so we first deal with this. Denote by $\mathcal{M}$ the set of all 2×2 hermitian traceless matrices. They are in correspondence with vectors in $\mathbb{R}^3$:

$$\boldsymbol{x} \in \mathbb{R}^3 \Leftrightarrow \boldsymbol{x} \cdot \boldsymbol{\sigma} = \begin{pmatrix} x_3 & x_1 - ix_2 \\ x_1 + ix_2 & -x_3 \end{pmatrix} \in \mathcal{M},$$

$$\det \boldsymbol{x} \cdot \boldsymbol{\sigma} = -\boldsymbol{x} \cdot \boldsymbol{x}. \tag{3.76}$$

Under conjugation by any $u \in SU(2)$, $\mathcal{M}$ is mapped onto itself:

$$u \in SU(2), \ \boldsymbol{x} \cdot \boldsymbol{\sigma} \in \mathcal{M} \to u\boldsymbol{x} \cdot \boldsymbol{\sigma} u^{-1} = \boldsymbol{x}' \cdot \boldsymbol{\sigma} \in \mathcal{M},$$

$$\boldsymbol{x}' = L(u)\boldsymbol{x}, \tag{3.77}$$

where $L(u)$ is some real 3×3 matrix. Since conjugation preserves the determinant, we have $L(u) \in O(3)$. Carrying out a second conjugation gives:

$$L(u')L(u) = L(u'u). \tag{3.78}$$

By continuity and since $SU(2)$ is (simply) connected, we have $\det L(u) = 1$, so $L(u) \in SO(3)$. Easy calculations show that every $A \in SO(3)$ arises as $L(u)$ for two choices of u, since $L(-u) = L(u)$. We now write $A(u)$ in place of $L(u)$, so we have the homomorphism condition:

$$u \in SU(2) \to A(u) \in SO(3), \ A(u')A(u) = A(u'u). \tag{3.79}$$

The explicit formulae which are useful are:

$$u\boldsymbol{x} \cdot \boldsymbol{\sigma} u^{-1} = (A(u)\boldsymbol{x}) \cdot \boldsymbol{\sigma},$$
$$u\sigma_j u^{-1} = A_{kj}(u)\sigma_k,$$
$$A_{kj}(u) = \frac{1}{2}\mathrm{Tr}(\sigma_k u \sigma_j u^{-1}). \tag{3.80}$$

Moreover the three parametrisations we have given for $SO(3)$ and $SU(2)$ match perfectly:

$$A(u(\boldsymbol{\alpha})) = A(\boldsymbol{\alpha}), \ A(u(a)) = A(a), \ A(u(\theta, \psi, \phi)) = A(\theta, \psi, \phi). \tag{3.81}$$

The kernel of the $SU(2) \to SO(3)$ homomorphism is the two element centre $\mathbb{Z}_2 = \{\mathbb{1}, -\mathbb{1}\}$ of $SU(2)$, so

$$SO(3) = SU(2)/\mathbb{Z}_2, \ A(u = \pm\mathbb{1}_{2\times2}) = \mathbb{1}_{3\times3}. \tag{3.82}$$

Thus $SU(2)$ is 'twice as big' as $SO(3)$, in each parametrisation its range is double that for $SO(3)$, but near the identity they are locally isomorphic and look the same.

Incidentally from Eq. (3.77) we have:

$$u \in SU(2) \to u\sigma_3 u^{-1} = \hat{\boldsymbol{n}} \cdot \boldsymbol{\sigma}, \ \hat{\boldsymbol{n}} \in \mathbb{S}^2.$$
$$\hat{\boldsymbol{n}} = (0,0,1) \Leftrightarrow u \in U(1, \hat{\boldsymbol{e}}_3), \tag{3.83}$$

so the result $SU(2)/U(1) = \mathbb{S}^2$ follows immediately.

Now we can discuss classes in $SU(2)$. Using axis-angle parameters we have from Eqs. (3.66, 3.80):

$$u, u(\hat{\boldsymbol{n}}, \alpha) \in SU(2):$$

$$u\, u(\hat{\boldsymbol{n}}, \alpha) u^{-1} = u(\cos\frac{\alpha}{2} - i\hat{\boldsymbol{n}} \cdot \boldsymbol{\sigma} \sin\frac{\alpha}{2}) u^{-1}$$

$$= \cos\frac{\alpha}{2} - i u \hat{\boldsymbol{n}} \cdot \boldsymbol{\sigma}\, u^{-1} \sin\frac{\alpha}{2}$$

$$= u(\hat{\boldsymbol{n}}', \alpha),$$

$$\hat{\boldsymbol{n}}' = A(u)\hat{\boldsymbol{n}}. \tag{3.84}$$

Thus over each class α is constant while $\hat{\boldsymbol{n}}$ varies over all of $\mathbb{S}^2$, so they can be listed:

$$\alpha = 0 \;:\; \mathcal{C}(0) = \{\mathbb{1}\}, \text{ single element;}$$

$$0 < \alpha < 2\pi \;:\; \mathcal{C}(\alpha) = \{u(\hat{\boldsymbol{n}}, \alpha)|\alpha \text{ fixed}, \hat{\boldsymbol{n}} \in \mathbb{S}^2 \text{ varying}\};$$

$$\alpha = 2\pi \;:\; \mathcal{C}(2\pi) = \{-\mathbb{1}\}, \text{ single element.} \tag{3.85}$$

As representative elements we can choose as in Eq. (3.30):

$$0 \le \alpha \le 2\pi \;:\; \mathcal{C}(\alpha) \to u(\hat{\boldsymbol{e}}_3, \alpha) = \begin{pmatrix} e^{-i\alpha/2} & 0 \\ 0 & e^{i\alpha/2} \end{pmatrix}. \tag{3.86}$$

A 'class function' is thus a (real or complex) function $f(\alpha)$, $0 \le \alpha \le 2\pi$.

3.5.2 Invariant integration over $SU(2)$

Let $f(u)$ be any function on $SU(2)$ belonging to a linear space. Its average over $SU(2)$ is expressible in various ways depending on the choice of parameters:

$$\int_{SU(2)} du\, f(u) = \int_{SU(2)} du\, (f(u_0 u) \text{ or } f(u u_0) \text{ or } f(u^{-1})), \text{ any fixed } u_0,$$

$$= \int_{0 \le \alpha \le 2\pi} d^3\alpha\, \frac{1 - \cos\alpha}{8\pi^2\alpha^2} f(\alpha)$$

$$= \frac{1}{\pi^2} \int_{\mathbb{S}^3} d^4 a\, \delta(a_0^2 + \boldsymbol{a}^2 - 1) f(a)$$

$$= \frac{1}{16\pi^2} \int_0^{4\pi} d\theta \int_0^{2\pi} d\phi \int_0^{\pi} d\psi\, \sin\psi f(\theta, \psi, \phi). \tag{3.87}$$

The Hilbert space $L^2(SU(2))$ which carries the (mutually commuting) left and right regular representations consists of complex valued $f(a)$ with finite norm:

$$L^2(SU(2)) = \{f(a) \in \mathbb{C} | \|f\|^2 = \int_{SU(2)} du \; |f(u)|^2 < \infty\}. \qquad (3.88)$$

For class functions we have a one dimensional integration:

$$f(u) = f(u_0 u u_0^{-1}), \text{ all } u, u_0 :$$

$$\int_{SU(2)} du \; f(u) = \frac{1}{\pi} \int_0^{2\pi} d\alpha \; \sin^2 \frac{\alpha}{2} f(\alpha). \qquad (3.89)$$

3.5.3 Some important properties of $SU(2)$ representations

This will be similar in treatment as with $SO(3)$ in 3.3.3. There will of course be tell-tale differences.

i) Every representation is equivalent to, and so may be assumed to be, a UR; so we need only find all UIR's up to equivalence.

ii) On account of the 2 to 1 homomorphism $SU(2) \rightarrow SO(3)$, every UR/UIR of $SO(3)$ is automatically also a UR/UIR of $SU(2)$ but not a faithful one as it does not distinguish u from $-u$. (Faithful $SU(2)$ UR's are not $SO(3)$ UR's). We just set $\mathcal{D}(u) = D(A(u))$.

iii) There is one UIR in any dimension $(2j+1)$ for $j = 0, \frac{1}{2}, 1, \cdots$. For integer j, in odd dimensions, these are the $SO(3)$ UIR's $D^{(l)}$; for $j = \frac{1}{2}, \frac{3}{2}, \cdots$, in even dimensions they are faithful $SU(2)$ UIR's $D^{(j)}$. The defining UIR (3.63) has $j = \frac{1}{2}$. The faithful UIR's of $SU(2)$ are called double valued or two-valued UIR's of $SO(3)$, even though strictly speaking they are not UIR's of $SO(3)$.

iv) For every j, $D^{(j)*}$ is equivalent to $D^{(j)}$. For integer j, as we know, they can be made real. However, for $j = \frac{1}{2}, \frac{3}{2}, \cdot$, they are pseudoreal and cannot be brought to real form.

In any UR of $SU(2)$ we write the operators as $D(u), D(\alpha), D(a)$ or $D(\theta, \psi, \phi)$. They obey

$$D(u')D(u) = D(u'u), \quad D(\alpha)D(\beta) = D(\alpha + \beta + \frac{1}{2}\alpha \wedge \beta + \cdots) \qquad (3.90)$$

The hermitian generators and commutation relations are similar to $SO(3)$:

$$J_j = i\frac{\partial}{\partial\alpha_j}D(\boldsymbol{\alpha})\bigg|_{\boldsymbol{\alpha}=0} \ , \ \ D(\boldsymbol{\alpha}) \simeq \mathbb{1} - i\boldsymbol{\alpha}\cdot\boldsymbol{J} + O(\alpha^2);$$

$$[J_j, J_k] = i\epsilon_{jkl}J_l. \tag{3.91}$$

So $SU(2)$ and $SO(3)$ have *the same* structure constants. In the defining UIR (3.63) we have $\boldsymbol{J} = \frac{1}{2}\boldsymbol{\sigma}$. In any UR, Eq. (3.91) integrates to

$$D(\boldsymbol{\alpha}) = e^{-i\boldsymbol{\alpha}\cdot\boldsymbol{J}}. \tag{3.92}$$

From general theory, any UR of $SU(2)$ arises from some hermitian representation of the commutation relations in (3.91), *with no added conditions*, by exponentiation. This is because $SU(2)$ is simply connected. In any UR, under conjugation the generators transform *as in the SO(3) case*:

$$D(u)e^{-i\boldsymbol{\alpha}\cdot\boldsymbol{J}}D(u)^{-1} = e^{-iA(u)\boldsymbol{\alpha}\cdot\boldsymbol{J}} = e^{-i\boldsymbol{\alpha}\cdot A(u)^T\boldsymbol{J}} \Rightarrow$$

$$D(u)J_jD(u)^{-1} = A_{kj}(u)J_k. \tag{3.93}$$

Thus the $SO(3)$ defining representation is the $SU(2)$ adjoint representation.

3.5.4 The $SU(2)$ UIR's

Now we accept all hermitian irreducible representations of the commutation relations (3.39, 3.91), for $j = 0, \frac{1}{2}, 1, \frac{3}{2}, \cdots$. For given j, $J^2 = j(j+1)$. $\mathbb{1}$ and J_3 has eigenvalues $m = j, j-1, \cdots, -j+1, -j$; and an orthonormal basis is written as $|j, m\rangle$. The generator actions are well known from quantum mechanics:

$$J^2|j, m\rangle = j(j+1)|j, m\rangle, \ J_3|j, m\rangle = m|j, m\rangle, \ \langle j, m'|j, m\rangle = \delta_{m'm};$$

$$(J_1 \pm iJ_2)|j, m\rangle = \sqrt{(j\mp m)(j\pm m+1)} \ |j, m\pm 1\rangle. \tag{3.94}$$

Again J_1 and J_3 are real symmetric, J_2 is pure imaginary antisymmetric.

The spectrum of J_3 tells us the reduction of the $SU(2)$ UIR under $U(1, \hat{\boldsymbol{e}}_3)$, say:

$$\text{UIR } D^{(j)}(\,\cdot\,) \text{ of } SU(2)\bigg|_{U(1,\hat{\boldsymbol{e}}_3)} = \sum_{m=-j}^{j}{}_{\oplus} D^{(m)}(\,\cdot\,) \text{ of } U(1, \hat{\boldsymbol{e}}_3). \tag{3.95}$$

So the jth irreducible character is

$$\chi^{(j)}(\alpha) = \sum_{m=-j}^{j} e^{-im\alpha} = \frac{\sin\left(j + \frac{1}{2}\right)\alpha}{\sin\frac{\alpha}{2}}, \quad 0 \leq \alpha \leq 2\pi. \tag{3.96}$$

One easily verifies their orthogonality and the expansion for class functions:

$$\frac{1}{\pi} \int_0^{2\pi} d\alpha \, \sin^2 \frac{\alpha}{2} \chi^{(j')}(\alpha)^* \chi^{(j)}(\alpha) = \delta_{j'j};$$

$$f(\alpha) = \sum_{j=0,\frac{1}{2},1\cdots} f_j \chi^{(j)}(\alpha) : \frac{1}{\pi} \int_0^{2\pi} d\alpha \, \sin^2 \frac{\alpha}{2} |f(\alpha)|^2 = \sum_j |f_j|^2. \tag{3.97}$$

3.5.5 The D-matrices, orthogonality and completeness

The structures are similar to the $SO(3)$ case except for including $j = \frac{1}{2}, \frac{3}{2}, \cdots$, and doubling the parameter ranges:

D-matrices

$$\langle j, m' | D(\theta, \psi, \phi) | j, m \rangle = D^{(j)}_{m'm}(\theta, \psi, \phi) = e^{-im'\theta - im\phi} d^{(j)}_{m'm}(\psi),$$

$$d^{(j)}_{m'm}(\psi) = \langle j, m' | e^{-i\psi J_2} | j, m \rangle = \text{real orthogonal}$$

$$= \sqrt{\frac{(j+m')!(j-m')!}{(j+m)!(j-m)!}} \sum_k (-1)^{j-m'-k} \binom{j+m}{j-m'-k} \binom{j-m}{k}$$

$$\times \left(\cos\frac{\psi}{2}\right)^{2k+m'+m} \left(\sin\frac{\psi}{2}\right)^{2j-2k-m'-m}; \tag{3.98a}$$

Orthonormality

$$\int_{SU(2)} du \, D^{(j')}_{m'_1 m'_2}(u)^* D^{(j)}_{m_1 m_2}(u) = \frac{1}{(2j+1)} \delta_{j'j} \delta_{m'_1 m_1} \delta_{m'_2 m_2}; \tag{3.98b}$$

Completeness

$$f \in L^2(SU(2)) : f(u) = \sum_{j=0,\frac{1}{2},1,\cdots} \sum_{m_1,m_2} (2j+1)^{1/2} f^{(j)}_{m_1 m_2} D^{(j)}_{m_1 m_2}(u),$$

$$f^{(j)}_{m_1 m_2} = (2j+1)^{1/2} \int_{SU(2)} du \, D^{(j)}_{m_1 m_2}(u)^* f(u);$$

$$\|f\|^2 = \int_{SU(2)} du \, |f(u)|^2 = \sum_{j \, m_1 m_2} |f^{(j)}_{m_1 m_2}|^2. \tag{3.98c}$$

We leave the details of the descent to $L^2(\mathbb{S}^2)$, analogous to Eq. (3.55) as an exercise.

Let us summarise the key features of $SU(2)$ and $SO(3)$ UIR's in a table:

Table 3.1 Key features of UIR's of $SU(2)$ and $SO(3)$

j	$\dim(2j+1)$	$SU(2)$ Rep.	$SO(3)$ Rep.	Reality	Remarks
0	1	Trivial	Trivial	Real	$D(\boldsymbol{\alpha}) = \mathbb{1}$
$\frac{1}{2}$	2	Defining, faithful, spinor	No	Pseudoreal	$D(\boldsymbol{\alpha})|_{2\pi} = -\mathbb{1}$
1	3	Adjoint, non Faithful	Defining, adjoint	Real	$D(\boldsymbol{\alpha})|_{2\pi} = \mathbb{1}$
$\frac{3}{2}$	4	Faithful	No	Pseudoreal	$D(\boldsymbol{\alpha})|_{2\pi} = -\mathbb{1}$
2	5	Nonfaithful	Faithful	Real	$D(\boldsymbol{\alpha})|_{2\pi} = \mathbb{1}$

$SU(2)$ has, up to equivalence, one UIR in each dimension, unlike any other compact Lie group.

For integer $j = 0, 1, 2, \cdots$, there is a symmetric bilinear invariant, like $\boldsymbol{a} \cdot \boldsymbol{b}$ in the case of 3-vectors. For $j = \frac{1}{2}, \frac{3}{2}, \cdots$, there is an antisymmetric bilinear invariant.

3.5.6 $SU(2)$ multispinors

This is similar in concept to Cartesian tensors for $SO(3)$. Use $\alpha, \beta, \cdots = 1, 2$ as indices for components of two component spinors. Then a rank $2j$ $SU(2)$ multispinor is a collection of 2^{2j} complex (in general) components with the transformation rule:

$$u \in SU(2): \quad \psi_{\alpha_1\alpha_2\cdots\alpha_{2j}} \rightarrow \psi'_{\alpha_1\alpha_2\cdots\alpha_{2j}} = u_{\alpha_1\beta_1} u_{\alpha_2\beta_2} \cdots u_{\alpha_{2j}\beta_{2j}} \psi_{\beta_1\beta_2\cdots\beta_{2j}}. \quad (3.99)$$

In the general case, ψ has 2^{2j} independent complex components. But if it is completely symmetric, which is consistent with (3.99), the number of components drops to $(2j + 1)$ – we get immediately the UIR $D^{(j)}$! There is no need to remove traces.

3.5.7 Weyl, Jordan and Schwinger constructions

These are closely linked to the symmetric multispinor idea. Let ξ, η be the two complex components of a spin $\frac{1}{2}$ object or 'wave function', with the $SU(2)$ transformation law:

$$u \in SU(2) : \quad \begin{pmatrix} \xi \\ \eta \end{pmatrix} \rightarrow \begin{pmatrix} \xi' \\ \eta' \end{pmatrix} = u \begin{pmatrix} \xi \\ \eta \end{pmatrix}. \quad (3.100)$$

Then for any given j the $(2j+1)$ monomials $\xi^{j+m}\eta^{j-m}/\sqrt{(j+m)!(j-m)!}$ transform unitarily among themselves according to the UIR $D^{(j)}$ of $SU(2)$. This is the Weyl result:

$$\frac{\xi'^{j+m'}\eta'^{j-m'}}{\sqrt{(j+m')!(j-m')!}} = \sum_m D^{(j)}_{m'm}(u) \frac{\xi^{j+m}\eta^{j-m}}{\sqrt{(j+m)!(j-m)!}} . \tag{3.101}$$

By considering all possible j we cover all polynomials in ξ, η and get all the $SU(2)$ UIR's, once each.

A quantum mechanical operator form of this is the Jordan–Schwinger construction. Take two commuting boson operator sets $a_\alpha, a_\alpha^\dagger$ obeying

$$[a_\alpha, a_\beta^\dagger] = \delta_{\alpha\beta}, \ [a_\alpha, a_\beta] = [a_\alpha^\dagger, a_\beta^\dagger] = 0. \tag{3.102}$$

The construction

$$\boldsymbol{J} = \frac{1}{2} a^\dagger \boldsymbol{\sigma} a, \ a = \begin{pmatrix} a_1 \\ a_2 \end{pmatrix} \tag{3.103}$$

yields a hermitian solution to the $SU(2)$ commutation relations. Note that the total number operator commutes with $\boldsymbol{J}$:

$$N = a^\dagger a = a_1^\dagger a_1 + a_2^\dagger a_2 \ : \ [N, \boldsymbol{J}] = 0. \tag{3.104}$$

We obtain an infinite dimensional UR of $SU(2)$ using the $\boldsymbol{J}$'s:

$$U(\boldsymbol{\alpha}) = e^{-i\boldsymbol{\alpha}\cdot\boldsymbol{J}}, \ U(\theta, \psi, \phi) = e^{-i\theta J_3} e^{-i\psi J_2} e^{-i\phi J_3}, \tag{3.105}$$

acting on the Hilbert space of the two identical oscillators. Then if we define the orthonormal basis

$$|j, m\rangle = (a_1^\dagger)^{j+m}(a_2^\dagger)^{j-m}|0, 0\rangle/\sqrt{(j+m)!(j-m)!},$$

$$\langle j'm'|jm\rangle = \delta_{j'j}\delta_{m'm}, \tag{3.106}$$

for $j = 0, \frac{1}{2}, 1, \cdots, \ m = j, j-1, \cdots, -j$, we have:

$$U(\boldsymbol{\alpha})|jm\rangle = \sum_{m'} D^{(j)}_{m'm}(\boldsymbol{\alpha})|jm'\rangle. \tag{3.107}$$

This is called the Schwinger representation of $SU(2)$, containing each $SU(2)$ UIR once. It has applications in many quantum mechanical problems. The Schwinger representation concept will be reexamined and generalised in Chapter 7.

Problems

P3.1　For the nonabelian group $O(2)$,

　　(i)　define a normalised integration process for complex functions on $O(2)$;
　　(ii)　use it to define the Hilbert space $L^2(O(2))$;
　　(iii)　Set up the right and left regular representations of $O(2)$ on $L^2(O(2))$;
　　(iv)　work out the details of the orthogonality and completeness of the matrix elements of
　　　　UIR D-matrices in Eqs. (3.11) and (3.12), in this Hilbert space.

P3.2　For two $(2l+1)$-component complex vectors

$$\{\psi_m^{(a)}, m = l, l-1, \cdots, -l; a = 1, 2\}$$

in the space of the UIR $D^{(l)}(A)$ of $SO(3)$, obeying the transformation rule

$$A \in SO(3) : \psi_m^{(a)} \to \psi_m'^{(a)} = \sum_{m'} D_{mm'}^{(l)}(A)\psi_{m'}^{(a)}, \; a = 1, 2,$$

show that the expression

$$\sum_{m=-l}^{l} (-1)^{l-m} \, \psi_m^{(1)} \psi_{-m}^{(2)}$$

is a symmetric bilinear invariant.

P3.3　For two $(2j+1)$-component complex vectors

$$\{\psi_m^{(a)}, m = j, j-1, \cdots, -j; a = 1, 2\}$$

in the space of the faithful UIR $D^{(j)}(u)$ of $SU(2)$, $j = 1/2, 3/2, 5/2, \cdots$, obeying the transformation rule

$$u \in SU(2) : \psi_m^{(a)} \to \psi_m'^{(a)} = \sum_{m'} D_{mm'}^{(j)}(u)\psi_{m'}^{(a)}, \; a = 1, 2,$$

show that the expression

$$\sum_{m=-j}^{j} (-1)^{j-m} \, \psi_m^{(1)} \psi_{-m}^{(2)}$$

is an antisymmetric bilinear invariant.

Bibliography

Edmonds, A. R. (1957). *Angular Momentum in Quantum Mechanics*. Princeton, New Jersey:
　　Princeton University Press.
Gelfand, I. M., Minlos, R. A. and Shapiro, Z. Y. (1963). *Representations of the Rotation and
　　Lorentz Groups and Their Applications*. New York: MacMillan.

Pauli, W. (1965). Continuous Groups in Quantum Mechanics. CERN Lectures 1956. *Ergebnisse der Exakten Naturwissenschaften*, 37: 85–104.

Schwinger, J. (1952). On Angular Momentum, USAEC Report NYO–3071. Reprint (1965) *Quantum Theory of Angular Momentum*. Biedenharn, L. C. and van Dam, H., eds. New York: Academic Press.

4

General Theory of Lie Groups and Lie Algebras

The study of $SO(3)$ and $SU(2)$ has shown how elements of a continuous group can be labelled by (a certain number of) real independent continuous coordinates or parameters; how the composition law can be expressed using these coordinates; how in a representation we encounter generators, commutation relations, structure constants; the representation of finite group elements by (products of) exponentials in the generators; and so on. Now we will try to understand all this in a more basic manner and in a general situation.

The work of this chapter and the next one will lead us to a vast generalisation of $SO(3)$ and $SU(2)$ resulting in the so-called classical families of continuous groups which are all, like $SO(3)$ and $SU(2)$, compact. (The concept of compactness will be briefly described in a heuristic manner in Chapter 5.) These are mathematical results from the late nineteenth and early twentieth centuries, associated with the names of Killing, Cartan and Weyl and are truly beautiful.

4.1 Local Coordinates, Group Composition, Inverses

Let a Lie group G be given. The dimension of G, also called its order, will hereafter be denoted by r rather than n. (In the development of the theory of compact simple Lie groups, the order is traditionally denoted by r; and another important property called the rank, which we will come to in Chapter 5, by ℓ. These are the notations used, for instance, in the classic 1951 Princeton lectures by Giulio Racah on Group Theory and Spectroscopy.) In some neighborhood $\mathcal{N}$ of $e \in G$, we use r essential real independent parameters to label group elements:

$$a, b, \cdots \in G \longmapsto \alpha = \{\alpha^j\},\ \beta = \{\beta^j\}, \cdots,\ j = 1, 2, \cdots, r. \tag{4.1}$$

It is understood that $a, b, \cdots \in \mathcal{N} \subset G$. As a convention we always assume

$$e \in G \longrightarrow \alpha^j = 0. \tag{4.2}$$

As $a \in G$ runs over $\mathcal{N}$, α runs over some open set around the origin in r-dimensional Euclidean space. So in this region and in $\mathcal{N}$, coordinates and group elements determine one another uniquely.

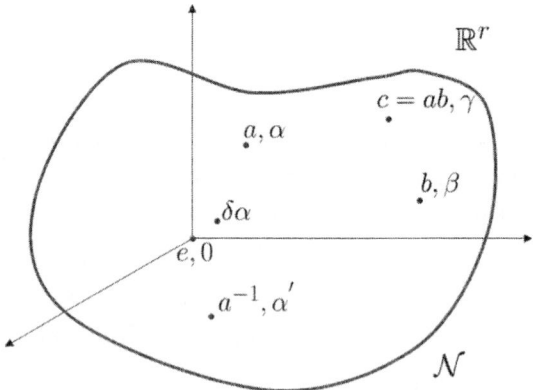

Figure 4.1   Local coordinates in the neighbourhood $\mathcal{N} \subset G$

The freedom to alter coordinates smoothly is always available, and we will use it later. We can imagine replacing α^j by $\bar{\alpha}^j$ in a smooth invertible way. But we always obey the convention (4.2):

$$\alpha^j = 0 \Leftrightarrow \bar{\alpha}^j = 0. \tag{4.3}$$

For given elements $a, b \in \mathcal{N}$ with coordinates α, β let the product $c = ab$ (also assumed to be in $\mathcal{N}$) have coordinates γ as in the figure. They are given by a system of r real functions of $2r$ arguments each:

$$c = ab \Leftrightarrow \gamma^j = f^j(\alpha; \beta),\ j = 1, 2, \cdots, r. \tag{4.4}$$

The composition law is expressed in coordinates by this system of functions. The convention (4.2) implies:

$$f^j(\alpha;0) = f^j(0;\alpha) = \alpha^j. \tag{4.5}$$

The existence of a unique inverse a^{-1} to each $a \in \mathcal{N}$, assuming also $a^{-1} \in \mathcal{N}$, is expressible as follows: given the coordinates α of a, the coordinates α' of a^{-1} are the unique solutions to

$$f^j(\alpha;\alpha') = 0 \text{ or } f^j(\alpha';\alpha) = 0. \tag{4.6}$$

If we switch to a new coordinate system $\overline{\alpha}$, then the group composition functions will be $\overline{f}^j(\overline{\alpha};\overline{\beta})$ having new functional forms. Symbolically:

$$\overline{\alpha} = \varphi(\alpha), \; \overline{\beta} = \varphi(\beta) \to \overline{f}(\overline{\alpha};\overline{\beta}) = \varphi(f(\alpha;\beta))$$
$$= \varphi(f(\varphi^{-1}(\overline{\alpha});\varphi^{-1}(\overline{\beta}))). \tag{4.7}$$

From the functions $f(\alpha;\beta)$ we obtain systems of more elementary or basic functions as follows. Consider functions $\Phi(g)$ defined on G. In coordinates we simply write $\Phi(\alpha)$. Let $\delta\alpha$ be the coordinates of some element very close to e. Multiplying a on the left or on the right by this element gives results near to a, with coordinates $f(\delta\alpha;\alpha)$ and $f(\alpha;\delta\alpha)$, respectively. The values of $\Phi(\alpha)$ at these nearby points are, respectively:

$$\Phi(f(\delta\alpha;\alpha)) \simeq \Phi(\alpha) + \delta\alpha^j \eta_j^k(\alpha)\frac{\partial}{\partial\alpha^k}\Phi(\alpha),$$

$$\eta_j^k(\alpha) = \left.\frac{\partial f^k}{\partial\beta^j}(\beta;\alpha)\right|_{\beta=0} \; ;$$

$$\Phi(f(\alpha;\delta\alpha)) \simeq \Phi(\alpha) + \delta\alpha^j \hat{\eta}_j^k(\alpha)\frac{\partial}{\partial\alpha^k}\Phi(\alpha),$$

$$\hat{\eta}_j^k(\alpha) = \left.\frac{\partial f^k}{\partial\beta^j}(\alpha;\beta)\right|_{\beta=0} \; ;$$

$$f^j(\delta\alpha;\alpha) \simeq \alpha^j + \eta_k^j(\alpha)\delta\alpha^k, \;\; f^j(\alpha;\delta\alpha) \simeq \alpha^j + \hat{\eta}_k^j(\alpha)\delta\alpha^k. \tag{4.8}$$

(Here the functions f are assumed to be sufficiently often differentiable.) As a temporary notation based on the above let us write:

$$X_j = \eta_j^k(\alpha)\frac{\partial}{\partial\alpha^k}, \;\; \hat{X}_j = \hat{\eta}_j^k(\alpha)\frac{\partial}{\partial\alpha^k}. \tag{4.9}$$

These are important first order partial differential operators applicable to functions on G.

Since by (4.5) at $\alpha = 0$

$$\eta_k^j(0) = \hat{\eta}_k^j(0) = \delta_k^j, \tag{4.10}$$

by continuity we suppose that the matrices $(\eta_k^j(\alpha)), (\hat{\eta}_k^j(\alpha))$ are both nonsingular over $\mathcal{N}$, and we write their inverses as ξ and $\hat{\xi}$:

$$\xi_k^j(\alpha)\eta_l^k(\alpha) = \hat{\xi}_k^j(\alpha)\hat{\eta}_l^k(\alpha) = \delta_l^j, \quad \xi_k^j(0) = \hat{\xi}_k^j(0) = \delta_k^j. \tag{4.11}$$

A notation for these matrices which is sometimes useful is

$$H(\alpha) = (\eta_k^j(\alpha)), \ \Xi(\alpha) = (\xi_k^j(\alpha)) \ : \ H(\alpha)\Xi(\alpha) = \Xi(\alpha)H(\alpha) = \mathbb{1} \tag{4.12}$$

For $SO(3)$ and $SU(2)$, using subscripts throughout, these functions turn out in axis-angle parameters to be

$$\eta_{jk}(\alpha) = \delta_{jk}\frac{\alpha}{2}\cot\frac{\alpha}{2} + \frac{\alpha_j\alpha_k}{\alpha^2}\left(1 - \frac{\alpha}{2}\cot\frac{\alpha}{2}\right) + \epsilon_{jkl}\frac{\alpha_l}{2},$$

$$\xi_{jk}(\alpha) = \delta_{jk}\frac{\sin\alpha}{\alpha} + \frac{\alpha_j\alpha_k}{\alpha^2}\left(1 - \frac{\sin\alpha}{\alpha}\right) - \epsilon_{jkl}\alpha_l\frac{1 - \cos\alpha}{\alpha^2}. \tag{4.13}$$

(More precisely, the superscripts in ξ_k^j, η_k^j appear as the first subscripts.)

For much of the following we use the functions η and ξ; parallel statements involving $\hat{\eta}, \hat{\xi}$ can be made. The η and ξ are each functions of r independent real variables, while the f are functions of $2r$ independent variables. Intrinsically, $f(a; b)$ depends on two group elements, $\eta(a)$ and $\xi(a)$ on one alone. This is why the latter are more elementary than the former.

Now we study the properties of f, η, ξ carefully. There are four stages involved:

1. Analysis of associativity

2. One parameter subgroups

3. Integrability and the Lie algebra concept

4. Reconstruction of the Lie group from its Lie algebra

4.2 Associativity as a System of (Nonlinear) PDE's

For three elements $a, b, c \in \mathcal{N} \subset G$, the associativity of group composition means that the functions f obey certain functional relations:

$$c(ab) = (ca)b \Leftrightarrow f^j(\gamma; f(\alpha; \beta)) = f^j(f(\gamma; \alpha); \beta). \tag{4.14}$$

Differentiate these with respect to γ^k and then set $\gamma = 0$, i.e., $c = e$:

$$\left.\frac{\partial f^j(\gamma; f(\alpha; \beta))}{\partial \gamma^k}\right|_{\gamma=0} = \frac{\partial f^j(\alpha; \beta)}{\partial \alpha^l} \left.\frac{\partial f^l(\gamma; \alpha)}{\partial \gamma^k}\right|_{\gamma=0}. \tag{4.15}$$

The left hand side and the second factor on the right are the η's defined in (4.8), so (4.15) is:

$$\eta_k^j(f(\alpha; \beta)) = \frac{\partial f^j(\alpha; \beta)}{\partial \alpha^l} \eta_k^l(\alpha),$$

i.e., $$\frac{\partial f^j(\alpha; \beta)}{\partial \alpha^l} = \eta_k^j(f(\alpha; \beta))\xi_l^k(\alpha), \quad f^j(0; \beta) = \beta^j \tag{4.16}$$

Using the notations (4.12), these equations are:

$$\frac{\partial f^j(\alpha; \beta)}{\partial \alpha^l} = (H(f(\alpha; \beta))\varXi(\alpha))_l^j. \tag{4.17}$$

(If we had instead differentiated with respect to the β's we would obtain

$$\frac{\partial f^j(\alpha; \beta)}{\partial \beta^l} = \hat{\eta}_k^j(f(\alpha; \beta))\hat{\xi}_l^k(\beta) = (\hat{H}(f(\alpha; \beta))\hat{\varXi}(\beta))_l^j. \tag{4.18}$$

Then if we think of the α as 'running variables' and the β as 'parameters', the group composition functions obey the (nonlinear) partial differential equations (4.16 or 4.17), plus initial conditions, with respect to the α. This is a consequence of associativity. We explore its meaning in this way.

If a Lie group G is given, and some (local) coordinates α are chosen: the functions $f(\alpha; \beta)$, and so the functions $\eta(\alpha)$ and $\xi(\alpha)$, are known; and the PDE's (4.16) with initial conditions are guaranteed to be obeyed. We now ask: if conversely the functions $\eta(\alpha)$, and hence $\xi(\alpha)$, are given: can we solve (4.16) for $f(\alpha; \beta)$? Can we then use them to define a group composition law, and will associativity hold? Here, there are two situations:

(a) Someone else knows some Lie group G, and has the $f(\alpha;\beta)$. She gives you only the resulting $\eta(\alpha)$. Then the PDE's (4.16) can be solved uniquely *by you* to get $f(\alpha;\beta)$; and if you use them to define group composition, you will find associativity.

(b) She does not know if a group G exists, but just gives you some matrix of functions $\eta(\alpha)$ obeying (4.10) and with an inverse matrix $\xi(\alpha)$. Can you solve the PDE's (4.16), find the $f(\alpha;\beta)$, use them to define group composition? Will it work, will associativity hold, will you get a group G at the end?

The answer in the situation (b) is, in general, **No**. For 'arbitrary' choices of η's, the PDE's (4.16) are internally inconsistent and have no solutions. For them to be consistent and solvable, the η's should obey important *integrability conditions*. If they do, then we can (hope to) solve (4.16) for $f(\alpha;\beta)$, define associative group composition, find inverses, and so (re)construct a group (apart from global aspects).

We now discuss situation (a), and turn to (b) later. Treat $\eta(\alpha)$, $\xi(\alpha)$ as given, assured that they 'belong' to some Lie group G. In the PDE's (4.16) regard $f(\alpha;\beta)$ as the *unknowns* and β as parameters appearing in the boundary conditions. We are assured then that a unique solution $f(\alpha;\beta)$ exists. For this solution we can now prove, appealing again to (4.16), that

$$f(\gamma;f(\alpha;\beta)) = f(f(\gamma;\alpha);\beta), \quad \text{independent } \gamma,\alpha,\beta. \tag{4.19}$$

To see this, suppress α and β and write $L(\gamma)$ and $R(\gamma)$ for the two sides here. With respect to γ, $L(\gamma)$ obeys a system of PDE's and boundary conditions

$$\frac{\partial L^{j}(\gamma)}{\partial \gamma^{k}} = (\eta(L)\xi(\gamma))^{j}_{k}, \quad L^{j}(0) = f^{j}(\alpha;\beta). \tag{4.20}$$

On the other hand, $R(\gamma)$ obeys:

$$\frac{\partial R^{j}(\gamma)}{\partial \gamma^{k}} = \frac{\partial f^{j}(\delta;\beta)}{\partial \delta^{l}}\bigg|_{\delta=f(\gamma;\alpha)} \frac{\partial f^{l}(\gamma;\alpha)}{\partial \gamma^{k}}$$

$$= [\eta(R)\xi(\delta)\eta(\delta)\xi(\gamma)]^{j}_{k} = (\eta(R)\xi(\gamma))^{j}_{k},$$

$$R^{j}(0) = f^{j}(\alpha;\beta). \tag{4.21}$$

Comparing, we see that $L(\gamma)$ and $R(\gamma)$ obey the same (nonlinear) PDE's (4.16) with the same boundary conditions at $\gamma = 0$. Hence, by the uniqueness of solutions to such systems of PDE's, $L(\gamma) = R(\gamma)$ and associativity holds.

What about the existence and uniqueness of inverses? All is well. Since $H(\alpha)$, $\Xi(\alpha)$ are nonsingular, so is $(\partial f^{j}(\alpha;\beta)/\partial\alpha^{k})$. Then, using the implicit function theorem and properties of Jacobians, given α there is a unique α' such that

$$f^j(\alpha; \alpha') = 0 : \alpha \to a \in \mathcal{N} \Rightarrow \alpha' \to a^{-1} \in \mathcal{N} :$$

$$aa^{-1} = e. \tag{4.22}$$

Will this a^{-1} also fulfill $a^{-1}a = e$? Yes, and the argument is this: use Eq. (4.22) twice and exploit the associativity (4.18) just established!

$$a \sim \alpha^j \to f^j(\alpha; \alpha') = 0 \to \alpha'^j \sim a^{-1};$$

$$a^{-1} \sim \alpha'^j \to f^j(\alpha'; \alpha'') = 0 \to \alpha''^j \sim (a^{-1})^{-1};$$

$$f(\alpha; f(\alpha'; \alpha'')) = f(f(\alpha; \alpha'); \alpha'') \Rightarrow \alpha = \alpha'',$$

$$\text{i.e. } a = (a^{-1})^{-1},$$

$$\text{i.e., } f(\alpha'; \alpha) = 0 \quad \text{or} \quad a^{-1}a = e. \tag{4.23}$$

(One can simplify these equations and expressions somewhat by writing b in place of a^{-1} throughout; so α'^j are the coordinates of b, and α''^j those of b^{-1}.) Therefore, for acceptable or permissible $\eta(\alpha)$, the PDE's (4.16) capture (locally) all the aspects of group structure: associative composition, and inverses.

4.3 One Parameter Subgroups, Canonical Coordinates of First Kind

A one parameter subgroup in G, an OPS, is a continuous differentiable family of elements $a(\sigma) \in G$, σ a real parameter, forming a subgroup:

$$a(\sigma)a(\tau) = a(\sigma + \tau), \quad a(0) = e, \quad a(\sigma)^{-1} = a(-\sigma). \tag{4.24}$$

By definition an OPS is abelian, the parameters just add. We have in hand G, a coordinate system, the functions $f, \eta, \xi, \hat{\eta}$ and $\hat{\xi}$. If $a(\sigma)$ has coordinates $\alpha^j(\sigma)$, they obey a system of ODE's. Of course $\alpha^j(0) = 0$. Then considering a small change σ to $\sigma + \delta\sigma$:

$$\alpha^j(\sigma + \delta\sigma) = \text{coordinates of } a(\delta\sigma + \sigma) = \text{coordinates of } a(\delta\sigma)a(\sigma)$$

$$= f^j(\alpha(\delta\sigma); \alpha(\sigma)). \tag{4.25}$$

At and near the identity, write

$$t^j = \left.\frac{d\alpha^j(\sigma)}{d\sigma}\right|_{\sigma=0}, \quad \alpha^j(\delta\sigma) \simeq \delta\sigma\, t^j. \tag{4.26}$$

Then (4.25) becomes

$$\alpha^j(\sigma + \delta\sigma) = f^j(t\,\delta\sigma; \alpha(\sigma))$$

$$= \alpha^j(\sigma) + \eta_k^j(\alpha(\sigma))t^k\delta\sigma + 0(\delta\sigma^2),$$

$$\text{i.e.,} \quad \frac{d\alpha^j(\sigma)}{d\sigma} = \eta_k^j(\alpha(\sigma))t^k, \quad \alpha^j(0) = 0. \tag{4.27}$$

The t^j are the components of the tangent vector to the OPS at the identity. In a similar way, we find that the $\alpha^j(\sigma)$ also obey

$$\frac{d\alpha^j(\sigma)}{d\sigma} = \hat{\eta}_k^j(\alpha(\sigma))t^k, \quad \alpha^j(0) = 0. \tag{4.28}$$

Now this can be reversed. Suppose we *choose* a real r-component vector t^j, then solve the ODE's (4.27) for $\alpha^j(\sigma)$. These will be the coordinates for elements on a OPS. The proof is similar to that of associativity of the f's. To show that

$$\alpha^j(\sigma + \tau) = f^j(\alpha(\sigma);\, \alpha(\tau)), \tag{4.29}$$

use σ as the running variable, suppress τ, and write $L(\sigma), R(\sigma)$ for the two sides of the equation. Then

$$\frac{dL^j(\sigma)}{d\sigma} = \eta_k^j(L)t^k, \quad L(0) = \alpha(\tau);$$

$$\frac{dR^j(\sigma)}{d\sigma} = \eta_k^j(f(\alpha(\sigma);\alpha(\tau)))\xi_l^k(\alpha(\sigma))\eta_m^l(\alpha(\sigma))t^m$$

$$= \eta_k^j(R)t^k, \quad R(0) = \alpha(\tau). \tag{4.30}$$

Since the (generally nonlinear) ODE's and initial conditions are the same, (4.29) is established; we do have an OPS. Thus there is a one-to-one correspondence between OPS's and tangent vectors t at $e \in G$.

Exponential notation, canonical coordinates

To begin with, denote the solution to the ODE's (4.27) for given t^j as $\alpha^j(\sigma; t)$: r functions of $(r+1)$ real arguments. But clearly only the product σt is relevant, so

$$\alpha^j(\sigma; t) = \chi^j(\sigma t) = r \text{ functions of } r \text{ real independent arguments.} \qquad (4.31)$$

From the properties of $\alpha(\sigma; t)$ we get those of $\chi(\sigma t)$:

$$\chi^j(0) = 0; \quad \chi^j(\sigma t) \simeq \sigma t^j + O(\sigma^2);$$

$$t^k \frac{\partial}{\partial t^k} \chi^j(t) = \eta_k^j(\chi) t^k;$$

$$\left. \frac{\partial \chi^j(t)}{\partial t^k} \right|_{t=0} = \delta_k^j. \qquad (4.32)$$

In a parallel fashion we also find:

$$t^k \frac{\partial}{\partial t^k} \chi^j(t) = \hat{\eta}_k^j(\chi) t^k. \qquad (4.33)$$

The last Jacobian property in (4.32) means that in some neighbourhood of e, say $\mathcal{N}$ itself, if some element a with coordinates α^j is given, we can find a vector t^j such that

$$\chi^j(t) = \alpha^j. \qquad (4.34)$$

That is: for *any* $a \in \mathcal{N}$, we have a *unique* tangent vector t such that the OPS determined by t passes through a at $\sigma = 1$. We can solve (4.34) for t in terms of α as say:

$$t^j = \zeta^j(\alpha), \qquad (4.35)$$

so also

$$\left. \frac{\partial \zeta^j(\alpha)}{\partial \alpha^k} \right|_{\alpha=0} = \delta_k^j. \qquad (4.36)$$

We now invent an expression or notation – admittedly formal at the moment – to express all this:

$$\chi^j(t) = \alpha^j \Leftrightarrow t^j = \zeta^j(\alpha) \Leftrightarrow a = \exp(t). \qquad (4.37)$$

Thus in this notation the elements on the OPS determined by t are symbolically written as $\exp(\sigma t)$ and obey, again formally:

$$\exp(\sigma t) \exp(\tau t) = \exp((\sigma + \tau)t). \qquad (4.38)$$

The identity $e = \exp(0)$, and all $a \in \mathcal{N}$ are obtained as $\exp(t)$ for t in a suitable region around 0.

We now exploit the one-to-one correspondence (4.37) between the given coordinates α^j and tangent vectors t^j to make a transformation of coordinates – assign to $a = \exp(t)$ the t^j themselves as new coordinates! Retaining the general notation 'α' for coordinates:

$$a \in \mathcal{N} \rightarrow \text{ old coordinates } \alpha^j \rightarrow \text{ new coordinates } \bar{\alpha}^j = t^j = \zeta^j(\alpha). \qquad (4.39)$$

In principle in the new system $\bar{\alpha}$ we have functions $\bar{f}, \bar{\eta}, \bar{\xi}$, etc. As the OPS $\{a(\sigma)\}$ with tangent vector t (at e) has (by definition!) the new coordinates $\bar{\alpha}^j(\sigma) = \sigma t^j$, the ODE's (4.27) become:

$$\frac{d\bar{\alpha}^j(\sigma)}{d\sigma} = \bar{\eta}^j_k(\bar{\alpha}(\sigma))t^k \Rightarrow \bar{\eta}(\bar{\alpha})\bar{\alpha} = \bar{\alpha}. \qquad (4.40)$$

These are the defining characteristics of these so-called canonical coordinates of the 1st kind. Dropping the overhead bar, we repeat: a system of coordinates $\{\alpha^j\}$ is a canonical coordinate system of the 1st kind if

$$\eta^j_k(\alpha)\alpha^k = \alpha^j \Leftrightarrow \text{ elements } a(\sigma) \text{ on an OPS have}$$

$$\alpha^j(\sigma) = \sigma t^j. \qquad (4.41)$$

Equally well such coordinates are characterised by

$$\hat{\eta}^j_k(\alpha)\alpha^k = \alpha^j. \qquad (4.42)$$

We can say:

Infinitely many general coordinate systems $\alpha, \bar{\alpha}, \bar{\bar{\alpha}}, \cdots$, related to one another by invertible *functional* transformations	Infinitely many but far fewer canonical coordinate systems of the 1st kind $t, \bar{t}, \bar{\bar{t}}, \cdots$, related to one another by invertible *linear* transformations

$$\qquad (4.43)$$

4.4 Integrability Conditions, Passage to the Lie Algebra

We now turn to situation (b) following Eq. (4.18): what are the necessary and sufficient conditions for the PDE's (4.16) to be solvable? As the running variables are α, and β

are parameters entering through initial conditions, these integrability conditions are:

$$\frac{\partial}{\partial \alpha^m} \frac{\partial f^j}{\partial \alpha^k} = \frac{\partial}{\partial \alpha^k} \frac{\partial f^j}{\partial \alpha^m},$$

$$\text{i.e.,} \quad \frac{\partial}{\partial \alpha^m}(\eta_l^j(f)\xi_k^l(\alpha)) = \frac{\partial}{\partial \alpha^k}(\eta_l^j(f)\xi_m^l(\alpha)). \tag{4.44}$$

We work out the terms using Leibnitz rule, use (4.16) when needed, and then regroup the terms so that on one side of the equation we have quantities evaluated at f, on the other side at α:

$$\xi_m^l(f)[\eta_j^n(f)\eta_{k,n}^m(f) - \eta_k^n(f)\eta_{j,n}^m(f)] = \eta_j^m(\alpha)\eta_k^n(\alpha)[\xi_{m,n}^l(\alpha) - \xi_{n,m}^l(\alpha)]. \tag{4.45}$$

The free indices are jkl. Since the two sides are quantities calculated at two independent points (remember the β's) in $\mathbb{R}^r$, each side must be a constant carrying indices jkl. So we have two sets of integrability conditions involving a suitable set of constants:

$$\xi_m^l(\alpha)[\eta_j^n(\alpha)\eta_{k,n}^m(\alpha) - \eta_k^n(\alpha)\eta_{j,n}^m(\alpha)] = -C_{jk}^l,$$

$$[\xi_{m,n}^l(\alpha) - \xi_{n,m}^l(\alpha)]\eta_j^m(\alpha)\eta_k^n(\alpha) = -C_{jk}^l. \tag{4.46}$$

These two systems coincide, and the integrability conditions can be conveniently expressed in two equivalent forms:

$$\eta_j^n(\alpha)\eta_{k,n}^m(\alpha) - \eta_k^n(\alpha)\eta_{j,n}^m(\alpha) = -C_{jk}^l \eta_l^m(\alpha),$$

$$\Delta_{mn}^l(\alpha) \equiv \xi_{m,n}^l(\alpha) - \xi_{n,m}^l(\alpha) + C_{jk}^l \xi_m^j(\alpha)\xi_n^k(\alpha) = 0. \tag{4.47}$$

An 'acceptable' set of $\eta_k^j(\alpha)$ must obey these PDE's for some constants C_{jk}^l. If they do, we can solve (4.16) (locally!) for $f(\alpha;\beta)$ and get the group structure in some neighbourhood $\mathcal{N}$ of e.

The C_{jk}^l are called 'structure constants of G'. They cannot be chosen freely but must obey some conditions which we will derive. If G were given, the C_{jk}^l can be computed and will obey these conditions. Later we see that all sets of 'acceptable' C_{jk}^l obeying these conditions lead to acceptable η's, hence to f's with all desirable properties.

We first study the C_{jk}^l, later look at the solutions for the f's. The properties of the C_{jk}^l lead to the Lie algebra concept.

Expressions for and properties of structure constants

The structure constants C^l_{jk} are a collection of r^3 real numbers subject to many conditions:

(i) Antisymmetry in the subscripts which is obvious:

$$C^l_{jk} = -C^l_{kj}. \tag{4.48}$$

(ii) Setting $\alpha = 0$ in the first of Eq. (4.47),

$$
\begin{aligned}
C^l_{jk} &= \eta^l_{j,k}(0) - \eta^l_{k,j}(0) \\
&= \frac{\partial}{\partial \alpha^k} \eta^l_j(\alpha) - \frac{\partial}{\partial \alpha^j} \eta^l_k(\alpha) \Big|_{\alpha=0} \\
&= \left(\frac{\partial}{\partial \alpha^k} \frac{\partial}{\partial \beta^j} - \frac{\partial}{\partial \alpha^j} \frac{\partial}{\partial \beta^k} \right) f^l(\beta;\alpha) \Big|_{\alpha=\beta=0}.
\end{aligned}
\tag{4.49}
$$

(So the integrability conditions in terms of $\hat{\eta}$ and $\hat{\xi}$ are similar to Eqs. (4.46, 4.47) with $\hat{C}^l_{jk} = -C^l_{jk}$ in place of C^l_{jk}.) If in a double Taylor series expansion of $f(\alpha;\beta)$ we write:

$$f^l(\alpha;\beta) = \alpha^l + \beta^l + a^l_{jk}\alpha^j \beta^k + \cdots, \tag{4.50}$$

then

$$C^l_{jk} = a^l_{jk} - a^l_{kj}. \tag{4.51}$$

(iii) Relation to commutators of group elements: Let a, b be both near e, with coordinates α^j, β^j which are retained to lowest nontrivial order. Let γ^j be the coordinates of $q(a, b) = aba^{-1}b^{-1}$. Then:

$$ab = q(a, b)ba \Rightarrow$$

$$\alpha^j + \beta^j + a^j_{kl}\alpha^k \beta^l \cdots = \gamma^j + f^j(\beta;\alpha) + a^j_{kl}\gamma^k f^l(\beta;\alpha) \cdots$$

$$= \gamma^j + \beta^j + \alpha^j + a^j_{kl}\beta^k \alpha^l + a^j_{kl}\gamma^k f^l(\beta;\alpha) \cdots,$$

i.e., $$\gamma^j = C^j_{kl}\alpha^k \beta^l + \cdots. \tag{4.52}$$

We can conclude then that

$$G \text{ abelian} \Leftrightarrow q(a, b) = e \Rightarrow C^l_{jk} = 0. \tag{4.53}$$

We will prove the converse later.

(iv) The Jacobi condition: Recall the first order partial differential operators X_j in Eq. (4.9). A simple calculation shows that the integrability condition (4.47), the η-version, is expressible in terms of commutators of the X's:

$$[X_j, X_k] = -C_{jk}^l X_l. \tag{4.54}$$

In an equivalent form we can say: if to each tangent vector t^j we associate the differential operator

$$\eta_{[t]} = -t^j X_j = -t^j \eta_j^m(\alpha) \frac{\partial}{\partial \alpha^m}, \tag{4.55}$$

then the integrability conditions are

$$[\eta_{[t]}, \eta_{[u]}] \equiv \eta_{[t]} \eta_{[u]} - \eta_{[u]} \eta_{[t]}$$

$$= \eta_{[v]},$$

$$v^j = C_{kl}^j t^k u^l \equiv [t, u]^j. \tag{4.56}$$

We say that $v \equiv [t, u]$ is the *Lie bracket* of the two vectors t and u. Since, however, operator commutators obey the Jacobi identity, so must Lie brackets and hence the structure constants:

$$[\eta_{[t]}, \eta_{[u]}] = \eta_{[t,u]} :$$

$$\sum_{\text{cyclic}} \left[[\eta_{[t]}, \eta_{[u]}], \eta_{[w]} \right] = \sum_{\text{cyclic}} [\eta_{[t,u]}, \eta_{[w]}]$$

$$= \sum_{\text{cyclic}} \eta_{[[t,u],w]} = \eta_{\sum_{\text{cyclic}}[[t,u],w]} = 0 \Rightarrow$$

$$\sum_{\text{cyclic}} [[t, u], w] = 0 \Rightarrow$$

$$C_{kl}^j C_{mn}^l + C_{ml}^j C_{nk}^l + C_{nl}^j C_{km}^l = 0, \quad \text{all } jkmn. \tag{4.57}$$

Thus the complete set of conditions on structure constants making them acceptable are: reality, antisymmetry and the Jacobi identities, *nothing more!* This will be seen later.

These properties of the C_{jk}^l lead to the concept of the *Lie algebra*.

4.5 Lie Algebras

This concept formalises the properties of, and operations with, tangent vectors and OPS's in a Lie group. A (real) Lie algebra L is a real linear vector space with elements (vectors) $u, v, t, w, \cdots$ among which a *Lie bracket* $[,]$ is defined with these properties:

(i) $u, v \in L \Rightarrow t = [u, v] \in L$; $\qquad\qquad\qquad\qquad\qquad\qquad\qquad$ (4.58a)

(ii) $[u, v] = -[v, u]$, antisymmetry; $\qquad\qquad\qquad\qquad\qquad\qquad$ (4.58b)

(iii) $[u, v]$ linear in u, v; $\qquad\qquad\qquad\qquad\qquad\qquad\qquad\qquad$ (4.58c)

(iv) Jacobi identity: $[[u, v], w] + [[v, w], u] + [[w, u], v] = 0$ all u, v, w. $\quad$ (4.58d)

If we choose a basis $\{e_j\}$ for $L, j = 1, 2, \cdots, r =$ dimension of L, then:

$$u = u^j e_j \; v = v^j e_j \to [u, v] = C_{jk}^l u^j v^k e_l,$$

$$[e_j, e_k] = C_{jk}^l e_l. \qquad\qquad\qquad (4.59)$$

We recover the structure constants and all their properties. Since the C_{jk}^l arose from the Lie group G, they and the above Lie algebra 'belong' to G; so we denote it specifically as $\boldsymbol{G}$.

Repeating for emphasis: given an r-parameter Lie group G, the tangent vectors $t, u, \cdots$ at e to all OPS's in G together constitute the Lie algebra $\boldsymbol{G}$ of G. The Lie bracket $[u, v]$ is determined by G 'in the small', say via $q(a, b)$. Alternatively, the structure of G near e leads to its Lie algebra $\boldsymbol{G}$, an r-dimensional real vector space with a Lie bracket determined by G. So if two Lie groups G, G' are *locally isomorphic*, as we saw with $SO(3)$ and $SU(2)$, they have the *same* Lie algebra: $\boldsymbol{G} = \boldsymbol{G}'$.

Changing coordinates $\alpha \to \bar{\alpha}$ over $\mathcal{N} \subset G$ causes a change of basis, a linear transformation in $\boldsymbol{G}$.

A more 'advanced' view is to say that $\boldsymbol{G}$ is a collection of vector fields X_j (or $\hat{X}_j$) defined globally over G, possessing certain invariances and closed under commutation.

4.6 Local Reconstruction of G from $\boldsymbol{G}$

Suppose an acceptable set of C_{jk}^l, obeying the antisymmetry and Jacobi conditions, is given. Can we then say we will be able to compute the functions $\eta_k^j(\alpha)$? If the answer is 'yes', then we know that we can go on to compute $f(\alpha; \beta)$ too, the associativity property and inverses will all be secured, and so from C^l_{jk} we would have built up G locally.

But *we know in advance* that we can not find $\eta^j_k(\alpha)$ if only C^l_{jk} are given. Why? Suppose you already have a group G in hand. Clearly there is infinite freedom to change the coordinates over $\mathcal{N}$, leaving them unchanged near e. Such changes of coordinates would definitely change $\eta(\alpha)$ but not C^l_{jk}. Unless we restrict the coordinate system in some way, we cannot hope to find $\eta(\alpha)$ starting from C^l_{jk}. The key is to use canonical coordinates of the 1st kind. It *is then* possible, given an acceptable set of C^l_{jk}, to solve uniquely the combined system of PDE's, algebraic relations and initial conditions:

$$\Delta^l_{mn}(\alpha) \equiv \xi^l_{m,n}(\alpha) - \xi^l_{n,m}(\alpha) + C^l_{jk}\xi^j_m(\alpha)\xi^k_n(\alpha) = 0,$$

$$\xi^j_k(\alpha)\alpha^k = \alpha^j, \quad \xi^j_k(0) = \delta^j_k. \tag{4.60}$$

This is done in two steps: first we find $\xi(\alpha), \eta(\alpha)$ using a subset of the equations $\Delta^l_{mn}(\alpha) = 0$, then we show that this solution obeys all of $\Delta^l_{mn}(\alpha) = 0$. After this we proceed to solve (4.16) for $f(\alpha, \beta)$.

Step 1: The system of equations

$$\alpha^m \Delta^l_{mn}(\alpha) = 0, \quad \xi^j_k(\alpha)\alpha^k = \alpha^j, \quad \xi^j_k(0) = \delta^j_k \tag{4.61}$$

is a consequence of, but not equivalent to, (4.60); but it suffices to find $\xi(\alpha)$. We develop $\alpha^m \Delta^l_{mn}(\alpha) = 0$ as follows:

$$\alpha^m \Delta^l_{mn}(\alpha) = \alpha^m \xi^l_{m,n}(\alpha) - \alpha^m \xi^l_{n,m}(\alpha) + \alpha^m C^l_{jk}\xi^j_m(\alpha)\xi^k_n(\alpha)$$

$$= \delta^l_n - \xi^l_n(\alpha) - \alpha^m \frac{\partial}{\partial \alpha^m}\xi^l_n(\alpha) + C^l_{jk}\alpha^j\xi^k_n(\alpha)$$

$$\text{(using } \xi(\alpha)\alpha = \alpha)$$

$$= 0. \tag{4.62}$$

Define an $r \times r$ 'antisymmetric' matrix $C(\alpha)$ by

$$C^j_k(\alpha) = C^j_{lk}\alpha^l, \quad C(\alpha)\alpha = 0. \tag{4.63}$$

Then (4.62) becomes:

$$\left(\alpha^m \frac{\partial}{\partial \alpha^m} + 1\right)\xi^l_n(\alpha) = \delta^l_n + C^l_k(\alpha)\xi^k_n(\alpha),$$

i.e., $\left(\alpha^m \frac{\partial}{\partial \alpha^m} + 1\right)\xi(\alpha) = \mathbb{1} + C(\alpha)\xi(\alpha). \tag{4.64}$

This determines the 'radial dependence': setting $\alpha = \tau\alpha_0$ for any fixed α_0, we have:

$$\frac{\partial}{\partial \tau}(\tau\xi(\tau\alpha_0)) = \mathbb{1} + C(\alpha_0)\tau\xi(\tau\alpha_0),$$

i.e., $\quad \dfrac{\partial^n}{\partial \tau^n}(\tau\xi(\tau\alpha_0)) = C(\alpha_0)\dfrac{\partial^{n-1}}{\partial \tau^{n-1}}(\tau\xi(\tau\alpha_0)), \ n \geq 2$

$$= C(\alpha_0)^{n-1}(\mathbb{1} + C(\alpha_0)\tau\xi(\tau\alpha_0))$$

$$= C(\alpha_0)^{n-1} \text{ at } \tau = 0,$$

i.e., $\quad \tau\xi(\tau\alpha_0) = \tau \cdot \mathbb{1} + \displaystyle\sum_{n=2}^{\infty} \frac{\tau^n}{n!}C(\alpha_0)^{n-1}$

$$= \tau \cdot \left.\frac{e^z - 1}{z}\right|_{z=\tau C(\alpha_0)},$$

i.e., $\quad \xi(\alpha) = \left.\dfrac{e^z - 1}{z}\right|_{z=C(\alpha)} = \mathbb{1} + \displaystyle\sum_{n=1}^{\infty} \frac{C(\alpha)^n}{(n+1)!}.$ \hfill (4.65)

We see immediately that, thanks to (4.63), all of Eqs. (4.61) are obeyed.

Step 2: Next we should check that (4.65) indeed obeys all of (4.60). We exploit certain differential identities obeyed by $\Delta^l_{mn}(\alpha)$, following directly from the definition in (4.60). Indicating cyclic sums by $(\cdots)$, and as with Maxwell's equations:

$$\sum_{(mnp)} \Delta^l_{mn,p}(\alpha) = \sum_{(mnp)} C^l_{jk}\left(\xi^j_{m,p}\xi^k_n + \xi^j_m\xi^k_{n,p}\right)$$

$$= \sum_{(mnp)} C^l_{jk}\left(\xi^j_{m,p} - \xi^j_{p,m}\right)\xi^k_n$$

$$= \sum_{(mnp)} C^l_{jk}\left(\Delta^j_{mp} - C^j_{uv}\xi^u_m\xi^v_p\right)\xi^k_n$$

$$= \sum_{(mnp)} C^l_{jk}\xi^k_n\Delta^j_{mp} + \left(\sum_{(kuv)} C^l_{jk}C^j_{uv}\right)\xi^u_m\xi^k_n\xi^v_p. \hfill (4.66)$$

These are identically obeyed. Now we multiply both sides by α^p, appeal to the Jacobi identity for the structure constants and to the fact that Δ already obeys (4.61), to obtain:

$$\alpha^p \Delta^l_{mn,p} + \alpha^p \Delta^l_{np,m} + \alpha^p \Delta^l_{pm,n} = C^l_{jk} \alpha^k \Delta^j_{nm} = C(\alpha)^l_j \Delta^j_{mn},$$

$$\text{i.e., } \left(\alpha^p \frac{\partial}{\partial \alpha^p} + 2 \right) \Delta^l_{mn}(\alpha) = C(\alpha)^l_j \Delta^j_{mn}(\alpha);$$

therefore with $\alpha = \tau \alpha_0$ again,

$$\frac{\partial}{\partial \tau} (\tau^2 \Delta^l_{mn}(\tau(\alpha_0))) = C(\alpha_0)^l_j (\tau^2 \Delta^j_{mn}(\tau \alpha_0)). \tag{4.67}$$

We see that at $\tau = 0$, all the derivatives of $\tau^2 \Delta(\tau \alpha_0)$ vanish, so we have secured $\Delta^l_{mn}(\alpha) = 0$, which was our aim.

It is to be emphasised that the integrability conditions (4.60) for the solubility of the PDE's (4.16) for the group composition functions $f(\alpha, \beta)$ can be solved to obtain $\xi(\alpha)$ in canonical coordinates of the 1st kind, provided only that the structure constants obey the antisymmetry and Jacobi conditions. Nothing more is to be required of them.

Now we invert $\xi(\alpha)$ to obtain $\eta(\alpha)$, then solve (4.16) for $f(\alpha; \beta)$ knowing in advance that the solution exists. Actually we use a 'double barreled' approach to computing $f(\alpha; \beta)$, combining Eq. (4.16) and (4.18). So, using canonical coordinates of the 1st kind, the full list of ingredients for the calculation is:

$$\frac{\partial f^j}{\partial \alpha^l}(\alpha; \beta) = (\eta(f)\xi(\alpha))^j_l, \quad \frac{\partial f^j}{\partial \beta^l}(\alpha; \beta) = (\hat\eta(f)\hat\xi(\beta))^j_l;$$

$$\xi(\alpha) = \left(\frac{e^z - 1}{z} \right)_{z=C(\alpha)}, \quad \eta(\alpha) = \left(\frac{z}{e^z - 1} \right)_{z=C(\alpha)};$$

$$\hat\xi(\alpha) = \left(\frac{e^z - 1}{z} \right)_{z=-C(\alpha)}, \quad \hat\eta(\alpha) = \left(\frac{z}{e^z - 1} \right)_{z=-C(\alpha)}. \tag{4.68}$$

We also use the following facts and notations:

$$\frac{z}{e^z - 1} = -\frac{z}{2} + \psi(z),$$

$$\psi(z) = \psi(-z) = \frac{z}{2} \coth \frac{z}{2} = 1 + \frac{z^2}{12} + \cdots;$$

$$C(\alpha)\beta = [\alpha, \beta] = C_\alpha \beta,$$

$$C_\alpha = \text{linear operator on } \mathbf{G}, \text{ linear in } \alpha. \tag{4.69}$$

Then we set up an ODE for $f(\alpha; \beta)$ which treats both arguments on the same footing: We know $f(\alpha; \beta)$ exists, and we may use all available information to find it as conveniently as possible. So we work with $f(\tau\alpha; \tau\beta)$:

$$\frac{d}{d\tau}f^j(\tau\alpha; \tau\beta) = (\eta(f)\xi(\tau\alpha))^j_l\alpha^l + (\hat{\eta}(f)\hat{\xi}(\tau\beta))^j_l\beta^l$$

$$= \eta(f)^j_l\alpha^l + \hat{\eta}(f)^j_l\beta^l$$

$$= \left\{\left(\left(\frac{-z}{2} + \psi(z)\right)_{z=C(f)}\alpha + \left(\frac{z}{2} + \psi(z)\right)_{z=C(f)}\beta\right)\right\}^j$$

$$= \left\{\frac{1}{2}C_f(\beta - \alpha) + \psi(C_f)(\alpha + \beta)\right\}^j,$$

i.e., $\quad \dfrac{d}{d\tau}f(\tau\alpha; \tau\beta) = \dfrac{1}{2}[\alpha - \beta, f] + \psi(C_f)(\alpha + \beta).$ $\qquad$ (4.70)

Now we substitute a power series for $f(\tau\alpha; \tau\beta)$, develop both sides and compare terms:

$$f(\tau\alpha; \tau\beta) = \sum_{n=1}^{\infty}\tau^n f_n(\alpha; \beta) = \tau f_1(\alpha; \beta) + \tau^2 f_2(\alpha; \beta) + \cdots :$$

$$f_1 + 2\tau f_2 + 3\tau^2 f_3 + \cdots = \frac{1}{2}[\alpha - \beta, \tau f_1 + \tau^2 f_2 + \cdots]$$

$$+ \alpha + \beta + \frac{1}{12}[\tau f_1 + \tau^2 f_2 + \cdots, [\tau f_1 + \tau^2 f_2 + \cdots, \alpha + \beta]]\cdots ;$$

$$f_1 = \alpha + \beta; \quad f_2 = \frac{1}{2}[\alpha, \beta]; \quad f_3 = \frac{1}{12}[\alpha - \beta, [\alpha, \beta]]; \cdots . \qquad (4.71)$$

So the group composition functions are obtained as an infinite series:

$$f(\alpha; \beta) = \alpha + \beta + \frac{1}{2}[\alpha, \beta] + \frac{1}{12}[\alpha - \beta, [\alpha, \beta]] + \cdots . \qquad (4.72)$$

Remembering that we are using canonical coordinates in which, by (4.37, 4.39), the element $a = \exp(t)$ has coordinates t^j, i.e., t^j are simultaneously components of a tangent vector and coordinates of a finite group element, we can rewrite (4.72) in intrinsic form as:

$$u, v \in \boldsymbol{G} \to \exp(u), \exp(v) \in G:$$

$$\exp(u)\exp(v) = \exp\left(u + v + \frac{1}{2}[u, v] + \frac{1}{12}[u - v, [u, v]] + \cdots\right). \qquad (4.73)$$

This is called the Baker–Campbell–Hausdorff formula, and we have here the exponential map from some neighbourhood of the zero vector in $\boldsymbol{G}$ to some neighbourhood $\mathcal{N}$ of e in G.

We see here, as we promised after Eq. (4.53): G is abelian if and only if all $C^l_{jk} = 0$.

To summarise: Given a Lie group G and some coordinates α over $\mathcal{N} \subset G$, we can find the structure constants C^l_{jk}, then set up the Lie algebra $\boldsymbol{G}$ of G. Conversely, given an acceptable set of C^l_{jk}, working in a canonical coordinate system of the 1st kind, we can build up $\xi(\alpha), \eta(\alpha), \hat{\xi}(\alpha), \hat{\eta}(\alpha), f(\alpha; \beta)$ for small enough α, β and so determine the structure of G locally.

4.7 General Remarks on the $G \to \boldsymbol{G}$ Relationship, Some Definitions Concerning Lie Algebras

The problem of dealing with a Lie group G has resolved itself into two parts: determining the structure of G 'in the small', then a study of its global properties. So if two Lie groups G, G' are 'locally isomorphic', i.e., 'look the same' near e, e' but possibly differ globally, this will not show up in $\boldsymbol{G}, \boldsymbol{G'}$ at all. G and G' will give rise to, or possess, the *same* Lie algebra, $\boldsymbol{G'} = \boldsymbol{G}$. Such is the case with $SO(3)$ and $SU(2)$, also with $SO(3,1), SL(2,\mathbb{C})$ and $SO(3,\mathbb{C})$ as we will see later.

In general, then, several Lie groups $G, G', G'' \cdots$ may well have the same Lie algebra $\boldsymbol{L} = \boldsymbol{G} = \boldsymbol{G'} = \boldsymbol{G''} \cdots$. Then, in the reconstruction of a group from $\boldsymbol{L}$, which group do we get? Of all the groups $G, G', \cdots$ having the same $\boldsymbol{L}$, *only one* is simply connected; it is called the universal covering group of all the others. Writing $\overline{G}$ for it, what $\boldsymbol{L}$ or $\boldsymbol{G}$ leads to unambiguously is this $\overline{G}$. Recall that $SU(2)$ is simply connected, $SO(3)$ is not.

Homomorphisms, automorphisms and isomorphisms among Lie algebras are easily defined. For example: $\boldsymbol{L}$ and $\boldsymbol{L'}$ are *isomorphic* if we have a linear one-to-one onto invertible map $\varphi : \boldsymbol{L} \to \boldsymbol{L'}$ preserving Lie brackets:

$$x, y \in \boldsymbol{L} \to \varphi(x), \varphi(y) \in \boldsymbol{L'} : \varphi([x, y]) = [\varphi(x), \varphi(y)] \qquad (4.74)$$

Then we may regard $\boldsymbol{L}$ and $\boldsymbol{L'}$ as being 'the same'. So

$$G' \text{ and } G \text{ locally isomorphic} \iff \boldsymbol{G'} \text{ and } \boldsymbol{G} \text{ isomorphic}. \qquad (4.75)$$

The various concepts developed for groups, as applied to Lie groups, naturally lead to corresponding Lie algebra concepts. Let us run through them briefly.

1. A Lie algebra $\boldsymbol{L}$ is abelian or commutative if all Lie brackets vanish:

$$x, y \in \boldsymbol{L} \to [x, y] = 0 : \text{abelian} \qquad (4.76)$$

2. A subset $L' \subset L$ is a sub-Lie algebra, or more simply a subalgebra if, first, it is a linear subspace in L, and second it is closed under the Lie bracket operation:

$$x, y \in L', \ a, b \in \mathbb{R} \Rightarrow ax + by, \ [x, y] \in L'. \qquad (4.77)$$

So we have

$$H \subset G \text{ is a Lie subgroup of } G \Leftrightarrow H \subset G \text{ is a Lie subalgebra.} \qquad (4.78)$$

3. A subalgebra $L' \subset L$ is a *normal* or *invariant* subalgebra if

$$x \in L', \ y \in L \Rightarrow [x, y] \in L'. \qquad (4.79)$$

As a refinement of (4.78) we then have:

$$H \subset G \text{ is an invariant Lie subgroup of } G \Leftrightarrow$$

$$H \subset G \text{ is an invariant Lie subalgebra.} \qquad (4.80)$$

Going back to (4.79), projecting L as a vector space with respect to L', the quotient space L/L' is also a Lie algebra, the factor algebra.

4. Given a Lie algebra L, its commutator subalgebra L_1 consists of all linear combinations of all Lie brackets $[x, y]$ for $x, y \in L$.
We write this suggestively as

$$L_1 = [L, L] = \text{derived algebra of } L. \qquad (4.81)$$

It is trivial to check that L_1 is an invariant subalgebra of L, and the factor L/L_1 is abelian. One sees that if L was itself abelian, $L_1 = 0$ and conversely. As an example of the opposite, for $SO(3)$, $L = L_1$.

5. Now we look at two concepts which we did not explore in the group context, but which are quite basic for general theory. Given L of dimension r, form its commutator subalgebra $L_1 = [L, L]$; then form $L_2 = [L_1, L_1]$, and so on. So we have a sequence of Lie algebras

$$L_0 = L, \ L_1 = [L_0, L_0], \ L_2 = [L_1, L_1], \cdots, \ L_{j+1} = [L_j, L_j], \cdots. \qquad (4.82)$$

At each stage L_{j+1} is an invariant subalgebra of L_j, with abelian quotient L_j/L_{j+1}. Now at every stage where L_{j+1} is a proper subspace of L_j the dimension gets reduced by at least one. Since $\dim L = r$, before we reach L_r we must either reach dimension zero or stabilise at some L_j, namely $L_{j+1} = L_j$ and so thereafter $L_{j+2} = L_{j+3} = \cdots = L_j$ as well. Then we define:

$$L \text{ is } solvable \equiv \text{ for some } n \le r \text{ we have } L_n = 0. \tag{4.83}$$

6. Another series of subalgebras we can form is this: we construct $L_1 = [L, L]$ as before, but now write it as L^1. Next we define $L^2 = [L, L^1] =$ all linear combinations of $[x, y]$ for all $x \in L, y \in L^1$. This is *not* the commutator subalgebra of L^1 but is typically larger. Since L^1 is an invariant subalgebra in L, we see that L^2 is a subspace of L^1 and also an invariant subalgebra of L^1. In fact we even have $L^2 =$ invariant subalgebra in L. So we have:

$$L^1 = L_1 = [L, L] = \text{ invariant (commutator) subalgebra in } L;$$

$$L^2 = [L, L^1] = \text{ invariant subalgebra in } L^1 \text{ and in } L. \tag{4.84}$$

We continue this process with $L^3 = [L, L^2], \cdots, L^{j+1} = [L, L^j] \cdots$, so L^{j+1} is an invariant subalgebra in $L^j, L^{j-1}, \cdots, L$. So we generate the series

$$L^0 = L_0 = L, \; L^1 = L_1 = [L, L^0], \; L^2 = [L, L^1], \cdots, \; L^{j+1} = [L, L^j], \cdots. \tag{4.85}$$

If this becomes zero at some n, i.e., $L^n = 0$, we say L is *nilpotent*.

7. The connection between solvability and nilpotency is this. For the former we had the sequence (4.82) of commutator subalgebras

$$L = L_0, \; L_{j+1} = [L_j, L_j] \text{ for } j = 0, 1, 2, \cdots; \tag{4.86}$$

for the latter we have the sequence (4.85)

$$L^0 = L_0 = L, \; L^{j+1} = [L, L^j] \text{ for } j = 0, 1, 2, \cdots. \tag{4.87}$$

So $L_1 = L^1$. How about L_2 versus L^2, L_3 versus $L^3, \cdots$? We see easily by induction that

$$L_j \subseteq L^j. \tag{4.88}$$

So a *nilpotent* Lie algebra is *solvable* but not conversely.

8. A Lie algebra L is simple (semisimple) if it has no proper invariant (abelian invariant) subalgebra. Therefore

$$L \text{ simple} \quad \Rightarrow \quad L \text{ semisimple but not conversely:}$$

$$L \text{ solvable} \quad \Rightarrow \quad L_1 = \text{ invariant proper subalgebra of } L$$

$$\Rightarrow \quad L \text{ not simple;}$$

$$L \text{ simple} \quad \Rightarrow \quad L_1 = L \Rightarrow L \text{ not solvable.} \tag{4.89}$$

Thus being simple and being solvable are mutually exclusive properties.

9. Let L' be an invariant subalgebra of L. If we can find another invariant subalgebra L'' of L such that as vector spaces $L = L' \oplus L''$ as a direct sum, then we say L is the *direct sum Lie algebra* of L' and L''. Since $L' \cap L''$ vanishes, we have in this case:

$$[L', L'] \subseteq L', \quad [L', L''] = 0, \quad [L'', L''] \subseteq L''. \tag{4.90}$$

10. Lastly we define semidirect sums: L is the semidirect sum of subalgebras L' and L'' if

$$L = L' \oplus L'', \quad [L', L'] \subseteq L', \quad [L', L''] \subseteq L'', \quad [L'', L''] \subseteq L''. \tag{4.91}$$

So in this case L'' is an *invariant subalgebra* in L, but L' is not necessarily so.

4.8 Representations of Lie Algebras – A Brief Look

We have studied representations of (finite and some continuous) groups in some detail. Now to get a representation of the (real) Lie algebra G associated with the Lie group G, we must have a (real or complex) linear vector space V, of dimension n say, so that

$$\text{elements in } G \quad \rightarrow \quad \text{linear transformations on } V,$$

$$\text{Lie bracket in } G \quad \rightarrow \quad \text{commutator of transformations on } V. \tag{4.92}$$

Why is this so, why is the abstract Lie bracket in G 'realised' as the *commutator* $XY - YX$ in a linear representation on V? The reason is interesting and of fundamental importance. Basically to each element $a \simeq \{\alpha^j\} \in G$ we have a representation matrix

$\mathcal{D}(\alpha)$:

$$\mathcal{D}(\alpha)\mathcal{D}(\beta) = \mathcal{D}(f(\alpha,\beta)), \; \mathcal{D}(0) = \mathbb{1}. \tag{4.93}$$

In canonical coordinates of the 1st kind we have for elements on an OPS:

$$\mathcal{D}(\alpha) = \mathbb{1} + \alpha^j X_j + 0(\alpha^2):$$
$$\mathcal{D}(\exp(u)) = \exp(u^j X_j). \tag{4.94}$$

Formal exponentials turn into actual exponentials! Then from Eq. (4.52):

$$c = q(a,b) = aba^{-1}b^{-1} \to \gamma^j = C^j_{kl}\alpha^k\beta^l + \cdots,$$
$$\mathcal{D}(aba^{-1}b^{-1}) \simeq \mathbb{1} + \gamma^j X_j + \cdots;$$
$$\text{i.e., } (\mathbb{1} + \boldsymbol{\alpha} \cdot \boldsymbol{X} + \cdots)(\mathbb{1} + \boldsymbol{\beta} \cdot \boldsymbol{X} + \cdots)(\mathbb{1} - \boldsymbol{\alpha} \cdot \boldsymbol{X} \cdots)(\mathbb{1} - \boldsymbol{\beta} \cdot \boldsymbol{X} \cdots)$$
$$\simeq \mathbb{1} + C^j_{kl}\alpha^k\beta^l X_j \cdots,$$
$$\text{i.e., } \boldsymbol{\alpha} \cdot \boldsymbol{X} \boldsymbol{\beta} \cdot \boldsymbol{X} - \boldsymbol{\beta} \cdot \boldsymbol{X} \boldsymbol{\alpha} \cdot \boldsymbol{X} = C^j_{kl}\alpha^k\beta^l X_j. \tag{4.95}$$

This is the reason! In detail we have: if e_j, $j = 1, 2, \cdots, r$ is a basis for $\boldsymbol{G}$, with general element $x = x^j e_j \in \boldsymbol{G}$, in a representation of $\boldsymbol{G}$ we map $e_j \in \boldsymbol{G} \to$ linear transformation X_j on V, the *generators*:

$$x = x^j e_j \in \boldsymbol{G} \to X = x^j X_j \text{ on } V;$$
$$[e_j, e_k] = C^l_{jk} e_l \text{ in } \boldsymbol{G} \to [X_j, X_k] \equiv X_j X_k - X_k X_j = C^l_{jk} X_l. \tag{4.96}$$

Then, apart from global aspects, by exponentiation we go to a representation of G:

$$x = x^j e_j \in \boldsymbol{G} \to X = x^j X_j \text{ on } V,$$
$$\exp(x) \in G \to \mathcal{D}(\exp(x)) = e^X \text{ on } V. \tag{4.97}$$

Here too 'formal' exponentials turn into 'honest' exponentials in a representation.

We can see also that the structure constants C^l_{jk} can be computed once for all from a defining faithful matrix representation of G, and then used for constructing all other representations.

To get a unitary representation of G, V must carry or be endowed with a hermitian scalar product, and then the generators X_j must be antihermitian. However, in quantum mechanics we usually extract a factor of i to deal with hermitian generators. So hereafter:

Hermitian representation of $G : e_j \rightarrow X_j = X_j^\dagger$,

$$[e_j, e_k] = C_{jk}^l e_l \rightarrow [X_j, X_k] = iC_{jk}^l X_l,$$

$$\mathcal{D}(\exp(u)) = \exp(-iu^j X_j). \tag{4.98}$$

(More properly we may say: $e_j \rightarrow$ hermitian X_j, Lie bracket $\rightarrow -i$ times commutator.) For a real orthogonal representation of G, we need a real representation space V with inner product, and pure imaginary antisymmetric generators X_j, so that finally $\mathcal{D}(\exp(u))$ are real orthogonal.

The meanings of reducible, irreducible, decomposable, indecomposable representation are the same as with group representations. For the special processes (1.38) defined in Chapter 1 to get new representations from a given one, in terms of generators we have:

Representation of G		Representation of G
$\mathcal{D} \rightarrow (\mathcal{D}^T)^{-1}$	———	$X_j \rightarrow -X_j^T$
$\mathcal{D} \rightarrow (\mathcal{D}^\dagger)^{-1}$	———	$X_j \rightarrow X_j^\dagger$
$\mathcal{D} \rightarrow (\mathcal{D}^*)$	———	$X_j \rightarrow -X_j^*$
$\mathcal{D}^\dagger \mathcal{D} = \mathbb{1}$	———	$X_j^\dagger = X_j$
$\mathcal{D}^T \mathcal{D} = \mathbb{1}$	———	$X_j^T = -X_j$
$\mathcal{D}^* = \mathcal{D}$	———	$X_j^* = -X_j$

$$\tag{4.99}$$

In these rules the quantum mechanical factor of i has been taken into account.

4.9 The Adjoint Representation

This is a special and important representation of G (and of G). The Lie algebra G itself is the (real) vector space on which G and then G are both represented. For any $x \in G$, define a linear transformation ad $\cdot x$ on G in this way:

$$\text{ad} \cdot x : y \in G \rightarrow y' = (\text{ad} \cdot x)y = [x, y] \in G. \tag{4.100}$$

This is linear in y, so it is a linear transformation on G. Is it a representation of G? Yes, because of the Jacobi identity:

$$[\text{ad} \cdot x, \text{ad} \cdot y]z = (\text{ad} \cdot x)(\text{ad} \cdot y)z - (\text{ad} \cdot y)(\text{ad} \cdot x)z$$

$$= [x, [y, z]] - [y, [x, z]]$$

$$= -[z, [x, y]] = (\text{ad} \cdot [x, y])z,$$

i.e., $[\text{ad} \cdot x, \text{ad} \cdot y] = \text{commutator}$

$$= \text{ad} \cdot [x, y], \quad \text{Lie bracket.} \tag{4.101}$$

The generator matrices are made up of the structure constants themselves, with the i factor:

$$(\text{ad} \cdot \boldsymbol{e}_j)\boldsymbol{e}_k = [\boldsymbol{e}_j, \boldsymbol{e}_k] = C_{jk}^l \boldsymbol{e}_l \Rightarrow \text{in adjoint representation}$$

$$(X_j)_k^l = i \times \text{ coefficient of } \boldsymbol{e}_l \text{ in } (\text{ad} \cdot \boldsymbol{e}_j)\boldsymbol{e}_k$$

$$= iC_{jk}^l,$$

$$[X_j, X_k] = iC_{jk}^l X_l, \text{ just the Jacobi identities (4.57).} \tag{4.102}$$

Upon exponentiation we get the matrices of the (real) adjoint representation of G.

4.10 Summary

Since several important concepts have been dealt with in this chapter, a summary of the sequence of steps we have taken is now presented: (i) description of group elements by systems of real independent coordinates, and the freedom in the choice of such coordinate systems; (ii) associativity of group composition expressed via a system of PDE's, involving particular sets of functions on the group; (iii) one parameter subgroups, OPS's, and special canonical coordinates of the first kind they lead to; (iv) integrability conditions for the PDE's expressing associativity, leading to the Lie algebra concept; (v) local reconstruction of the Lie group from its Lie algebra, exploiting properties of canonical coordinates.

The general theory of Lie groups and Lie algebras will be used in the next chapter to look at the compact case. In the later Chapters 8 and 10 we deal with certain noncompact groups important in physics, namely the real symplectic groups, and groups associated with spacetime in nonrelativistic and special relativistic situations.

Problems

P4.1   Obtain the expressions (4.13) for the functions $\eta_{jk}(\alpha)$, $\xi_{jk}(\alpha)$ for the group $SO(3)$.

P4.2   Show that among three dimensional real vectors $\boldsymbol{a}, \boldsymbol{b}, \cdots$, the vector product operation

$$\boldsymbol{a}, \boldsymbol{b} \to \boldsymbol{a} \wedge \boldsymbol{b}$$

is a realisation of the Lie bracket concept. Similarly show that among functions $f(\boldsymbol{q},\boldsymbol{p})$, $g(\boldsymbol{q},\boldsymbol{p})$, $\cdots$, on a classical mechanical $2n$-dimensional phase space, the Poisson bracket

$$\{f,g\} = \sum_{j=1}^{n} \left(\frac{\partial f}{\partial q_j} \frac{\partial g}{\partial p_j} - \frac{\partial f}{\partial p_j} \frac{\partial g}{\partial q_j} \right)$$

is also a realisation of this concept.

Bibliography

Gilmore, R. (1974). *Lie Groups, Lie Algebras and Some of Their Applications.* New York: Wiley.

Pontrjagin, L. (1958). *Topological Groups.* Princeton, New Jersey: Princeton University Press.

Racah, G. (1965). Group Theory and Spectroscopy, Princeton Lectures 1951. *Ergebnisse der Exakten Naturwissenschaften*, 37: 28.

Sudarshan, E. C. G. and Mukunda, N. (1974). *Classical Dynamics – A Modern Perspective.* New York: Wiley. Reprint (2015); New Delhi: Hindustan Book Agency.

5

Compact Simple Lie Algebras – Classification and Irreducible Representations

In the previous chapter we dealt with real Lie algebras built on real linear vector spaces, as these are what arise when analysing Lie groups whose elements are labelled by certain numbers of real essential independent coordinates. The further study of 'all possible Lie algebras and Lie groups', their survey and classification, is quite intricate. We will only try to convey the flavour of the concepts and methods involved. Two crucial steps are: complexification of real Lie algebras, and the direct definition of complex Lie algebras; and the exploration of the properties of solvability, semisimplicity and simplicity. Our final goal is to arrive at the compact simple Lie algebras – CSLA – and their associated Lie groups. We will only aim to give a 'physical account' of arguments and strategies in this field. It will be largely descriptive with limited derivations.

A few words about the concept of compactness may be useful at this point. This is a topological property – a given topological space $\mathcal{T}$ may either be compact or noncompact. A topology on $\mathcal{T}$ is defined through the notion and properties of a class of subsets of $\mathcal{T}$ that are called 'open sets'. Familiar examples are bounded connected open intervals on the real line; bounded simply connected open regions in the plane, etc. The space $\mathcal{T}$ is compact if every covering of it by a family of open sets contains a finite subset of these open sets which also cover $\mathcal{T}$: every open cover contains a finite subcover. Otherwise it is noncompact. In an intuitive sense both the real line and the plane are noncompact. On the other hand, both the groups $SO(3)$ and $SU(2)$ are compact. An intuitive indication of this is that with respect to the invariant volume elements for them given in Eqs. (3.34, 3.87) they both have finite total volumes.

5.1 From a Real Lie Algebra to Its Complexification

Let L be an n-dimensional real Lie algebra. We first complexify it as a vector space to arrive at $\tilde{L}$ consisting of formal expressions of the form

$$z = x + iy \in \tilde{L}, \quad x, y \in L. \tag{5.1}$$

The equality of such expressions, multiplication by complex numbers, are obvious:

$$x' + iy' = x + iy \Leftrightarrow x' = x, \ y' = y;$$
$$(a + ib)(x + iy) = ax - by + i(ay + bx);$$
$$(a + ib)(x + iy) = 0 \Leftrightarrow a + ib = 0 \text{ and/or } x + iy = 0. \tag{5.2}$$

The complex dimension of $\tilde{L}$ is also n. Lie brackets in $\tilde{L}$ are defined by

$$[x + iy, \ u + iv] = [x, u] - [y, v] + i([x, v] + [y, u]). \tag{5.3}$$

A basis $\{e_j\}$ for L remains a basis for $\tilde{L}$; in such a basis, for both L and $\tilde{L}$ the structure constants C_{jk}^l are real. One can of course change to a general (complex) basis for $\tilde{L}$, when the structure constants become complex; however, they would have arisen from a real set. $\tilde{L}$ is the unique complex form of the real L.

We can next define a complex Lie algebra $\tilde{L}$ starting with a complex vector space and setting up Lie brackets obeying the antisymmetry and Jacobi conditions. In a general basis for $\tilde{L}$, the structure constants, $\tilde{C}_{jk}^l$ say, will be complex. A priori there may be no basis in which they all become real. On the other hand, it can happen that $\tilde{L}$ has several real forms, with corresponding bases in which all $\tilde{C}_{jk}^l$ are real. Allowing for some amount of repetition, we can say:

(i) A real Lie algebra L of dimension n leads to a unique complex Lie algebra $\tilde{L}$, its complex extension or form, with the same dimension n. If a basis for L is used as is for $\tilde{L}$, the structure constants remain unchanged and real.

(ii) Many distinct real L's may lead to the same complex $\tilde{L}$. For instance, $SO(3)$ and $SO(2, 1)$ have a common complex extension.

(iii) Given directly a complex $\tilde{L}$ of dimension n, it may have no, or several distinct, real forms. In the former case there is no basis for $\tilde{L}$ in which all $\tilde{C}^l_{jk}$ are real.

The use of complex $\tilde{L}$ is unavoidable for general theory, the 'return to real L's' may be difficult because of (ii), (iii) above. The notions of: subalgebra, invariant subalgebra, factor algebra, abelian case, commutator subalgebra, solvable, nilpotent, semisimple, simple cases, all generalise easily to the complex situation. One finds:

$$L \text{ abelian} \iff \tilde{L} \text{ abelian,}$$

$$L \text{ solvable} \iff \tilde{L} \text{ solvable,}$$

$$\text{but } L \text{ simple} \not\Rightarrow \tilde{L} \text{ simple.} \tag{5.4}$$

Solvability is a crucial generalisation of abelianness. For both real and complex cases, omitting tildes:

L solvable $\rightarrow$ every subalgebra solvable;

$\qquad\qquad$ factor algebras with respect to invariant subalgebras solvable;

L has an invariant subalgebra L' such that $L', L/L'$ both solvable

$$\rightarrow L \text{ solvable.} \tag{5.5}$$

Now comes the important *Levi splitting theorem:* Any (real or complex) Lie algebra L can be split into the semidirect sum of a semisimple subalgebra S and a maximal solvable invariant subalgebra T:

$$L = S \oplus T, \quad [S, S] \subseteq S, \quad [L, T] \subseteq T. \tag{5.6}$$

Does this help us classify all possible Lie algebras? The answer is – No! All semisimple ones can be classified, but not all solvable ones. The theory of the semisimple case is based on (magnificent!) work by Killing, Cartan, Weyl... . The first step is Cartan's 1894 theorem: A (real or complex) semisimple L is the direct sum of mutually commuting simple nonabelian subalgebras:

$$L \text{ semisimple} \iff L = L' \oplus L'' \oplus L''' \oplus \cdots,$$

$$\text{each simple, nonabelian;}$$

$$[L', L'' \text{ or } L''' \cdots] = [L'', L''' \text{ or } \cdots] = \cdots = 0. \tag{5.7}$$

Next, the complex simple Lie algebras are all known and classified. There are four great families, each infinite; and five exceptional cases.

All this is based on powerful theorems of Cartan and others, and repeated very skillful exploitation of the Jacobi identities (which amounts to working within the adjoint representation). We do not go into any detail but only mention two very interesting results:

1. In terms of the (possibly complex) structure constants C^l_{jk} of L in some basis define a 'metric tensor'

$$g_{jk} = g_{kj} = C^m_{jl} C^l_{km}. \qquad (5.8)$$

Then (Cartan):

$$L \text{ is semisimple} \quad \Leftrightarrow \quad \det(g_{jk}) \neq 0. \qquad (5.9)$$

2. In this case, from the structure constants C^l_{jk} we can construct a totally antisymmetric object:

$$C_{jkl} = C^m_{jk} g_{ml} = -C_{jlk} = -C_{lkj},$$
$$C^l_{jk} = g^{lm} C_{jkm}. \qquad (5.10)$$

Here is the proof:

$$\begin{aligned}
C_{jkl} &= C^m_{jk} g_{ml} = C^m_{jk} C^p_{mq} C^q_{lp} \\
&= -(C^m_{kq} C^p_{mj} + C^m_{qj} C^p_{mk}) C^q_{lp} \text{ by Jacobi on first two factors} \\
&= -C^m_{kq} C^q_{lp} C^p_{mj} - C^q_{lp} C^p_{km} C^m_{jq} \\
&= C^m_{kq} C^q_{lp} C^p_{jm} - C^m_{lq} C^q_{kp} C^p_{jm} \text{ by } p \to q \to m \to p \text{ in } 2^{\text{nd}} \text{ term} \\
&= (C^m_{kq} C^q_{lp} - C^m_{lq} C^q_{kp}) C^p_{jm} = -C_{jlk}. \qquad (5.11)
\end{aligned}$$

As mentioned, using Eqs. (5.9, 5.10) and the Jacobi identities repeatedly, a great deal of information on the structure constants can be obtained. We, however, switch attention at this point to the CSLA and describe them in some detail. In the simple case the complex to real transition is based on three fundamental results:

(I)   Every complex simple L has a basis in which all C^l_{jk} are real.

(II) (Weyl) Every complex simple L has a *unique* compact simple real form.

(III) (Weyl) If G is a (real) simple compact Lie group, and G' is locally isomorphic to G, then G' is also compact. So compactness can be 'read off' from the Lie algebra itself!

We now describe various properties of CSLA, the so-called Cartan–Weyl structure. To begin with, for a CSLA the metric tensor g_{jk} of (5.8) is positive definite (and of course real), so we can choose a basis to achieve

$$g_{jk} = \delta_{jk}, \qquad j, k = 1, 2, \cdots, r. \tag{5.12}$$

The usual notation for the dimension or order of L is r; also as the letter l is reserved for 'rank', we will avoid using it hereafter as either superscript or subscript. We recall that in the $SO(3)$ or $SU(2)$ case it is convenient to use complex combinations of real Lie algebra elements, such that:

$$J_\pm = J_1 \pm iJ_2 : \ [J_3, J_\pm] = \pm J_\pm, \ \ [J_+, J_-] = 2J_3. \tag{5.13}$$

This generalises in a very elegant way. It is also helpful to always keep in mind some generic or unspecified irreducible hermitian representation of the real Lie algebra L, so Lie brackets appear as i times commutators. Of course complex combinations like $J_\pm$ will be nonhermitian.

There is a certain maximal set of mutually commuting hermitian generators (real elements of L) written as H_a with $a = 1, 2, \cdots, l$, where l is called the rank of L:

$$H_a^\dagger = H_a, \ \ [H_a, H_b] = 0, \ \ a, b = 1, 2, \cdots, l < r. \tag{5.14}$$

In a general representation the H_a form part of a complete commuting set. In an irrep we can assume all the H_a are diagonal. They span the Cartan subalgebra of L, a maximal abelian subalgebra. In the $SU(2)$ case we have $r = 3, l = 1$.

The 'remaining' basis elements of L are $(r-l)$ in number in the real sense. However, analogous to $J_\pm$ in the $SU(2)$ case they can be arranged into $\frac{1}{2}(r-l)$ pairs of mutually complex, or better hermitian conjugate, elements. (Always, $r - l$ is an even integer; in the case of $SU(2)$ this difference is 2.) We need to count and label them suitably.

There is a collection of $r - l$ distinct real non-zero l-dimensional or l-component root vectors $\{\boldsymbol{\alpha}\}$, $\boldsymbol{\alpha} = \{\alpha_a, a = 1, 2, \cdots, l\}$, forming the *root space* $\mathcal{R}$, with many properties:

$$\mathcal{R} = \{\boldsymbol{\alpha}\} = \text{ set of } r - l \text{ roots};$$

$$\boldsymbol{\alpha} \in \mathcal{R} \Rightarrow -\boldsymbol{\alpha} \in \mathcal{R}; \ \ \pm 2\boldsymbol{\alpha}, \ \pm 3\boldsymbol{\alpha}, \cdots, \notin \mathcal{R};$$

$$g_{ab} = \sum_{\boldsymbol{\alpha} \in \mathcal{R}} \alpha_a \alpha_b = \delta_{ab}. \tag{5.15}$$

There are enough independent root vectors to span l-dimensional space, so $\frac{1}{2}(r-l) \geq l$ or $\frac{1}{2}(r - 3l)$ is a nonnegative integer. The number of linearly independent root vectors is l, the rank. For each $\boldsymbol{\alpha} \in \mathcal{R}$, we have a (one) complex element $E_{\boldsymbol{\alpha}}$ formed like $J_{\pm}$ out of the real elements of $\boldsymbol{L}$, and then the basic Lie brackets or commutation relations of $\boldsymbol{L}$ take the canonical Cartan–Weyl form:

$$[H_a, H_b] = 0;$$

$$[H_a, E_{\boldsymbol{\alpha}}] = \alpha_a E_{\boldsymbol{\alpha}};$$

$$[E_{\boldsymbol{\alpha}}, E_{-\boldsymbol{\alpha}}] = \alpha_a H_a;$$

$$\boldsymbol{\alpha}, \boldsymbol{\beta}, \boldsymbol{\alpha} + \boldsymbol{\beta} \in \mathcal{R} : [E_{\boldsymbol{\alpha}}, E_{\boldsymbol{\beta}}] = N_{\alpha\beta} E_{\boldsymbol{\alpha}+\boldsymbol{\beta}}, N_{\alpha\beta} \neq 0;$$

$$\boldsymbol{\alpha}, \boldsymbol{\beta} \in \mathcal{R}, \boldsymbol{\alpha} + \boldsymbol{\beta} \notin \mathcal{R} : [E_{\boldsymbol{\alpha}}, E_{\boldsymbol{\beta}}] = 0. \tag{5.16}$$

However, 'vectors' in $\boldsymbol{L}$ in the strict sense are linear combinations

$$\sum_{a=1}^{l} \lambda_a H_a + \sum_{\mathcal{R}} \left(\lambda_{\boldsymbol{\alpha}} E_{\boldsymbol{\alpha}} + \lambda_{\boldsymbol{\alpha}}^* E_{-\boldsymbol{\alpha}} \right), \lambda_a \text{ real.} \tag{5.17}$$

In a hermitian representation of $\boldsymbol{L}$ we have:

$$H_a^{\dagger} = H_a, \quad E_{\boldsymbol{\alpha}}^{\dagger} = E_{-\boldsymbol{\alpha}}. \tag{5.18}$$

For each root $\boldsymbol{\alpha} \in \mathcal{R}$, we have a corresponding $SU(2)^{(\alpha)}$ subalgebra in $\boldsymbol{L}$: the identifications of $J_3, J_{\pm}$ of Eq. (5.13) are

$$J_3 = \frac{\alpha_a H_a}{|\boldsymbol{\alpha}|^2}, J_+ = \frac{\sqrt{2}}{|\boldsymbol{\alpha}|} E_{\boldsymbol{\alpha}}, J_- = \frac{\sqrt{2}}{|\boldsymbol{\alpha}|} E_{-\boldsymbol{\alpha}},$$

$$|\boldsymbol{\alpha}|^2 = \alpha_a \alpha_a. \tag{5.19}$$

Thus there are $\frac{1}{2}(r - l)$ generally mutually noncommuting $SU(2)$'s within $\boldsymbol{L}$. Many of the properties so far described (and more to come!) arise from the known properties of the UIR's of $SU(2)$.

5.2 Properties of Roots and Root Space

For a rank l CSLA, root space is a set of distinct nonzero real vectors, $r - l$ in number, in a real Euclidean l-dimensional space. The roots come in pairs $\pm\boldsymbol{\alpha}$, and all together constitute the set $\mathcal{R}$. The angles between roots and the ratios of their lengths are quantised in a correlated manner:

Table 5.1   Properties of roots $\boldsymbol{\alpha}$ in $\mathcal{R}$

θ=angle between $\boldsymbol{\alpha}$ and $\boldsymbol{\beta}$	Ratio $|\boldsymbol{\alpha}|/|\boldsymbol{\beta}|$
90^0	free
60^0 or 120^0	1
45^0 or 135^0	$\sqrt{2}$ or $\frac{1}{\sqrt{2}}$
30^0 or 150^0	$\sqrt{3}$ or $\frac{1}{\sqrt{3}}$

No other values of θ are possible.

Having chosen the Cartan subalgebra elements H_a in a definite sequence, a root $\boldsymbol{\alpha} = \{\alpha_1, \alpha_2, \cdots, \alpha_l\}$ is *positive* if its first nonzero component is positive, otherwise it is *negative*. Then we separate $\mathcal{R}$ into two subsets:

$$\mathcal{R}_+ = \text{set of } \frac{1}{2}(r - l) \text{ positive roots,}$$

$$\mathcal{R}_- = \text{set of } \frac{1}{2}(r - l) \text{ negative roots,}$$

$$\boldsymbol{\alpha} \in \mathcal{R}_+ \Leftrightarrow -\boldsymbol{\alpha} \in \mathcal{R}_-;$$

$$\mathcal{R} = \mathcal{R}_+ \cup \mathcal{R}_-. \tag{5.20}$$

The *simple* roots form a subset $\mathfrak{s} \subset \mathcal{R}_+$: a positive root $\boldsymbol{\alpha} \in \mathcal{R}_+$ is a simple root if and only if it *cannot* be expressed as a nonnegative integral linear combination of other positive roots. So we have:

$$\mathfrak{s} \subset \mathcal{R}_+ \subset \mathcal{R}. \tag{5.21}$$

The properties of the simple roots are really remarkable and elegant:

1. The allowed angles and length ratios between simple roots are a subset of the possibilities in Table 5.1. This is shown in Table 5.2.

Table 5.2   Properties of simple roots $\boldsymbol{\alpha}$ in $\mathfrak{s}$

θ=angle between $\boldsymbol{\alpha}$ and $\boldsymbol{\beta}$	Ratio $|\boldsymbol{\alpha}|/|\boldsymbol{\beta}|$
90^0	free
120^0	1
135^0	$\sqrt{2}$ or $\frac{1}{\sqrt{2}}$
150^0	$\sqrt{3}$ or $\frac{1}{\sqrt{3}}$

2. The simple root vectors are linearly independent and there are exactly l of them, written as $\boldsymbol{\alpha}^{(a)}$:

$$\mathfrak{s} = \{\boldsymbol{\alpha}^{(1)}, \boldsymbol{\alpha}^{(2)}, \cdots, \boldsymbol{\alpha}^{(l)}\} \subset \mathcal{R}_+, \text{ set of } l \text{ simple roots}$$

$$\text{a subset of } \frac{1}{2}(r-l) \text{ positive roots.} \tag{5.22}$$

3. Each positive root is uniquely expressible as a linear combination of simple roots with nonnegative integer coefficients:

$$\boldsymbol{\alpha} \in \mathcal{R}_+ \Rightarrow \boldsymbol{\alpha} = \sum_{a=1}^{l} n_a \boldsymbol{\alpha}^{(a)}, \ n_a \geq 0, \text{ integers, unique for given } \boldsymbol{\alpha}. \tag{5.23}$$

Therefore also for negative roots:

$$\boldsymbol{\alpha} \in \mathcal{R}_- \Rightarrow \boldsymbol{\alpha} = \sum_{a=1}^{l} n_a \boldsymbol{\alpha}^{(a)}, \ n_a \leq 0, \text{ integers, unique for given } \boldsymbol{\alpha}. \tag{5.24}$$

4. The lengths and angles of simple roots allow the unique reconstruction of $\mathcal{R}_+$ and $\mathcal{R}_-$, and thereafter of $\mathcal{R}$.

$$\text{Geometry of } \mathfrak{s} \rightarrow \text{ unique reconstruction of } \mathcal{R}_+, \mathcal{R}_-, \mathcal{R}. \tag{5.25}$$

5. In this way the problem of classifying all possible CSLA becomes that of building up all allowed simple root systems. The information on angles and length ratios of the set of simple roots $\mathfrak{s}$ can be depicted in a two dimensional *Dynkin diagram*. Thus each CSLA corresponds uniquely to some permitted Dynkin diagram. Indeed the complete classification of all possible CSLA's can be carried out by looking for all possible root systems obeying all the geometrical conditions described earlier, all flowing from the Jacobi conditions. Since we will not make specific use of Dynkin diagrams in the rest of this chapter or in later ones, we forego a detailed study of them.

We said earlier that all complex simple Lie algebras are known, and that each of them has a unique compact real form. So the CSLA classification follows the same pattern: four infinite classical families, plus five exceptional ones. We now study the significant aspects of the former, beginning in each case with what we call a 'covariant' description, and then connecting to the canonical Cartan–Weyl description. By 'covariant' description we mean the description of the Lie algebra with real vectors and basis elements, corresponding to hermitian generators in any unitary representation.

The names given by Cartan to the four infinite families are A_l, B_l, C_l and D_l, with l the rank. For practical reasons, corresponding to the order of increasing algebra, we will study them in the sequence $D_l - B_l - C_l - A_l$. In each case there is a defining matrix representation denoted by $\mathcal{D}$; the hermitian (or other convenient) generators in $\mathcal{D}$; their commutation and other algebraic relations; order and rank; Cartan subalgebra basis $\{H_a\}$; roots $\mathcal{R}$ and complex combinations E_α of generators; positive roots $\mathcal{R}_+$ and simple roots $\mathfrak{s}$. In some cases, split index notation is useful.

5.3 The $SO(2l)$ Family D_ℓ

This is the family of real proper orthogonal rotations in real even dimensional spaces. The defining representation $\mathcal{D}$ is

$$SO(2l) = \{A = 2l\text{-dimensional real matrix} \,|\, A^T A = \mathbb{1}, \ \det A = 1\}. \tag{5.26}$$

These matrices act on $2l$-component vectors in $2l$-dimensional real Euclidean space. For $l = 1$ we have an abelian group; for $l = 2$ we find the group is nonsimple; so we start at $l = 3$.

Let indices $A, B, C, \cdots$ range over $1, 2, \cdots, 2l$. (Overuse of some letters is unavoidable!) The generators for $\mathcal{D}$ are all $2l \times 2l$ pure imaginary hermitian matrices, i.e., i times all $2l \times 2l$ real antisymmetric matrices. The number of such independent matrices is $\frac{1}{2} \cdot 2l \cdot (2l - 1) = l(2l - 1)$, which is the order of D_l. In covariant form, a basis for such matrices and their commutation relations are:

$$M_{AB} = -M_{BA} = M_{AB}^\dagger \ : (M_{AB})_{CD} = i(\delta_{AC}\delta_{BD} - \delta_{AD}\delta_{BC});$$

$$[M_{AB}, M_{CD}] = i(\delta_{BC}M_{AD} - \delta_{AC}M_{BD} + \delta_{BD}M_{CA} - \delta_{AD}M_{CB}). \tag{5.27}$$

These hermiticity relations and commutation relations hold for the generators in *every* UIR.

It is useful to introduce *split index* notation: divide the $2l$ values of $A, B, \cdots$ into l pairs; the ath pair is $2a - 1, 2a$ for $a = 1, 2, \cdots, l$; within each pair we have indices r, s taking values $1, 2$:

$$A = 1, 2, \cdots, 2l \to ar, \ a = 1, 2, \cdots, l; \ r = 1, 2;$$

$$A = 2(a - 1) + r. \tag{5.28}$$

From (5.27) it is obvious that $M_{12}, M_{34}, \cdots , M_{2l-1,2l}$ are l mutually commuting elements – they span the Cartan subalgebra, so the rank is l. So we write:

$$H_1 = M_{12}, \ H_2 = M_{34}, \cdots , \ H_a = M_{2a-1,2a}, \ a = 1,2,\cdots ,l;$$

$$[H_a, H_b] = 0. \tag{5.29}$$

It is easy to check that $\{H_a\}$ is a maximal set of abelian generators. In the defining representation $\mathcal{D}$ they are block diagonal, not diagonal as $\mathcal{D}$ is real:

$$(H_a)_{AB} = i \left(\delta_{A,2a-1} \delta_{B,2a} - \delta_{A,2a} \delta_{B,2a-1} \right);$$

$$H_a = i \begin{pmatrix} 0 \cdots\cdots\cdots 0 \\ \vdots \ \ddots \ \ \ \ \vdots \\ \vdots \ \ \ i\sigma_2 \ \ \ \vdots \\ \vdots \ \ \ \ \ \ \ddots \ \vdots \\ 0 \cdots\cdots\cdots 0 \end{pmatrix} \tag{5.30}$$

In H_a, the ath 2×2 block 'on the diagonal' is $i\sigma_2 = \left(\begin{smallmatrix} 0 & 1 \\ -1 & 0 \end{smallmatrix} \right)$. So, to repeat, in $\mathcal{D}$ the H_a are not diagonal as given, but are easily diagonalisable. And they obey

$$\mathrm{Tr}(H_a H_b) = 2\delta_{ab}. \tag{5.31}$$

With easy algebra, from (5.27) we can find all roots $\boldsymbol{\alpha} \in \mathcal{R}$. Since these are real l-component vectors, we introduce the standard unit vectors in l-dimensional space:

$$\boldsymbol{e}_a = (0,0,\cdots , 1,0,\cdots ,0), \ 1 \text{ in } a^{\text{th}} \text{ position}, \ a = 1,2,\cdots ,l. \tag{5.32}$$

The roots can be found by constructing suitable $E_{\boldsymbol{\alpha}}$'s out of M_{AB}. In split index notation we write:

$$\begin{aligned} H_a &= M_{a1,a2}, \ a = 1,2,\cdots ,l; \\ X_{ab} &= M_{a1,b1} = -X_{ba}; \\ Y_{ab} &= M_{a1,b2}; \\ Z_{ab} &= M_{a2,b1} = -Y_{ba}; \\ W_{ab} &= M_{a2,b2} = -W_{ba}. \end{aligned} \tag{5.33}$$

All $X_{ab}, Y_{ab}, Z_{ab}, W_{ab}$ are hermitian. Then the only nonzero Lie brackets with one factor being H_a are (no sums!):

$$[H_a, X_{ab} \text{ or } Y_{ab} \text{ or } Z_{ab} \text{ or } W_{ab}] = i(-Z_{ab} \text{ or } - W_{ab} \text{ or } X_{ab} \text{ or } Y_{ab}), \ a < b;$$

$$[H_b, X_{ab} \text{ or } Y_{ab} \text{ or } Z_{ab} \text{ or } W_{ab}] = i(-Y_{ab} \text{ or } X_{ab} \text{ or } - W_{ab} \text{ or } Z_{ab}), \ a < b. \tag{5.34}$$

Other such brackets vanish:

$$[H_a, X_{bc} \text{ or } Y_{bc} \text{ or } Z_{bc} \text{ or } W_{bc}] = 0, \quad a \neq b, a \neq c, \ b < c. \tag{5.35}$$

The evaluation of commutators among X, Y, Z, W is left as an exercise.

Based on (5.34) we see how to form E_α's (again no sums!):

$$[H_a \text{ or } H_b, \ X_{ab} - i\epsilon' \, Y_{ab} - i\epsilon \, Z_{ab} - \epsilon'\epsilon \, W_{ab}] =$$
$$(\epsilon \text{ or } \epsilon')(X_{ab} - i\epsilon' Y_{ab} - i\epsilon Z_{ab} - \epsilon'\epsilon W_{ab}),$$
$$\text{all } \ a < b; \ \epsilon', \epsilon = \pm 1. \tag{5.36}$$

Thus we have the possible roots and E_α's (up to normalisation but obeying $E_\alpha^+ = E_{-\alpha}$):

$$\boldsymbol{\alpha} = \epsilon \boldsymbol{e}_a + \epsilon' \boldsymbol{e}_b, \ a < b = 2, \cdots, l :$$
$$E_\alpha = X_{ab} - i\epsilon' Y_{ab} - i\epsilon Z_{ab} - \epsilon'\epsilon W_{ab}. \tag{5.37}$$

The set of all roots is

$$\mathcal{R} = \{\pm \boldsymbol{e}_a \pm \boldsymbol{e}_b, \ a < b = 2, \cdots, l\}, \tag{5.38}$$

$2l(l-1)$ in number. There are no roots like $\pm 2\boldsymbol{e}_a, \pm 3\boldsymbol{e}_a, \cdots$, etc. With a little algebra beginning with low values of l, we find:

$$\mathcal{R}_+ = \text{positive roots} = \{\boldsymbol{e}_a \pm \boldsymbol{e}_b, \ a < b\}, \ l(l-1) \text{ in number};$$
$$\mathfrak{s} = \text{simple roots} = \{\boldsymbol{\alpha}^{(1)} = \boldsymbol{e}_1 - \boldsymbol{e}_2, \ \boldsymbol{\alpha}^{(2)} = \boldsymbol{e}_2 - \boldsymbol{e}_3, \cdots, \boldsymbol{\alpha}^{(l-1)} = \boldsymbol{e}_{l-1} - \boldsymbol{e}_l,$$
$$\boldsymbol{\alpha}^{(l)} = \boldsymbol{e}_{l-1} + \boldsymbol{e}_l\}, \ l \text{ in number}. \tag{5.39}$$

It is good to verify that the restrictions in Tables 5.1 and 5.2 are respected, and also to find the nonzero $N_{\alpha\beta}$.

In summary for $SO(2l) = D_l$ we have:

Order $l(2l - 1)$, rank l

$$H_a = M_{2a-1,2a}, \quad a = 1, 2, \cdots, l;$$

$$\mathcal{R} = \{\epsilon \boldsymbol{e}_a + \epsilon' \boldsymbol{e}_b, \ \epsilon' = \pm 1, \ \epsilon = \pm 1, a < b\}, \ 2l(l-1) \text{ roots};$$

$$\mathcal{R}_+ = \{\boldsymbol{e}_a + \epsilon' \boldsymbol{e}_b, \ \epsilon' = \pm 1, a < b\}, \ l(l-1) \text{ positive roots};$$

$$\mathfrak{s} = \{\boldsymbol{e}_1 - \boldsymbol{e}_2, \ \boldsymbol{e}_2 - \boldsymbol{e}_3, \cdots, \ \boldsymbol{e}_{l-1} - \boldsymbol{e}_l, \ \boldsymbol{e}_{l-1} + \boldsymbol{e}_l\}, l \text{ simple roots};$$

$$E_{\epsilon \boldsymbol{e}_a + \epsilon' \boldsymbol{e}_b} = X_{ab} - i\epsilon' \Upsilon_{ab} - i\epsilon Z_{ab} - \epsilon'\epsilon W_{ab}. \tag{5.40}$$

5.4 The $SO(2l + 1)$ Family B_l

This is the family of real proper orthogonal rotations in real odd dimensional Euclidean spaces. The defining representation $\mathcal{D}$ is:

$$SO(2l + 1) = \{A = (2l+1)\text{-dimensional real matrix} \,|A^T A = \mathbb{1}, \ \det A = 1\}. \tag{5.41}$$

These matrices act on $(2l+1)$-component real vectors in Euclidean $(2l+1)$-dimensional real space. We can take $l = 1, 2, \cdots$ for the present.

Now we let the indices $A, B, \cdots$ go over $1, 2, \cdots, 2l+1$: the $SO(2l)$ range plus one more value, $2l+1$. The independent generators for $\mathcal{D}$, and their commutation relations, read exactly as before, namely (5.27) with the extended range for $A, B, C, \cdots$. The order is $l(2l + 1)$.

The $\{H_a\}$ of $SO(2l)$ remain good as a maximal abelian subalgebra, with $a = 1, 2, \cdots, l$. Thus both $SO(2l)$ and $SO(2l + 1)$ have rank l. Again in $\mathcal{D}$ we have

$$\text{Tr}(H_a H_b) = 2\delta_{ab}. \tag{5.42}$$

All the roots of $SO(2l)$, namely $\pm \boldsymbol{e}_a \pm \boldsymbol{e}_b$ for $a < b$, are again present as roots, as the previous calculations remain valid. The added roots now come from the extra generators $M_{A,2l+1}$ for $A = 1, 2, \cdots, 2l$. For them we have under commutation with H_a (no sums!):

$$[H_a, M_{br,2l+1}] = i\delta_{ab}(\delta_{r2} M_{b1,2l+1} - \delta_{r1} M_{b2,2l+1}),$$

$$\therefore \ [H_a, M_{b1,2l+1} - i\epsilon M_{b2,2l+1}] = \epsilon \delta_{ab}(M_{b1,2l+1} - i\epsilon M_{b2,2l+1}), \ \epsilon = \pm 1 \tag{5.43}$$

(As before the commutators among $X, Y, Z, W, M_{A,2l+1}$ are left as exercises.) Therefore the new roots are $\epsilon \boldsymbol{e}_b$ for $b = 1, 2, \cdots, l$:

$$\boldsymbol{\alpha} = \epsilon \boldsymbol{e}_b: \quad E_{\boldsymbol{\alpha}} = M_{b1,2l+1} - i\epsilon M_{b2,2l+1}. \tag{5.44}$$

With sample calculations for small values of l one can find $\mathcal{R}_+$ and $\mathfrak{s}$. All put together the $SO(2l+1)$ summary, analogous to (5.40), reads:

Order $l(2l+1)$, rank l;

$$H_a = M_{2a-1,2a}, \ a = 1, 2, \cdots, l;$$
$$\mathcal{R} = \{\epsilon \boldsymbol{e}_a + \epsilon' \boldsymbol{e}_b, \ \epsilon' = \pm 1, \epsilon = \pm 1, \ a < b; \ \epsilon \boldsymbol{e}_a, \ \epsilon = \pm 1\}, 2l^2 \text{ roots};$$
$$\mathcal{R}_+ = \{\boldsymbol{e}_a + \epsilon' \boldsymbol{e}_b, \ \epsilon' = \pm 1, \ a < b; \ \boldsymbol{e}_a\}, \ l^2 \text{ positive roots};$$
$$\mathfrak{s} = \{\boldsymbol{e}_1 - \boldsymbol{e}_2, \ \boldsymbol{e}_2 - \boldsymbol{e}_3, \cdots, \ \boldsymbol{e}_{l-1} - \boldsymbol{e}_l, \ \boldsymbol{e}_l\}, \ l \text{ simple roots};$$
$$E_{\epsilon \boldsymbol{e}_a + \epsilon' \boldsymbol{e}_b} = X_{ab} - i\epsilon' Y_{ab} - i\epsilon Z_{ab} - \epsilon' \epsilon W_{ab},$$
$$E_{\epsilon \boldsymbol{e}_a} = M_{a1,2l+1} - i\epsilon M_{a2,2l+1}. \tag{5.45}$$

5.5 The $USp(2l)$ Family C_l

In the two previous families the defining representations were made up of real matrices, and the algebra was fairly easy. The next two families have defining representations using complex matrices, and the algebra is somewhat involved. The groups C_l and their Lie algebras are rather unfamiliar. They share some properties with $SO(2l)$ and some with $SO(2l+1)$, so they stand somewhere in between. Even the defining representation is unfamiliar to many. We study the various aspects in some detail.

The defining representation $\mathcal{D}$ is via certain complex matrices, on a $2l$-dimensional complex linear space. As with $SO(2l)$, let indices $A, B, \cdots = 1, 2, \cdots, 2l$. First define a *‘symplectic metric’* matrix

$$\eta = (\eta_{AB}) = -\eta^\dagger = -\eta^T = \begin{pmatrix} \epsilon & & & 0 \\ & \ddots & & \\ & & \epsilon & \\ & & & \ddots \\ 0 & & & \epsilon \end{pmatrix}, \quad \epsilon = i\sigma_2 = \begin{pmatrix} 0 & 1 \\ -1 & 0 \end{pmatrix},$$

$$\eta^2 = -\mathbb{1}. \tag{5.46}$$

This is real antihermitian antisymmetric and block diagonal, with the 2×2 block $\epsilon = i\sigma_2$ appearing l times. In split index notation we have

$$\eta_{ar,bs} = \delta_{ab}\epsilon_{rs}, \quad a, b = 1, \cdots, l; \quad r, s = 1, 2. \tag{5.47}$$

Then the defining representation $\mathcal{D}$ of C_l, the unitary symplectic group $USp(2l)$ in $2l$ complex dimensions, is:

$$USp(2l) = \{ U = 2l\text{-dimensional complex matrix} \mid U^\dagger U = \mathbb{1}, \ U^T \eta U = \eta \}. \tag{5.48}$$

Thus in addition to being unitary, these transformations preserve an antisymmetric nondegenerate bilinear form in $2l$ complex dimensions.

The properties of the generator matrices are found by setting $U \simeq \mathbb{1} - i\epsilon J$ for infinitesimal ϵ in the defining conditions:

$$J^\dagger = J,$$
$$J^T \eta + \eta J = 0 \ \text{ or } \ (\eta J)^T = \eta J = \text{symmetric}. \tag{5.49}$$

We need to analyse these conditions carefully and find a complete independent set of hermitian generators in terms of which a general J can be expanded as a *real* linear combination. This is quite intricate.

From the symplectic condition in (5.49) we see that ηJ is a complex symmetric matrix, so J itself is a complex linear combination of a set of $l(2l + 1)$ independent $2l \times 2l$ matrices defined as follows:

$$J = \frac{1}{2}\omega_{AB}J_{AB}, \quad J_{AB} = J_{BA}, \quad \omega_{AB} = \omega_{BA},$$
$$(J_{AB})_{CD} = i(\eta_{AC}\delta_{BD} + \eta_{BC}\delta_{AD}). \tag{5.50}$$

These matrices J_{AB}, which like M_{AB} in (5.27) are pure imaginary, have the following hermitian conjugation and commutation properties:

$$J_{AB}^\dagger = \eta_{AC}\eta_{BD}J_{CD}, \tag{5.51a}$$

$$[J_{AB}, J_{CD}] = i(\eta_{CA}J_{BD} + \eta_{CB}J_{AD} + \eta_{DA}J_{BC} + \eta_{DB}J_{AC}). \tag{5.51b}$$

Therefore these are the covariant commutation relations and hermiticity conditions obeyed by the generators in any UR of $USp(2l)$. The conditions on ω_{AB} ensuring

$J^\dagger = J$ are

$$\omega^*_{AB} = \eta_{AC}\eta_{BD}\omega_{CD}. \tag{5.52}$$

Now we analyse (5.51a) in detail, to pick out the independent hermitian generators, their number and expressions in the defining representation $\mathcal{D}$ in Eq. (5.48). This will allow us to identify the order, the Cartan subalgebra generators $\{H_a\}$, the rank, and the root system $\mathcal{R}$. At this point it helps to switch to split index notation: $A, B, C, D, \cdots \rightarrow ar, bs, cu, dv, \cdots$. Then (5.51a) reads:

$$J_{AB} \equiv J_{ar,bs} \equiv J_{bs,ar} \equiv J_{rs}^{(a,b)} :$$

$$J_{11}^{(a,b)\dagger} = J_{22}^{(a,b)}, \quad J_{12}^{(a,b)\dagger} = -J_{21}^{(a,b)}. \tag{5.53}$$

We separate the cases $a = b$ and $a < b$, and choose the following as independent hermitian generators:

$$J_1^{(a)} = \frac{i}{4}\left(J_{22}^{(a,a)} - J_{11}^{(a,a)}\right), J_2^{(a)} = \frac{1}{4}\left(J_{22}^{(a,a)} + J_{11}^{(a,a)}\right),$$

$$J_3^{(a)} = \frac{i}{2}J_{12}^{(a,a)}, \quad a = 1, 2, \cdots, l; \tag{5.54a}$$

$$K_1^{(a,b)} = \frac{i}{2}\left(J_{22}^{(a,b)} - J_{11}^{(a,b)}\right),$$

$$K_2^{(a,b)} = \frac{1}{2}\left(J_{22}^{(a,b)} + J_{11}^{(a,b)}\right),$$

$$K_3^{(a,b)} = \frac{i}{2}\left(J_{12}^{(a,b)} + J_{21}^{(a,b)}\right),$$

$$K_0^{(a,b)} = \frac{1}{2}\left(J_{12}^{(a,b)} - J_{21}^{(a,b)}\right), \quad a < b = 2, 3, \cdots, l. \tag{5.54b}$$

There are $3l$ generators in the first set and $2l(l-1)$ in the second, adding up to a total of $l(2l+1)$ independent hermitian generators. Thus $USp(2l)$ is of order $l(2l+1)$, the same as $SO(2l+1)$.

The motivation for the definitions (5.54) is that in the defining representation $\mathcal{D}$ we find the following 2×2 block forms:

$$J_j^{(a)} \rightarrow a: \begin{pmatrix} & \overset{\displaystyle a}{\ddots} & \\ 0 & & 0 \\ & \ddots & \\ & \frac{1}{2}\sigma_j & \\ & & \ddots \\ 0 & & 0 \end{pmatrix},$$

$$
K_j^{(a,b)} \rightarrow \quad
\begin{array}{c} a: \\ b: \end{array}
\overset{\overset{\displaystyle a \quad\quad\quad b}{\cdots \quad\quad \cdots}}{
\begin{pmatrix}
0 & & & & \\
 & \ddots & & \frac{i}{2}\sigma_j & \\
 & & & & \\
 & -\frac{i}{2}\sigma_j & & \ddots & \\
 & & & & 0
\end{pmatrix}}, a < b.
$$

$$
K_0^{(a,b)} \rightarrow \quad
\begin{array}{c} a: \\ b: \end{array}
\overset{\overset{\displaystyle a \quad\quad\quad b}{\cdots \quad\quad \cdots}}{
\begin{pmatrix}
0 & & & & \\
 & \ddots & & \frac{i}{2}\,\mathbb{1} & \\
 & & & & \\
 & -\frac{i}{2}\,\mathbb{1} & & \ddots & \\
 & & & & 0
\end{pmatrix}}, a < b. \tag{5.55}
$$

The general commutation relations among $J_j^{(a)}, K_0^{(a,b)}, K_k^{(a,b)}$ may be read off quite easily from these forms in $\mathcal{D}$, since in any event the general result $(5.51b)$ has been obtained from $\mathcal{D}$. We see that the $J_j^{(a)}$ generate a useful block-diagonal $SU(2) \times SU(2) \times \cdots \times SU(2)$ subgroup made up of l commuting factors acting, respectively, on the pairs of dimensions $1, 2; 3, 4; 5, 6; \cdots; 2l-1, 2l$. Each $SU(2)$ factor appears in $\mathcal{D}$ via its defining two dimensional representation given in Eq. (3.63). Note that since $SU(2) = USp(2)$,

$$
U = \text{ block diagonal } (u_1, u_2, \cdots, u_l),
$$
$$
u_a \in SU(2), \tag{5.56}
$$

does obey (5.48). So the commutation relations

$$
[J_j^{(a)}, J_k^{(b)}] = i\delta_{ab}\epsilon_{jkl}J_l^{(a)}, \text{ no sum on } a, \tag{5.57}
$$

immediately follow. The commutators between J's on the one hand and K's on the other are also easily found:

$$
[J_j^{(a)}, K_0^{(b,c)}] = \frac{i}{2}(\delta_{ab} - \delta_{ac})K_j^{(b,c)},
$$
$$
[J_j^{(a)}, K_k^{(b,c)}] = \frac{i}{2}(\delta_{ac} - \delta_{ab})\delta_{jk}K_0^{(b,c)} + \frac{i}{2}(\delta_{ab} + \delta_{ac})\epsilon_{jkl}K_l^{(b,c)}, \quad b < c,
$$

$$
\text{no sums on } a, b, c. \tag{5.58}
$$

The $[K, K]$ commutators are left as an exercise.

One can guess that $H_a = J_3^{(a)}$ for $a = 1, 2, \cdots, l$ span a Cartan subalgebra. They do, as one can check that they are a maximal abelian set. In $\mathcal{D}$, $H_a = \frac{1}{2}\sigma_3$ in the ath

diagonal 2×2 block and zero elsewhere, so they are already diagonal. We see that in $\mathcal{D}$ we have

$$\text{Tr}(H_a \, H_b) = \frac{1}{2}\delta_{ab}. \tag{5.59}$$

The rank of $USp(2l)$ is thus l.

The root system $\mathcal{R}$

Within each $SU(2)$ factor in the $SU(2) \times SU(2) \times \cdots \times SU(2)$ subgroup we find roots $\pm e_a$ corresponding to the raising and lowering operators:

$$[H_a, J_1^{(b)} \pm iJ_2^{(b)}] = \pm\delta_{ab}(J_1^{(b)} \pm iJ_2^{(b)}), \text{ no sum on } b. \tag{5.60}$$

Out of the K's we can form the remaining combinations:

$$[H_a, K_0^{(b,c)}] = \frac{i}{2}(\delta_{ab} - \delta_{ac})K_3^{(b,c)}, \ [H_a, K_3^{(b,c)}] = -\frac{i}{2}(\delta_{ab} - \delta_{ac})K_0^{(b,c)} \Rightarrow$$

$$[H_a, K_0^{(b,c)} \pm iK_3^{(b,c)}] = \pm\frac{1}{2}(\delta_{ab} - \delta_{ac})(K_0^{(b,c)} \pm iK_3^{(b,c)});$$

$$[H_a, K_1^{(b,c)} \text{ or } K_2^{(b,c)}] = \frac{i}{2}(\delta_{ab} + \delta_{ac})(K_2^{(b,c)} \text{ or } - K_1^{(b,c)}) \Rightarrow$$

$$[H_a, K_1^{(b,c)} \pm iK_2^{(b,c)}] = \pm\frac{1}{2}(\delta_{ab} + \delta_{ac})(K_1^{(b,c)} \pm iK_2^{(b,c)}). \tag{5.61}$$

Therefore the roots and corresponding E_α's are:

$$
\begin{array}{ccc}
\boldsymbol{\alpha} & E_\alpha & \\
\pm e_a & J_1^{(a)} \pm iJ_2^{(a)} & a = 1, 2, \cdots, l \\
\pm\frac{1}{2}(e_b + e_c) & K_1^{(b,c)} \pm iK_2^{(b,c)} & b < c \\
\pm\frac{1}{2}(e_b - e_c) & K_0^{(b,c)} \pm iK_3^{(b,c)} & b < c
\end{array} \tag{5.62}
$$

As expected these add up to a system of $2l^2$ roots:

$$\mathcal{R} = \{\pm e_a, \ a = 1, 2, \cdots, l; \ \pm\frac{1}{2}(e_b \pm e_c), \ b < c = 2, 3, \cdots, l\}. \tag{5.63}$$

The positive roots are just half of these, l^2 in number:

$$\mathcal{R}_+ = \{e_a, \ a = 1, 2, \cdots, l; \ \frac{1}{2}(e_a \pm e_b), \ a < b = 2, 3, \cdots, l\}. \tag{5.64}$$

The l simple roots turn out to be

$$\mathfrak{s} = \{\boldsymbol{\alpha}^{(1)} = \frac{1}{2}(\boldsymbol{e}_1 - \boldsymbol{e}_2), \; \boldsymbol{\alpha}^{(2)} = \frac{1}{2}(\boldsymbol{e}_2 - \boldsymbol{e}_3), \; \cdots, \; \boldsymbol{\alpha}^{(l-1)} = \frac{1}{2}(\boldsymbol{e}_{l-1} - \boldsymbol{e}_l), \; \boldsymbol{\alpha}^{(l)} = \boldsymbol{e}_l\}.$$

(5.65)

Therefore we have the $USp(2l)$ summary:

Order $l(2l+1)$, rank l;

$$H_a = J_3^{(a)}, \quad a = 1, 2, \cdots, l;$$

$$\mathcal{R} = \{\epsilon \boldsymbol{e}_a, \epsilon = \pm 1, a = 1, \cdots, l; \; \frac{1}{2}(\epsilon \boldsymbol{e}_a + \epsilon' \boldsymbol{e}_b),$$

$$\epsilon, \epsilon' = \pm 1, a < b = 2, \cdots, l\}, \; 2l^2 \text{ roots};$$

$$\mathcal{R}_+ = \{\boldsymbol{e}_a, a = 1, \cdots, l; \; \frac{1}{2}(\boldsymbol{e}_a + \epsilon' \boldsymbol{e}_b), \; \epsilon' = \pm 1, a < b = 2, \cdots, l\}, \; l^2 \text{ roots};$$

$$\mathfrak{s} = \{\frac{1}{2}(\boldsymbol{e}_1 - \boldsymbol{e}_2), \frac{1}{2}(\boldsymbol{e}_2 - \boldsymbol{e}_3), \cdots, \frac{1}{2}(\boldsymbol{e}_{l-1} - \boldsymbol{e}_l), \boldsymbol{e}_l\}, \; l \text{ roots};$$

$$E_{\pm \boldsymbol{e}_a} = J_1^{(a)} \pm i J_2^{(a)}, \; E_{\pm \frac{1}{2}(\boldsymbol{e}_a + \boldsymbol{e}_b)} = K_1^{(a,b)} \pm i K_2^{(a,b)},$$

$$E_{\pm \frac{1}{2}(\boldsymbol{e}_a - \boldsymbol{e}_b)} = K_0^{(a,b)} \pm i K_3^{(a,b)}.$$

(5.66)

5.6 The $SU(l+1)$ Family A_l

The work involved in analysing the defining representations $\mathcal{D}$ for $SO(2l)$, $SO(2l+1)$ and $USp(2l)$ has been fairly 'clean'. While $SU(l+1)$ is easier to define and picture than $USp(2l)$ because of familiarity with quantum mechanics, it turns out to be more 'messy' when we have to find H_a, roots, etc. It is easier to handle hermiticity than in the $USp(2l)$ case, but the roots are more complicated than in the previous three cases.

The defining representation $\mathcal{D}$ for $SU(l+1)$ consists of all unitary unimodular matrices in $(l+1)$ complex dimensions:

$$SU(l+1) = \{U = (l+1)\text{-dimensional complex matrix}| U^\dagger U = \mathbb{1}, \det \; U = 1\}.$$

(5.67)

To find the generators and their properties we set $U \simeq \mathbb{1} - i\epsilon H$ where ϵ is small and find:

$$H^\dagger = H, \quad \text{Tr } H = 0.$$

(5.68)

The number of real independent 'matrix elements' in H are: l diagonal ones, and $2 \times \frac{1}{2}l(l+1) = l(l+1)$ off diagonal ones, totalling $l(l+2) = (l+1)^2 - 1$, which is the order of $SU(l+1)$.

The covariant or tensor approach to the commutation relations expresses a general H, taking account of tracelessness, as

$$H = b^{\mu}_{\lambda} A^{\lambda}_{\mu}, \quad b^{\mu*}_{\lambda} = b^{\lambda}_{\mu}, \quad b^{\lambda}_{\lambda} = 0;$$

$$(A^{\lambda}_{\mu})^{\rho}_{\sigma} = \delta^{\lambda}_{\sigma}\delta^{\rho}_{\mu} - \frac{1}{l+1}\delta^{\lambda}_{\mu}\delta^{\rho}_{\sigma},$$

$$\lambda, \mu, \rho, \sigma = 1, 2, \cdots, l+1. \tag{5.69}$$

Here superscripts/subscripts function as row/column indices. Then the hermiticity, algebraic and commutation relations among A^{λ}_{μ} are:

$$A^{\lambda\dagger}_{\mu} = A^{\mu}_{\lambda}, \quad A^{\lambda}_{\lambda} = 0,$$

$$[A^{\lambda}_{\mu}, A^{\rho}_{\sigma}] = \delta^{\lambda}_{\sigma} A^{\rho}_{\mu} - \delta^{\rho}_{\mu} A^{\lambda}_{\sigma}. \tag{5.70}$$

These are the 'covariant' relations that have to be obeyed by the generators in any UR of $SU(l+1)$.

It is clear that the subset of real diagonal traceless matrices among the generators of $\mathcal{D}$ forms a Cartan subalgebra. A basis for these is made up, for instance, of $A^1_1, A^2_2, \cdots, A^l_l$ with $A^{l+1}_{l+1} = -(A^1_1 + A^2_2 + \cdots A^l_l)$. Thus the rank is l. We see that both order and rank for $SU(l+1)$ are found much more easily than for $USp(2l)$. However, these 'diagonal' A^{λ}_{λ} (no sum) fail to obey trace orthogonality in the defining representation $\mathcal{D}$, a property which we found easily for $SO(2l)$, $SO(2l+1)$ and $USp(2l)$. It is desirable to have this property again, even though this makes the ensuing algebra a little cumbersome. We choose an independent set of diagonal real traceless generators as follows:

$$H_a = \frac{1}{\sqrt{a(a+1)}}\text{diag}\,(1, 1, \cdots, 1, -a, 0, \cdots, 0), \quad a = 1, 2, \cdots, l,$$

$$\text{Tr}(H_a\,H_b) = \delta_{ab}, \tag{5.71}$$

and take them as the basis for the Cartan subalgebra. That this is maximal abelian is obvious.

We expect to find $l(l+1)$ distinct roots in $\mathcal{R}$, with corresponding (nonhermitian) E_{α}'s. These are in fact just the A^{λ}_{μ} for $\lambda \neq \mu$. To see this, it is simplest to use Dirac notation in the space of the defining representation $\mathcal{D}$. We use an orthonormal basis of vectors

$$|\lambda\rangle, \ \lambda = 1, 2, \cdots, l+1 : \ \langle\lambda'|\lambda\rangle = \delta_\lambda^{\lambda'}, \ \sum_{\lambda=1}^{l+1} |\lambda\rangle\langle\lambda| = \mathbb{1}. \qquad (5.72)$$

Then the matrices $A^\lambda_{\ \mu}$, H_a appear conveniently as operators:

$$\langle\rho|A^\lambda_{\ \mu}|\sigma\rangle \equiv (A^\lambda_{\ \mu})^\rho_{\ \sigma} = \delta^\lambda_{\ \sigma}\delta^\rho_{\ \mu} - \frac{1}{l+1}\delta^\lambda_{\ \mu}\delta^\rho_{\ \sigma} \Rightarrow$$

$$A^\lambda_{\ \mu} = |\mu\rangle\langle\lambda| - \frac{1}{l+1}\cdot\delta^\lambda_{\ \mu}\cdot\mathbb{1};$$

$$H_a = \frac{1}{\sqrt{a(a+1)}}\left(\sum_{\nu=1}^{a} |\nu\rangle\langle\nu| - a|a+1\rangle\langle a+1|\right), \ \ a = 1, 2, \cdots, l. \qquad (5.73)$$

When needed for clarity, certain summations will be indicated explicitly. The basis vectors $|\lambda\rangle$ are naturally simultaneous eigenvectors of H_a with specific eigenvalues:

$$H_a|\lambda\rangle = m_a^{(\lambda)}|\lambda\rangle, \ a = 1, 2, \cdots, l; \ \lambda = 1, 2, \cdots, l+1;$$

$$m_a^{(\lambda)} = \frac{1}{\sqrt{a(a+1)}}(\theta(a-\lambda) - a\delta_{\lambda,a+1}),$$

$$\theta(0 \text{ or } 1 \text{ or } 2 \cdots) = 1, \ \ \theta(-1 \text{ or } -2 \cdots) = 0. \qquad (5.74)$$

We will soon see as part of the general theory of representations that these l-component real vectors $\boldsymbol{m}^{(\lambda)}$ are the *weights* in the defining representation $\mathcal{D}$. In detail:

$$\boldsymbol{m}^{(1)} = \left(\frac{1}{\sqrt{1\cdot 2}}, \frac{1}{\sqrt{2\cdot 3}}, \frac{1}{\sqrt{3\cdot 4}}, \ \cdots, \frac{1}{\sqrt{l(l+1)}}\right);$$

$$\boldsymbol{m}^{(2)} = \left(\frac{-1}{\sqrt{1\cdot 2}}, \frac{1}{\sqrt{2\cdot 3}}, \frac{1}{\sqrt{3\cdot 4}}, \ \cdots, \frac{1}{\sqrt{l(l+1)}}\right);$$

$$\boldsymbol{m}^{(3)} = \left(0, \frac{-2}{\sqrt{2\cdot 3}}, \frac{1}{\sqrt{3\cdot 4}}, \ \cdots, \frac{1}{\sqrt{l(l+1)}}\right);$$

$$\boldsymbol{m}^{(4)} = \left(0, 0, \frac{-3}{\sqrt{3\cdot 4}}, \ \cdots, \frac{1}{\sqrt{l(l+1)}}\right);$$

. . . .

$$m^{(l)} = \left(0,\ 0,\ 0,\ \cdots,\ \frac{-(l-1)}{\sqrt{(l-1)l}},\ \frac{1}{\sqrt{l(l+1)}}\right);$$

$$m^{(l+1)} = \left(0,\ 0,\ 0,\ \cdots,0,\ \frac{-l}{\sqrt{l(l+1)}}\right). \tag{5.75}$$

The commutators $[H_a, A^\lambda{}_\mu]$ are now very simple:

$$[H_a, A^\lambda{}_\mu] = \left(m_a^{(\mu)} - m_a^{(\lambda)}\right) A^\lambda{}_\mu. \tag{5.76}$$

In one sense the finding of the roots $\boldsymbol{\alpha}$ and the E_α is very easy, but the expressions for the roots are complicated! We have:

$$\boldsymbol{\alpha} = m^{(\mu)} - m^{(\lambda)} :\ E_\alpha \doteq A^\lambda{}_\mu,\ \lambda, \mu = 1, 2, \cdots, l+1,\ \lambda \neq \mu. \tag{5.77}$$

The number of roots is exactly $l(l+1)$:

$$\mathcal{R} = \{m^{(\lambda)} - m^{(\mu)},\ \lambda \neq \mu,\ \lambda, \mu = 1, 2, \cdots, l+1\}. \tag{5.78}$$

By carefully checking the signs of the first nonvanishing components, we find the positive roots:

$$\mathcal{R}_+ = m^{(1)} - m^{(\lambda)},\ \lambda = 2, 3 \cdots, l+1;$$

$$m^{(3)} - m^{(\lambda)},\ \lambda = 2;$$

$$m^{(4)} - m^{(\lambda)},\ \lambda = 2, 3;$$

$$m^{(5)} - m^{(\lambda)},\ \lambda = 2, 3, 4;$$

$$\cdots\cdots\cdots\cdots\cdots$$

$$m^{(l)} - m^{(\lambda)},\ \lambda = 2, 3 \cdots, l-1;$$

$$m^{(l+1)} - m^{(\lambda)},\ \lambda = 2, 3 \cdots, l. \tag{5.79}$$

The total number is $\frac{1}{2}l(l+1)$. After some work the simple roots are found to be:

$$\mathfrak{s}:\ \boldsymbol{\alpha}^{(1)} = m^{(3)} - m^{(2)};\ \boldsymbol{\alpha}^{(2)} = m^{(4)} - m^{(3)};\ \boldsymbol{\alpha}^{(3)} = m^{(5)} - m^{(4)};$$

$$\cdots\cdots\ \boldsymbol{\alpha}^{(l-1)} = m^{(l+1)} - m^{(l)};\ \boldsymbol{\alpha}^{(l)} = m^{(1)} - m^{(l+1)}. \tag{5.80}$$

In Eq. (5.70) we have the covariant or tensor form of the $SU(l+1)$ Lie algebra relations. On account of the great importance of these groups, an explicitly hermitian form highlighting the $SO(l+1)$ subgroup is also useful:

$$J_{\lambda\mu} = i(A^\lambda{}_\mu - A^\mu{}_\lambda) = -J_{\mu\lambda},$$

$$Q_{\lambda\mu} = A^\lambda{}_\mu + A^\mu{}_\lambda = Q_{\mu\lambda}, \quad Q_{\lambda\lambda} = 0. \tag{5.81}$$

Both $J_{\lambda\mu}$ and $Q_{\lambda\mu}$ are hermitian, the number of independent components being $\frac{1}{2}l(l+1)$ and $\frac{1}{2}(l+1)(l+2) - 1$, respectively. The former constitutes a basis for the **$SO(l+1)$** subalgebra, while the latter are the quadrupole generators of $SU(l+1)$:

$$[J_{\lambda\mu}, J_{\rho\sigma}] = i(\delta_{\lambda\rho}J_{\mu\sigma} - \delta_{\mu\rho}J_{\lambda\sigma} + \delta_{\lambda\sigma}J_{\rho\mu} - \delta_{\mu\sigma}J_{\rho\lambda}),$$

$$[J_{\lambda\mu}, Q_{\rho\sigma}] = i(\delta_{\lambda\rho}Q_{\mu\sigma} - \delta_{\mu\rho}Q_{\lambda\sigma} + \delta_{\lambda\sigma}Q_{\rho\mu} - \delta_{\mu\sigma}Q_{\rho\lambda}),$$

$$[Q_{\lambda\mu}, Q_{\rho\sigma}] = i(\delta_{\lambda\rho}J_{\mu\sigma} + \delta_{\mu\rho}J_{\lambda\sigma} - \delta_{\lambda\sigma}J_{\rho\mu} - \delta_{\mu\sigma}J_{\rho\lambda}). \tag{5.82}$$

To summarise: for $SU(l+1)$ we have order $l(l+2)$ and rank l; the $\{H_a\}$ of the Cartan subalgebra are in Eqs. (5.71, 5.73); the roots $\mathcal{R}$ are in Eq. (5.78); positive roots $\mathcal{R}_+$ in Eq. (5.79); simple roots $\mathfrak{s}$ in Eq. (5.80); and the E_α in Eq. (5.77).

We make a final comment regarding the simple roots $\mathfrak{s}$. Even though the l-component vectors $\boldsymbol{m}^{(\lambda)}$ in Eq. (5.75) look complicated, their differences appearing in $\mathfrak{s}$ in Eq. (5.80) are fairly simple:

$$\mathfrak{s} : \boldsymbol{\alpha}^{(a)} = \frac{1}{\sqrt{a+1}}(\sqrt{a}\,\boldsymbol{e}_a - \sqrt{a+2}\,\boldsymbol{e}_{a+1}), \quad a = 1, 2, \cdots, l-1;$$

$$\boldsymbol{\alpha}^{(l)} = \sum_{a=1}^{l-1} \frac{\boldsymbol{e}_a}{\sqrt{a(a+1)}} + \sqrt{\frac{l+1}{l}}\,\boldsymbol{e}_l. \tag{5.83}$$

We will use these expressions later.

Among the members of the four classical families, at the Lie algebra level there are some coincidences in low dimensions:

$$\boldsymbol{A}_1 = \boldsymbol{B}_1 = \boldsymbol{C}_1; \quad \boldsymbol{B}_2 = \boldsymbol{C}_2; \quad \boldsymbol{A}_3 = \boldsymbol{D}_3. \tag{5.84}$$

Further, $\boldsymbol{D}_1$ is abelian; $\boldsymbol{D}_2$ is not simple as it is $\boldsymbol{B}_1 \oplus \boldsymbol{B}_1$. So a nonredundant list of CSLA's is:

$$\boldsymbol{A}_l = SU(l+1), \quad l \geq 1;$$

$$\boldsymbol{B}_l = SO(2l+1), \quad l \geq 2;$$

$$\boldsymbol{C}_l = USp(2l), \quad l \geq 3;$$

$$D_l = SO(2l), \quad l \geq 4;$$

$$G_2, F_4, E_6, E_7, E_8. \tag{5.85}$$

5.7 The Exceptional Groups

There are five of these 'unusual' groups, we just give their names, orders, ranks, dimension of the smallest or defining UIR's, and some comments. They are unusual in the sense of being isolated and not forming parts of infinite families like $A_\ell, \cdots, D_\ell$: The significance of the entries in the last column varies from case to case. The group G_2

Table 5.3   The exceptional groups

Name	Order	Rank	Dimension of $\mathcal{D}$	Related to
G_2	14	2	7	$SO(7) = B_3$
F_4	52	4	26	$SO(9) = B_4$
E_6	78	6	27,27*	$SU(6) = A_5$
E_7	133	7	56	$SU(8) = A_7$
E_8	248	8	248	$SO(16) = D_8$

is in fact a proper subgroup of $SO(7)$, defined by the property that it leaves invariant a certain trilinear expression in the components of real seven dimensional vectors. In the other four cases we do not have a subgroup relationship, but something more intricate. The root system of F_4 is obtained by starting with the 32 roots of $B_4 = SO(9)$ and adding 16 more roots in a particular manner. The situation is similar in the remaining three cases.

5.8 Representations of CSLA's

For each of $A_1,..., D_l$ we have seen one UIR, the defining representation $\mathcal{D}$. Now we describe some aspects of generic UIR's. We imagine a complex vector space V of some dimension, and an irrep D acting on it. Various properties will be made plausible but not proved.

Since the H_a are commuting hermitian operators on V, they can be assumed to be simultaneously diagonal. Denote the eigenvalues by m_a, and a common eigenvector by $|m \cdots \rangle$:

$$H_a|m \cdots \rangle = m_a|m \cdots \rangle, \quad |m \cdots \rangle \in V, a = 1, 2, \cdots, l. \tag{5.86}$$

The l-component real vector $\boldsymbol{m} = (m_a)$ is called a *weight vector* or simply a *weight*. It is a vector in the same l-dimensional Euclidean space to which roots belong. We can construct a basis for V with such $|\boldsymbol{m}\cdots\rangle$.

We assume each $|\boldsymbol{m}\cdots\rangle$ is normalised. Vectors with different weights are mutually orthogonal, since $H_a^\dagger = H_a$.

While the H_a do commute pairwise, they are in general not a complete commuting set. This explains the dots after $\boldsymbol{m}$ in $|\boldsymbol{m}\cdots\rangle$. For a given $\boldsymbol{m}$, there may be several independent simultaneous eigenvectors. This is the *multiplicity problem*, which we discuss later. A weight $\boldsymbol{m}$ for which there is only one $|\boldsymbol{m}\cdots\rangle$ is non degenerate: it is called a *simple* weight (of course for the UIR D on V).

The non hermitian combinations $E_{\boldsymbol{\alpha}}$ affect $|\boldsymbol{m}\cdots\rangle$ in a simple way – the weight changes in a definite manner. From (5.16) we find:

$$H_a E_{\boldsymbol{\alpha}}|\boldsymbol{m}\cdots\rangle = [H_a, E_{\boldsymbol{\alpha}}]|\boldsymbol{m}\cdots\rangle + E_{\boldsymbol{\alpha}} H_a|\boldsymbol{m}\cdots\rangle$$
$$= (\boldsymbol{m}+\boldsymbol{\alpha})_a E_{\boldsymbol{\alpha}}|\boldsymbol{m}\cdots\rangle. \tag{5.87}$$

So either $E_{\boldsymbol{\alpha}}|\boldsymbol{m}\cdots\rangle$ vanishes, or it is another simultaneous eigenvector of all H_a with shifted weight:

$$E_{\boldsymbol{\alpha}}|\boldsymbol{m}\cdots\rangle = \cdots|\boldsymbol{m}+\boldsymbol{\alpha}\cdots\rangle. \tag{5.88}$$

The multiplicities may be different, there will be numerical coefficients and in general a sum of terms. Thus the relationship is

$$\text{Roots } \boldsymbol{\alpha} \ \sim \text{ differences of weights } \boldsymbol{m}$$
$$\sim \text{ weights in the adjoint representation.} \tag{5.89}$$

(So one of Cartan's theorems is that in the adjoint representation all nonzero weights are nondegenerate, i.e., simple.) Given the UIR D on V, the set of all weights present is denoted by W. Here no attention is paid to multiplicities.

Now we look at intricate geometrical relationships among roots and weights. Take some root $\boldsymbol{\alpha} \in \mathcal{R}$, and its associated $SU(2)^{(\alpha)}$ subgroup with generators given in (5.19). For any weight $\boldsymbol{m} \in W$ we have:

$$J_3|\boldsymbol{m}\cdots\rangle = m|\boldsymbol{m}\cdots\rangle, \quad m = \boldsymbol{m}\cdot\boldsymbol{\alpha}/|\boldsymbol{\alpha}|^2. \tag{5.90}$$

From our knowledge of $SU(2)$ representations, m must be one of $0, \pm\frac{1}{2}, \pm 1, \cdots$. Precisely:

$$\alpha \in \mathcal{R}, \ m \in W \Rightarrow 2\alpha \cdot m/|\alpha|^2 = 0, \ \pm 1, \ \pm 2, \cdots. \tag{5.91}$$

Again from $SU(2)$ representation theory we can see by applying $E_{\pm\alpha}$ to $|m \cdots\rangle$ repeatedly that there must be some integers $p, q \geq 0$ such that

$$m + p\alpha, \ m + (p - 1)\alpha, \cdots, m + \alpha, m, \ m - \alpha, \ \cdots m - q\alpha \in W,$$

$$m + (p + 1)\alpha, \ m - (q + 1)\alpha \notin W,$$

$$\text{i.e., } E_\alpha |m + p\alpha \cdots\rangle = E_{-\alpha} |m - q\alpha \cdots\rangle = 0. \tag{5.92}$$

In particular there can be no gaps in this sequence of weights in W. Now we must consider the problem of multiplicity.

Consider the set of vectors in V with the above string of weights, each with some multiplicity:

$$|m + p\alpha \cdots\rangle, |m + (p - 1)\alpha \cdots\rangle, \cdots,$$

$$|m + \alpha, \cdots\rangle, |m, \cdots\rangle, |m - \alpha, \cdots\rangle, \cdots |m - q\alpha \cdots\rangle. \tag{5.93}$$

These are carried into one another by the $SU(2)^{(\alpha)}$ generators, so they must support a certain set of UIR's of $SU(2)$ – a certain spectrum of j values, each appearing a certain number of times. The string of J_3 eigenvalues is evidently

$$m + p, \ m + p - 1, \cdots, \ m + 1, m, m - 1, \cdots, m - q, \tag{5.94}$$

where m is given by (5.90). So in the spectrum of j-values present there is a maximum,

$$\mathcal{J}_{\max} = m + p = -(m - q),$$

$$\text{i.e., } \mathcal{J}_{\max} = \frac{1}{2}(p + q), \ m = \frac{1}{2}(q - p). \tag{5.95}$$

The UIR with $\mathcal{J} = \mathcal{J}_{\max}$ may appear one or more times, namely the multiplicity of the weights $m + p\alpha$ and $m - q\alpha$ *which must be the same.*

A characteristic feature of $SU(2)$ UR's is that if $J_3 = m$ occurs, so does $J_3 = -m$. In fact within $SU(2)$ they are unitarily related:

$$e^{i\pi J_2} |j, m, \cdots\rangle \sim |j, -m, \cdots\rangle. \tag{5.96}$$

So we see that

$$\alpha \in \mathcal{R}, \ m \in W \Rightarrow m - 2\alpha \, m \cdot \alpha/|\alpha|^2 \in W, \tag{5.97}$$

with the same multiplicity since $e^{i\pi J_2}$ is unitary. Summing up: for any $\boldsymbol{m} \in W$ and $\boldsymbol{\alpha} \in \mathcal{R}$,

(i) $\exists$ integers $p, q \geq 0 : \boldsymbol{m} + p\boldsymbol{\alpha},\, \boldsymbol{m} + (p-1)\boldsymbol{\alpha}, \cdots, \boldsymbol{m} - q\boldsymbol{\alpha} \in W,$

$\boldsymbol{m} + (p+1)\boldsymbol{\alpha},\, \boldsymbol{m} - (q+1)\boldsymbol{\alpha} \notin W;$

(ii) $2\boldsymbol{m} \cdot \boldsymbol{\alpha}/|\boldsymbol{\alpha}|^2 = q - p = 0,\ \pm 1, \pm 2, \cdots;$

(iii) $\boldsymbol{m} - 2\boldsymbol{\alpha}\, \boldsymbol{\alpha} \cdot \boldsymbol{m}/|\boldsymbol{\alpha}|^2 \in W,$ same multiplicity as $\boldsymbol{m}$. $\hspace{2em}$ (5.98)

The reflection operations $e^{i\pi J_2^{(\alpha)}}$ are important. We have one such operator for each $\boldsymbol{\alpha} \in \mathcal{R}$. In weight space this is reflection in a plane through the origin perpendicular to $\boldsymbol{\alpha}$. The set of all these reflections is a group, the finite *Weyl subgroup* of G. For each $\boldsymbol{m} \in W$, each element of the Weyl group applied to $\boldsymbol{m}$ gives some $\boldsymbol{m}' \in W$ with the same multiplicity. We then say $\boldsymbol{m}, \boldsymbol{m}'$ are *equivalent*, so

$$\text{Weights equivalent to } \boldsymbol{m} = \{\boldsymbol{m} - 2\boldsymbol{\alpha}\, \boldsymbol{\alpha} \cdot \boldsymbol{m}/|\boldsymbol{\alpha}|^2,\ \boldsymbol{\alpha} \in \mathcal{R}\}. \hspace{2em} (5.99)$$

Make a definite choice and sequence of the generators H_a. Then, a weight $\boldsymbol{m}$ is *higher* than a weight $\boldsymbol{m}'$ if in the difference $\boldsymbol{m} - \boldsymbol{m}'$ (which may not be a weight) the first nonvanishing component is *positive*. So for $\boldsymbol{m} \neq \boldsymbol{m}'$, one is definitely higher than the other. The highest weight in a set of equivalent weights is called a *dominant* weight.

Given the UIR D on V, with weights W: let $\boldsymbol{\Lambda} \in W$ be the *highest weight*. It is a fact that it is a simple weight, there is just one eigenvector $|\boldsymbol{\Lambda}\rangle$; and the UIR D is completely determined by $\boldsymbol{\Lambda}$. The fact that $\boldsymbol{\Lambda}$ cannot be 'raised' can be expressed by

$$\boldsymbol{\Lambda} + \boldsymbol{\alpha} \notin W,\quad E_{\boldsymbol{\alpha}}|\boldsymbol{\Lambda}\rangle = 0,\quad \text{all } \boldsymbol{\alpha} \in \mathcal{R}_+. \hspace{2em} (5.100)$$

More economically we can say:

$$\boldsymbol{\alpha}^{(a)} \in \mathfrak{s} :\ E_{\boldsymbol{\alpha}^{(a)}}|\boldsymbol{\Lambda}\rangle = 0. \hspace{2em} (5.101)$$

Thus $|\boldsymbol{\Lambda}\rangle$ is simultaneously the maximum magnetic quantum number ('$j = m$') state for all the $SU(2)^{(\alpha)}$'s, i.e.,

$$\boldsymbol{\alpha} \in \mathcal{R}_+ \Rightarrow 2\boldsymbol{\Lambda} \cdot \boldsymbol{\alpha}/|\boldsymbol{\alpha}|^2 = \text{integer} \geq 0. \hspace{2em} (5.102)$$

For $\boldsymbol{\alpha}^{(a)} \in \mathfrak{s}$ we use the notation

$$2\boldsymbol{\Lambda} \cdot \boldsymbol{\alpha}^{(a)}/|\boldsymbol{\alpha}^{(a)}|^2 = N_a \geq 0,\quad \boldsymbol{\Lambda} \leftrightarrow \{N_a\}. \hspace{2em} (5.103)$$

Each UIR is uniquely characterised by the set of l nonnegative integers $\{N_a\}$. So also two UIR's with the same highest weight are the same.

There are now two major questions to consider: can we survey the set of all possible UIR's in a systematic manner; and within a given UIR what can we say about multiplicities of various weights that are present? We take up these questions in this sequence.

5.9 Survey of UIR's, Fundamental UIR's, Elementary UIR's

We said above that each UIR, each highest weight Λ, is uniquely determined via Eq. (5.103) in terms of a set of nonnegative integers $\{N_a\}$. We can explicitly exhibit Λ. We use the fact that the simple roots $\{\boldsymbol{\alpha}^{(a)}\} = \mathfrak{s}$ are linearly independent and form an (oblique) basis for l-dimensional root and weight space. We expand Λ as

$$\Lambda = \sum_{a=1}^{l} \lambda_a \boldsymbol{\alpha}^{(a)}, \ \lambda_a \text{ real.} \tag{5.104}$$

We need to find the λ_a in terms of the N_a. From (5.103, 5.104) we have:

$$\sum_{a=1}^{l} \lambda_a 2 \boldsymbol{\alpha}^{(a)} \cdot \boldsymbol{\alpha}^{(b)} / |\boldsymbol{\alpha}^{(b)}|^2 = N_b, \ b = 1, 2, \cdots, l \tag{5.105}$$

To invert this system, we define the real $l \times l$ *Cartan matrix* A as

$$A = (A_{ab}), \ A_{ab} = 2\boldsymbol{\alpha}^{(a)} \cdot \boldsymbol{\alpha}^{(b)} / |\boldsymbol{\alpha}^{(b)}|^2. \tag{5.106}$$

The linear independence of the $\boldsymbol{\alpha}^{(a)}$'s ensures that A is nonsingular. Even though A is not symmetric, the properties in Table 5.2 ensure that its elements have simple values:

$$A_{aa} = 2, \ A_{ab} = 0 \text{ or } -1 \text{ or } -2 \text{ or } -3 \text{ for } a \neq b. \tag{5.107}$$

We write A as the product of a symmetric and a diagonal matrix:

$$A = \mathcal{S}\mathcal{D}, \ \mathcal{S}_{ab} = \mathcal{S}_{ba} = \boldsymbol{\alpha}^{(a)} \cdot \boldsymbol{\alpha}^{(b)}, \ \mathcal{D}_{ab} = 2\delta_{ab} / |\boldsymbol{\alpha}^{(a)}|^2. \tag{5.108}$$

Then the matrix

$$G = A^{-1}\mathcal{D}^{-1} = \mathcal{D}^{-1}\mathcal{S}^{-1}\mathcal{D}^{-1} \tag{5.109}$$

is also symmetric. The solution to (5.105), with λ_a and N_a written as l-component column vectors, is:

$$A^T \lambda = N \Rightarrow \lambda = (A^T)^{-1} N$$

$$= \mathcal{D} G N,$$

$$\lambda_a = \frac{1}{|\alpha^{(a)}|^2} \sum_a G_{ab} N_b. \tag{5.110}$$

This means that the highest weight vector Λ of the UIR $\{N_a\}$ is the following combination of simple roots:

$$\Lambda = \sum_{a,b} \frac{2}{|\alpha^{(a)}|^2} G_{ab} N_b \alpha^{(a)} = \sum_a N_a \Lambda^{(a)},$$

$$\Lambda^{(a)} = 2 \sum_b G_{ab} \alpha^{(b)} / |\alpha^{(b)}|^2. \tag{5.111}$$

Therefore a general highest weight vector Λ is a nonnegative integer linear combination of l special highest weight vectors $\Lambda^{(a)}$.

Each $\Lambda^{(a)}$ is itself the highest weight of a particular UIR. And there are l of them. These UIR's are called the *fundamental* UIR's, and the $\Lambda^{(a)}$'s are the *fundamental* highest or *dominant* weights. So if we know these UIR's, we 'know' all UIR's. We say:

$$\mathcal{D}^{(a)} = a^{\text{th}} \text{ fundamental UIR}$$

$$= \text{UIR } \textit{with highest weight } \Lambda^{(a)}, \; a = 1, 2, \cdots, l. \tag{5.112}$$

How do we find these $\Lambda^{(a)}$'s? They have a simple relationship to the simple roots. From (5.111)

$$2\Lambda^{(a)} \cdot \alpha^{(b)} / |\alpha^{(b)}|^2 = \delta_{ab}. \tag{5.113}$$

So, we begin with the simple roots $\alpha^{(a)} \in \mathfrak{s}$ and rescale each one to define

$$\hat{\alpha}^{(a)} = 2\alpha^{(a)} / |\alpha^{(a)}|^2, \; a = 1, 2, \cdots, l. \tag{5.114}$$

These are independent, but not unit, vectors. Then (5.113) which reads

$$\Lambda^{(a)} \cdot \hat{\alpha}^{(b)} = \delta_{ab} \tag{5.115}$$

means that the $\{\Lambda^{(a)}\}$ form the reciprocal basis to the $\{\hat{\alpha}^{(a)}\}$ in l-dimensional space.

Fundamental UIR's for the classical families

For each of the classical families $A_l, \ldots, D_l$ we know what $\mathfrak{s}$ looks like. So we can compute the fundamental highest weights $\{\Lambda^{(a)}\}$ geometrically. We do it in the same sequence as before – D_l, B_l, C_l, A_l. In each case we also describe, to the extent possible, how each fundamental UIR arises from the defining representation $\mathcal{D}$.

Case of $D_l = SO(2l)$

The simple roots are listed in Eq. (5.39). Since each of them has length $\sqrt{2}$, we have $\hat{\alpha}^{(a)} = \alpha^{(a)}$. So the fundamental dominant weights $\Lambda^{(a)}$ form a system reciprocal to the simple roots $\mathfrak{s}$. Simple geometry leads to the desired expressions. The corresponding UIR's, except for the last two which are surprises, are obtainable from the defining $2l$-dimensional UIR $\mathcal{D}$ in Eq. (5.26):

$$\Lambda^{(1)} = e_1 : \quad \mathcal{D}^{(1)} = \text{defining } 2l\text{-dimensional 'vector' representation } \mathcal{D};$$

$$\Lambda^{(2)} = e_1 + e_2 : \quad \mathcal{D}^{(2)} = \text{second rank antisymmetric tensors over } \mathcal{D};$$

$$\Lambda^{(3)} = e_1 + e_2 + e_3 : \quad \mathcal{D}^{(3)} = \text{third rank antisymmetric tensors over } \mathcal{D};$$

$$\cdots \cdots \cdots$$

$$\Lambda^{(l-2)} = e_1 + e_2 \cdots + e_{l-2} : \quad \mathcal{D}^{(l-2)}$$

$$= (l-2) \text{ rank antisymmetric tensors over } \mathcal{D};$$

$$\Lambda^{(l-1)} = \frac{1}{2}(e_1 + e_2 \cdots + e_{l-1} - e_l) : \quad \mathcal{D}^{(l-1)} = \Delta^{(1)}$$

$$= \text{ first spinor UIR of } SO(2l);$$

$$\Lambda^{(l)} = \frac{1}{2}(e_1 + e_2 \cdots + e_{l-1} + e_l) : \quad \mathcal{D}^{(l)} = \Delta^{(2)}$$

$$= \text{ second spinor UIR of } SO(2l). \tag{5.116}$$

So, the l fundamental UIR's are antisymmetric tensors of ranks $1, 2, \cdots, l-2$ over $\mathcal{D}$, then two inequivalent spinor UIR's $\Delta^{(1)}$ and $\Delta^{(2)}$ which, as we will see in the next chapter, are of dimension 2^{l-1} each.

Case of $B_l = SO(2l+1)$

Now there is a small difference. Since from Eq. (5.45),

$$\mathfrak{s} = \{e_1 - e_2, e_2 - e_3, \cdots, e_{l-1} - e_l, e_l\} \text{ with lengths } \sqrt{2}, \sqrt{2}, \cdots, \sqrt{2}, 1 \tag{5.117}$$

for $\hat{\alpha}^{(a)}$ we have:

$$\hat{\alpha}^{(1)} = \alpha^{(1)}, \hat{\alpha}^{(2)} = \alpha^{(2)}, \cdots, \hat{\alpha}^{(l-1)} = \alpha^{(l-1)}, \ \hat{\alpha}^{(l)} = 2e_l. \qquad (5.118)$$

Again simple geometry gives the reciprocal system of vectors and the situation is:

$$\Lambda^{(1)} = e_1 : \ \mathcal{D}^{(1)} = \text{defining } (2l+1)\text{-dimensional vector representation } \mathcal{D};$$

$$\Lambda^{(2)} = e_1 + e_2 : \ \mathcal{D}^{(2)} = \text{second rank antisymmetric tensors over } \mathcal{D};$$

$$\cdots\cdots\cdots\cdots$$

$$\Lambda^{(l-2)} = e_1 + e_2 + \cdots + e_{l-2} : \ \mathcal{D}^{(l-2)}$$

$$= (l-2) \text{ rank antisymmetric tensors over } \mathcal{D};$$

$$\Lambda^{(l-1)} = e_1 + e_2 + \cdots + e_{l-1} : \ \mathcal{D}^{(l-1)}$$

$$= (l-1) \text{ rank antisymmetric tensors over } \mathcal{D};$$

$$\Lambda^{(l)} = \frac{1}{2}(e_1 + e_2 + \cdots + e_l) : \ \mathcal{D}^{(l)}$$

$$= \Delta = \text{ spinor UIR of } SO(2l+1), \text{ dimension } 2^l. \qquad (5.119)$$

So the l fundamental UIR's are antisymmetric tensors of ranks $1, 2, \cdots, l-1$ over $\mathcal{D}$, plus one spinor UIR Δ.

Case of $C_l = USp(2l)$

Here the simple roots are given in Eq. (5.65). We can easily trace the passages $\alpha^{(a)} \rightarrow \hat{\alpha}^{(a)} \rightarrow \Lambda^{(a)}$:

$$\alpha^{(1)} = \frac{1}{2}(e_1 - e_2), \ \alpha^{(2)} = \frac{1}{2}(e_2 - e_3), \cdots, \ \alpha^{(l-1)} = \frac{1}{2}(e_{l-1} - e_l), \ \alpha^{(l)} = e_l \rightarrow$$

$$\hat{\alpha}^{(1)} = 2(e_1 - e_2), \ \hat{\alpha}^{(2)} = 2(e_2 - e_3), \cdots, \ \hat{\alpha}^{(l-1)} = 2(e_{l-1} - e_l), \ \hat{\alpha}^{(l)} = 2e_l \rightarrow$$

$$\Lambda^{(1)} = \frac{1}{2}e_1, \ \Lambda^{(2)} = \frac{1}{2}(e_1 + e_2), \cdots, \ \Lambda^{(l-1)} = \frac{1}{2}(e_1 + e_2 + \cdots + e_{l-1}),$$

$$\Lambda^{(l)} = \frac{1}{2}(e_1 + e_2 + \cdots + e_l). \qquad (5.120)$$

We see a uniform pattern in the $\Lambda^{(a)}$, with no surprises like spinors. It turns out:

$$\mathcal{D}^{(a)} = a^{\text{th}} \text{ fundamental UIR with highest weight } \Lambda^{(a)}$$

$$= \text{rank } a \text{ antisymmetric 'trace free' tensor}$$

$$\text{representation over the defining } 2l\text{-dimensional}$$

$$\text{representation } \mathcal{D}, \ a = 1, 2, \cdots, l. \tag{5.121}$$

One has to remove 'traces' from antisymmetric tensors using the symplectic metric matrix η of Eq. (5.46).

Case of $A_l = SU(l + 1)$

Some algebra (or geometry!) is needed in this case. The simple roots $\boldsymbol{\alpha}^{(a)}$, given in Eq. (5.80), are all of length $\sqrt{2}$, so $\hat{\boldsymbol{\alpha}}^{(a)} = \boldsymbol{\alpha}^{(a)}$. Using Eqs. (5.75, 5.77) we have:

$$\hat{\boldsymbol{\alpha}}^{(a)} = \frac{1}{\sqrt{a+1}} \left(\sqrt{a} \, \boldsymbol{e}_a - \sqrt{a+2} \, \boldsymbol{e}_{a+1} \right), \ a = 1, 2, \cdots, l-1;$$

$$\hat{\boldsymbol{\alpha}}^{(l)} = \sum_{a=1}^{l-1} \frac{\boldsymbol{e}_a}{\sqrt{a(a+1)}} + \sqrt{\frac{l+1}{l}} \, \boldsymbol{e}_l. \tag{5.122}$$

Before constructing the reciprocal system of $\Lambda^{(a)}$'s we look at the defining representation $\mathcal{D}$, the weights $\boldsymbol{m}$ of which are given in Eq. (5.75). We see that the highest weight in $\mathcal{D}$ is $\boldsymbol{m}^{(1)}$, and moreover

$$\boldsymbol{m}^{(1)} \cdot \hat{\boldsymbol{\alpha}}^{(a)} = 0, \ \ a = 1, 2, \cdots, l-1; \ \ \boldsymbol{m}^{(1)} \cdot \hat{\boldsymbol{\alpha}}^{(l)} = 1. \tag{5.123}$$

This means that $\mathcal{D}$ is the last of the fundamental UIR's:

$$\boldsymbol{m}^{(1)} = \Lambda^{(l)}, \ \mathcal{D} = \mathcal{D}^{(l)}. \tag{5.124}$$

This is a result of the way we have handled $SU(l + 1)$. The remaining fundamental dominant weights can be found after some work and the result is:

$$\Lambda^{(a)} = \sum_{b=1}^{a} \frac{\boldsymbol{e}_b}{\sqrt{b(b+1)}} - a \sum_{b=a+1}^{l} \frac{\boldsymbol{e}_b}{\sqrt{b(b+1)}}, \ a = 1, 2, \cdots, l-1. \tag{5.125}$$

Actually, looking at $\boldsymbol{m}^{(1)}$ in Eq. (5.75), we see that this formula works also for $a = l$, as then the second term in (5.125) is absent. So we can use the general expression (5.125) for all $a = 1, 2, \cdots, l$. We can check next how these arise from tensors over

$\mathcal{D}$, and the final picture is:

$$\Lambda^{(l)} = \text{highest weight of defining } (l+1)\text{-dimensional UIR } \mathcal{D};$$

$$\Lambda^{(l-1)} = \text{highest weight of } 2^{\text{nd}} \text{ rank antisymmetric tensor}$$
$$\text{representation over } \mathcal{D};$$

$$\Lambda^{(l-2)} = \text{highest weight of } 3^{\text{rd}} \text{ rank antisymmetric tensor}$$
$$\text{representation over } \mathcal{D};$$

$$\cdots\cdots\cdots\cdots\cdots\cdots\cdots\cdots\cdots\cdots\cdots\cdots$$

$$\Lambda^{(2)} = \text{highest weight of } (l-1)^{\text{th}} \text{ rank antisymmetric tensor}$$
$$\text{representation over } \mathcal{D};$$

$$\Lambda^{(1)} = \text{highest weight of } l^{\text{th}} \text{ rank antisymmetric tensor}$$
$$\text{representation over } \mathcal{D}. \tag{5.126}$$

So the l fundamental UIR's of $SU(l+1)$ are given by antisymmetric tensors over $\mathcal{D}$ of ranks $1, 2, \cdots, l$.

Recapitulating, the overall relationships of the fundamental UIR's to the defining UIR's $\mathcal{D}$ in the four families are:

Family	Fundamental UIR's via tensors over $\mathcal{D}$
$A_l = SU(l+1)$	a^{th} rank antisymmetric tensors over $\mathcal{D}$, $a = 1, 2, \cdots, l$.
$B_l = SO(2l+1)$	a^{th} rank antisymmetric tensors over $\mathcal{D}$, $a = 1, 2, \cdots,$ $l-1$; one spinor UIR Δ of dim 2^l.
$C_l = USp(2l)$	a^{th} rank antisymmetric 'traceless' tensors over $\mathcal{D}$, $a = 1, 2, \cdots, l$.
$D_l = SO(2l)$	a^{th} rank antisymmetric tensors over $\mathcal{D}$, $a = 1, 2, \cdots,$ $l-2$; two spinor UIR's $\Delta^{(1)}$, $\Delta^{(2)}$ of dim 2^{l-1} each.

$$\tag{5.127}$$

We do not discuss the fundamental UIR's for the exceptional groups as we have not studied their root systems.

Elementary UIR's for the classical families

We have seen in Eq. (5.127) that many of the fundamental UIR's, if not all, are carried by spaces of antisymmetric tensors over the defining representation $\mathcal{D}$. Thus if $\mathcal{D}$ is in hand, and may be one or two more UIR's, and we permit construction of tensor products plus complete antisymmetrisation, then all the $\mathcal{D}^{(a)}$ can be obtained. In this sense the elementary UIR's are far fewer than the rank l. The picture is:

Elementary UIR's :

$$A_l \to \mathcal{D}; \; B_l \to \mathcal{D}, \; \Delta; \; C_l \to \mathcal{D}; \; D_l \to \mathcal{D}, \Delta^{(1)}, \Delta^{(2)}. \qquad (5.128)$$

So the overall structure is:

Cartan subalgebra plus roots $\mathcal{R} = \{\alpha\} \to$ positive roots $\mathcal{R}_+$
$\to$ simple roots $\mathfrak{s} = \{\alpha^{(a)}\} \to$ reciprocal system
of fundamental dominant weights $\{\Lambda^{(a)}\} \to l$
fundamental UIR's $\mathcal{D}^{(a)} \to$ one or two or three elementary UIR's.

5.10 The General UIR $\{N_a\}$, Its Construction, Internal Structure, Reality

As mentioned after Eq. (5.103), a general UIR is determined by a set of l nonnegative integers $\{N_a\}$, with its highest weight Λ determined by Eq. (5.111). This UIR, as suggested by the latter equation, is obtained from the l fundamental UIR's as follows. Construct a direct product representation D with N_1 copies of $\mathcal{D}^{(1)}$, N_2 of $\mathcal{D}^{(2)}, \cdots$, N_l of $\mathcal{D}^{(l)}$:

$$D = \mathcal{D}^{(1)} \times \mathcal{D}^{(1)} \times \cdots \times \mathcal{D}^{(1)} \times \mathcal{D}^{(2)} \times \mathcal{D}^{(2)} \times \cdots \times \mathcal{D}^{(2)} \times \cdots$$

$$\longleftarrow \quad N_1 \quad \longrightarrow \quad \longleftarrow \quad N_2 \quad \longrightarrow$$

$$\cdots \times \mathcal{D}^{(l)} \times \mathcal{D}^{(l)} \times \cdots \times \mathcal{D}^{(l)} \qquad (5.129)$$

$$\longleftarrow \quad N_l \quad \longrightarrow$$

In this UR, the highest weight present is simple and is

$$\Lambda(\{N_a\}) = \sum_{a=1}^{l} N_a \Lambda^{(a)}, \; N_a \geq 0. \qquad (5.130)$$

Upon complete reduction of D, the 'leading' or 'largest' piece emanating from $|\Lambda(\{N_a\})\rangle$ by repeated actions of the E_α's is the general UIR $\{N_a\}$ we are after. We can summarise this survey of all possible UIR's through these Cartan rules:

(i) There are l fundamental dominant weights $\Lambda^{(a)}$, $a = 1, 2, \cdots, l$, determining the corresponding l fundamental UIR's $\mathcal{D}^{(a)}$ with these $\Lambda^{(a)}$ as highest weights.

(ii) Any UIR corresponds uniquely to a set of integers $\{N_a\}$, and has the highest weight $\Lambda(\{N_a\})$ given in Eq. (5.130).

So in this way the l fundamental UIR's are the basic building blocks for all UIR's. Evidently, referring to Eq. (5.129), we must work within the completely symmetric subspace in $\mathcal{D}^{(1)} \times \mathcal{D}^{(1)} \times \cdots \times \mathcal{D}^{(1)}$, similarly in $\mathcal{D}^{(2)} \times \mathcal{D}^{(2)} \times \cdots \times \mathcal{D}^{(2)}$, and so on.

For the dimensions of the UIR's $\{N_a\}$, there are formulae due to Weyl and others for each classical family and each exceptional group.

There are criteria to determine when the UIR $\{N_a\}$ is real, when pseudoreal, and when complex. Just as a UIR has a *unique simple highest weight* Λ, it also has a *unique simple lowest weight* ν, say. Then the UIR $\{N_a\}$ is equivalent to its complex conjugate if and only if $\Lambda = -\nu$. Under complex conjugation we have the switch $\Lambda \to -\nu$, $\nu \to -\Lambda$. Thus we see immediately why every UIR of $SU(2)$ is real or pseudoreal. But for $SU(3)$ we do have genuinely complex UIR's. It happens that the overall picture is this:

$$SU(n) \text{ for } n \geq 3, \ SO(4n+2) \text{ for } n \geq 2 \ , \ E_6 \text{ possess complex UIR's;}$$

$$\text{all other CSLA's: every UIR is real or pseudoreal.} \qquad (5.131)$$

The description of a general UIR by the set $\{N_a\}$ tells us graphically how to build it up from the fundamental UIR's. Can we specify the $\{N_a\}$ through the eigenvalues of some invariant operators? There are such operators, the Casimir polynomials formed out of the hermitian generators X_j. The first quadratic one is

$$C_2 = g^{jk} X_j X_k, \qquad (5.132)$$

and there are further cubic, quartic expressions, l independent ones in all. Each of these commutes with all the generators X_j, therefore with one another; each of them thus reduces in a UIR to a function of the N_a's and in principle these Casimirs determine all the N_a. B. Gruber and L. O'Raifeartaigh wrote an important paper on them [Gruber and O'Raifeartaigh, 1964].

Finally we look briefly at the structure of states 'inside' a generic UIR, the set of weights $\boldsymbol{m} \in W$ and their multiplicities. The highest weight $\Lambda \in W$ is simple but in general $\boldsymbol{m} \in W$ has multiplicity. In the general UIR $\{N_a\}$ it turns out that a basis vector in V needs $\frac{1}{2}(r - 3l)$ additional labels to supplement the eigenvalues m_a of the Cartan subalgebra generators H_a. So in general the basis vectors have the appearance (cf Eq. (5.86)):

$$|\boldsymbol{m}; \text{ eigenvalues of } \frac{1}{2}(r - 3l) \text{ operators; 'eigenvalues' of } l \text{ Casimirs }). \qquad (5.133)$$

Within the UIR, in general, a complete commuting set consists of l diagonal generators H_a, and $\frac{1}{2}(r - 3l)$ polynomials in generators involving the E_α's, a total of $\frac{1}{2}(r - l)$ hermitian operators. The last set of entries in (5.133) are constant within the UIR. In some simple UIR's, for instance each of the defining UIR's $\mathcal{D}$, there may be no multiplicity.

For $SU(l + 1)$, $SO(2l)$ and $SO(2l + 1)$ there are canonical ways of choosing the additional $\frac{1}{2}(r - 3l)$ diagonal operators forming with $\{H_a\}$ a complete commuting set. This is based on the fact that a UIR of $SU(l + 1)$ (or $SO(2l)$ or $SO(2l + 1)$) is simply reducible under the canonical subgroup $SU(l)$ (or $SO(2l-1)$ or $SO(2l)$). Therefore we can use the Casimir operators (basically the UIR labels $\{N\}$) of the chains of subgroups $SU(l+1) \supset SU(l) \supset SU(l-1) \cdots \supset SU(2)$ (or $SO(n) \supset SO(n-1) \supset \cdots \supset SO(3)$). This canonical subgroup chain approach solves the multiplicity and state labelling problem in these cases.

For a general $\boldsymbol{m} \in W$, we can make some statements similar to those for the highest weight Λ in Eqs. (5.103, 5.111), based on properties with respect to the $SU(2)^{(\alpha)}$:

$$\boldsymbol{m} \in W \Rightarrow 2\boldsymbol{m} \cdot \boldsymbol{\alpha}^{(a)}/|\boldsymbol{\alpha}^{(a)}|^2 \equiv \boldsymbol{m} \cdot \hat{\boldsymbol{\alpha}}^{(a)} = n_a = 0, \pm 1, \ \pm 2, \ \cdots ;$$

$$\boldsymbol{m} = \sum_{a=1}^{l} n_a \Lambda^{(a)}. \tag{5.134}$$

Within the UIR, what weights $\boldsymbol{m}$ are present? There is a nice answer: each $\boldsymbol{m} \in W$ is obtained from Λ by *subtracting* a unique nonnegative linear combination of the simple roots $\boldsymbol{\alpha}^{(a)}$ with integer coefficients:

$$\Lambda = \text{highest weight}, \ \boldsymbol{m} \in W \Rightarrow$$

$$\boldsymbol{m} = \Lambda - \sum_{a=1}^{l} v_a \boldsymbol{\alpha}^{(a)}, \ v_a = \text{unique, integer} \geq 0. \tag{5.135}$$

There is a recursive procedure to reach all possible $\boldsymbol{m} \in W$, and important formulae due to Freudenthal and Kostant. (Also, Brian C. Hall, 2015).

5.11 Orthogonality and Completeness of UIR Matrix Elements

In the cases of the order 3 rank 1 compact Lie groups $SO(3)$ and $SU(2)$, we have described in Chapter 3 the processes of invariant integration of functions defined on the manifolds of these groups, Eqs. (3.33, 3.34, 3.87). We have then described the properties of orthogonality and completeness of the matrix elements of all the UIR's of these groups, Eqs. (3.54, 3.98a–3.98c).

For a general compact simple Lie group G – any one of the four classical families or the five exceptional groups – we have seen that a general UIR can be labelled by a set of nonnegative integers $N_a, a = 1, 2, \cdots \ell$, with ℓ the rank of G. Thus the set of all UIR's is in one-to-one correspondence with such sets $\{N_a\}$. It turns out that the properties of orthogonality and completeness continue to be valid in all these cases. These are with respect to a process of invariant integration for functions on any of these groups. This is the content of the Peter–Weyl theory and Theorem [Brian C Hall, 2015], a beautiful generalisation of what we have seen in the $SO(3)$ and $SU(2)$ cases. We will describe these results (not prove them) in Chapter 7, using a compact notation for the UIR matrices, as they will be useful in the context of the topics dealt with there.

Problems

P5.1　The real three dimensional Lie algebras of the groups $SO(3)$ and $SO(2,1)$, the homogeneous Lorentz group in two space and one time dimensions, have the following Lie brackets among their basis elements:

$$SO(3) \quad [e_1, e_2] = e_3, \quad [e_2, e_3] = e_1, \quad [e_3, e_1] = e_2$$
$$SO(2,1) \quad [e'_1, e'_2] = -e'_3, \quad [e'_2, e'_3] = e'_1, \quad [e'_3, e'_1] = e'_2.$$

Show that their complex extensions are the same.

P5.2　Show from Eq. (5.27) that the Lie algebra $SO(4)$ splits into the sum of two commuting $SO(3)$ Lie algebras.

P5.3　Check explicitly the low dimensional coincidences (5.84) among Lie algebras.

P5.4　Is the adjoint UIR $(1, 1)$ of $SU(3)$ faithful? If not, why not?

Bibliography

Behrends, R. E., Dreitlein, J., Fronsdal C. and Lee, B. W. (1962). Simple Groups and Strong Interaction Symmetries. *Rev. Mod. Phys.* 34: 1.

Gruber, B. and O'Raifeartaigh, L. (1964). S-Theorem and Construction of the Invariants of the Semisimple Compact Lie Algebras. *J. Math. Phys.* 5: 1796.

Hall, B. C. (2015). *Lie Groups, Lie Algebras, and Representations. Graduate Texts in Mathematics*, 2nd ed, Vol: 222. Cham, Switzerland: Springer.

Racah, G. (1965). Group Theory and Spectroscopy, Princeton Lectures 1951. *Ergebnisse der Exakten Naturwissenschaften*, 37: 28.

Mukhi, S. and Mukunda, N. (2010). *Lectures on Advanced Mathematical Methods for Physicists*. New Delhi: Hindustan Book Agency.

Wybourne, B. G. (1974). *Classical Groups for Physicists*. New York: Wiley.

6

Spinor Representations of the Orthogonal Groups

In the pattern of fundamental UIR's for the four classical families, we found an unusual feature for the real orthogonal groups $D_l = SO(2l)$ and $B_l = SO(2l + 1)$. The fundamental UIR's are not all obtainable from the defining 'vector' representation $\mathcal{D}$ by tensorial constructions. At the level of elementary UIR's this is seen even more clearly. For $A_l = SU(l + 1)$ and $C_l = USp(2l)$, we have only one elementary UIR in each case, namely $\mathcal{D}$ itself. But for both B_l and D_l, the elementary UIR's consist of the 'vector' UIR $\mathcal{D}$, and either one or two spinor UIR's (These are actually double valued UIR's of the respective groups $SO(2l + 1), SO(2l)$.) We study them briefly in this chapter.

Cartan had found the spinor UIR's by around 1913. Independently, Dirac found them in 1928 for the Lorentz group, and then they entered physics. Of course even earlier spinors for $SU(2)$ and $SO(3)$ were used in Pauli's description of spin in the non relativistic framework. In 1935 Brauer and Weyl [Brauer and Weyl, 1935] gave an elegant account of Cartan's fundamental spinor UIR's for B_l and D_l using the Dirac approach, i.e., via the algebra of γ matrices which we develop below.

6.1 Spinor UIR's for $D_l = SO(2l)$

We see from the discussion in Section 5.3 that the defining vector representation $\mathcal{D}$ of $SO(2l)$ of dimension $2l$, has the following highest weight and other weights making up W:

$$\mathcal{D} \text{ of } SO(2l) : \Lambda = e_1 = (1,0,\cdots,0);$$

$$W = \{\pm e_a, \ a = 1,2,\cdots,l\}. \tag{6.1}$$

All these weights are simple. It is clear that in any UIR formed out of tensors over $\mathcal{D}$, all weights present will be (positive or negative) *integer* linear combinations of the e_a. We see this in Eq. (5.116) for the highest weights $\Lambda^{(2)}, \Lambda^{(3)}, \cdots, \Lambda^{(l-2)}$ of the respective fundamental UIR's. In contrast, the last two fundamental UIR's, the 'spinor' UIR's, have the highest weights

$$\Delta^{(1)} \ \to \ \Lambda^{(l-1)} = \frac{1}{2}\left(e_1 + e_2 + \cdots + e_{l-1} - e_l\right) = \frac{1}{2}(1,1,\cdots,1,-1),$$

$$\Delta^{(2)} \ \to \ \Lambda^{(l)} = \frac{1}{2}\left(e_1 + e_2 + \cdots + e_l\right) = \frac{1}{2}(1,1,\cdots,1,1), \tag{6.2}$$

making it clear that they cannot arise from $\mathcal{D}$ via tensors. Each of $\Delta^{(1)}, \Delta^{(2)}$ is of dimension 2^{l-1}, an exponential rather than a polynomial dependence on l.

We now sketch the Brauer–Weyl construction, omitting most of the proofs. We ask for a set of matrices γ_A, $A = 1,2,\cdots,2l$ or $A = ar$ with $a = 1,2,\cdots,l$ and $r = 1,2$, obeying these conditions:

$$\gamma_A^\dagger = \gamma_A, \ \ \{\gamma_A,\gamma_B\} = 2\delta_{AB}, \ \ \text{set } \{\gamma_A\} \text{ irreducible.} \tag{6.3}$$

Up to unitary equivalence, there is a unique solution, and it is of dimension 2^l. One builds the γ's out of l independent, i.e., mutually commuting, Pauli triplets each of dimension two:

$$\sigma^{(a)}, \ a = 1,2,\cdots,l :$$

$$\gamma_A = \gamma_{ar} = \sigma_r^{(a)}\sigma_3^{(a+1)}\cdots\sigma_3^{(l)}, \ a = 1,2,\cdots,l; \ r = 1,2. \tag{6.4}$$

(An alternative construction is

$$\gamma_{ar} = \sigma_3^{(1)}\sigma_3^{(2)}\cdots\sigma_3^{(a-1)}\sigma_r^{(a)}, \tag{6.5}$$

but we work with (6.4).) So in γ_{ar} there are no factors from $\sigma^{(1)}, \sigma^{(2)}, \cdots, \sigma^{(a-1)}$. Hermiticity of γ_A is obvious, the anticommutation relations are easy, irreducibility is a (nontrivial) exercise.

Denote by V the complex vector space, of dimension 2^l, on which the γ_A act. On V we have a (double-valued) UR of $SO(2l)$ with hermitian generators

$$M_{AB} = M_{AB}^\dagger = -M_{BA} = \frac{i}{4}[\gamma_A, \gamma_B]. \tag{6.6}$$

It is easy to verify that the commutation relations (5.27) are obeyed. The corresponding Cartan subalgebra generators H_a are very simple:

$$H_a = M_{a1,a2} = \frac{i}{2}\gamma_{a1}\gamma_{a2} = \frac{i}{2}\sigma_1^{(a)}\sigma_2^{(a)} = -\frac{1}{2}\sigma_3^{(a)}, \quad a = 1, 2, \cdots, l. \tag{6.7}$$

In the usual representation, the H_a are already all diagonal. Let us introduce an orthonormal basis for V as:

$$|\epsilon\rangle \equiv |\epsilon_1, \epsilon_2, \cdots, \epsilon_l\rangle \equiv |\{\epsilon_a\}\rangle, \quad \epsilon_a = \pm 1,$$

$$\sigma_3^{(a)}|\epsilon\rangle = -\epsilon_a|\epsilon\rangle,$$

$$H_a|\epsilon\rangle = \frac{1}{2}\epsilon_a|\epsilon\rangle,$$

$$\langle\epsilon'|\epsilon\rangle = \prod_{a=1}^{l} \delta_{\epsilon'_a, \epsilon_a}. \tag{6.8}$$

So the weights $\{m\} = W$ present in this hermitian representation of $SO(2l)$ are $\frac{1}{2}\epsilon$, 2^l in number, and all of them are nondegenerate, i.e., simple.

Now, while the set of γ_A's is irreducible, we ask whether the set of M_{AB}'s is reducible or irreducible. It turns out that we can find a nontrivial operator commuting with all the M_{AB}. Therefore they, and the (double-valued) UR of $SO(2l)$ they generate, are reducible. Remembering that

$$\gamma_1\gamma_2 = i\sigma_3^{(1)}, \quad \gamma_3\gamma_4 = i\sigma_3^{(2)}, \cdots, \gamma_{2l-1}\gamma_{2l} = i\sigma_3^{(l)}, \tag{6.9}$$

we define (with F for Five):

$$\gamma_F = i^l\gamma_1\gamma_2\cdots\gamma_{2l} = (-1)^l\sigma_3^{(1)}\sigma_3^{(2)}\cdots\sigma_3^{(l)}. \tag{6.10}$$

Then we have the easy consequences:

$$\gamma_F^\dagger = \gamma_F^{-1} = \gamma_F^T = \gamma_F, \quad \text{Tr } \gamma_F = 0;$$

$$\{\gamma_A, \gamma_F\} = [M_{AB}, \gamma_F] = 0. \tag{6.11}$$

Since γ_F is diagonal in the basis in (6.8), we call it a 'Weyl basis':

$$\gamma_F|\epsilon\rangle = \epsilon_1\epsilon_2\cdots\epsilon_l|\epsilon\rangle. \tag{6.12}$$

The eigenvalues of γ_F are ± 1, the corresponding eigenspaces are subspaces V_1, V_2 in V:

$$V = V_1 \oplus V_2, \ \dim V_1 = \dim V_2 = 2^{l-1};$$

$$V_1 = Sp\{|\epsilon\rangle : \gamma_F = -1, \ \epsilon_1\epsilon_2\cdots\epsilon_l = -1\},$$

$$V_2 = Sp\{|\epsilon\rangle : \gamma_F = +1, \ \epsilon_1\epsilon_2\cdots\epsilon_l = +1\}. \tag{6.13}$$

Clearly the highest weights in V_1 and V_2 are $\mathbf{\Lambda}^{(l-1)}$ and $\mathbf{\Lambda}^{(l)}$ of Eq. (6.2), respectively. We have here the two inequivalent spinor UIR's of $SO(2l)$:

$$\Delta^{(1)} \text{ on } V_1, \ \Delta^{(2)} \text{ on } V_2, \text{ dimension } 2^{l-1} \text{ each, fundamental UIR's of } D_l. \tag{6.14}$$

In the same sense in which faithful UIR's of $SU(2)$ were described as double-valued or two-valued UIR's of $SO(3)$ in Chapter 3, here too $\Delta^{(1)}$ and $\Delta^{(2)}$ are (as already mentioned) double valued UIR's of $SO(2l)$. The replacement for $SU(2)$ here is, in the notation of Chapter 4 Section 4.7, the simply connected universal covering group $\overline{SO(2l)} \equiv Spin(2l)$ of $SO(2l)$; so $\Delta^{(1)}$ and $\Delta^{(2)}$ are faithful UIR's of $Spin(2l)$. Since the eigenvalues of γ_F are ± 1, vectors in V_1 (respectively V_2) are called left handed (respectively right handed) spinors.

Here is a summary of facts related to $SO(2l)$ spinors:

γ_A : dim 2^l, irreducible, unique up to unitary equivalence;

M_{AB} : dim 2^l, reducible;

$[M_{AB}, \gamma_C] = i(\delta_{BC}\gamma_A - \delta_{AC}\gamma_B)$;

$\{\gamma_A, \gamma_F\} = [M_{AB}, \gamma_F] = 0$;

$V = V_1 \oplus V_2$, dim 2^{l-1} each;

$\gamma_F = -1$ on V_1, $+1$ on V_2;

$\gamma_A V_1 \subset V_2$, $\gamma_A V_2 \subset V_1$;

$M_{AB} V_1 \subset V_1$, $M_{AB} V_2 \subset V_2$;

V_1 carries $\Delta^{(1)}$ of dim 2^{l-1}, weights $W = \{\frac{1}{2}\epsilon | \ \epsilon_1\epsilon_2\cdots\epsilon_l = -1\}$,

V_2 carries $\Delta^{(2)}$ of dim 2^{l-1}, weights $W = \{\frac{1}{2}\epsilon | \ \epsilon_1\epsilon_2\cdots\epsilon_l = +1\}$. $\tag{6.15}$

6.2 Spinor UIR for $B_l = SO(2l+1)$

The situation now is a little easier than with D_l as there is only one UIR to deal with. Its highest weight and dimension are, in contrast to Eq. (6.2):

$$\Delta \to \Lambda^{(l)} = \frac{1}{2}(e_1 + e_2 + \cdots + e_l), \text{ dimension } 2^l. \tag{6.16}$$

The Dirac algebra now has $(2l+1)\gamma$'s, one more than before:

$$\gamma_A^\dagger = \gamma_A, \ \{\gamma_A, \gamma_B\} = 2\delta_{AB}, \text{ set } \{\gamma_A\} \text{ irreducible, } A, B = 1, 2, \cdots, 2l+1. \tag{6.17}$$

It turns out that there are two *inequivalent* solutions. We can take γ_A for $A = 1, 2, \cdots, 2l$ to be as before, given in Eq. (6.4); and for γ_{2l+1} we can take either γ_F or $-\gamma_F$:

$$A = 1, 2. \cdots, 2l : \ \gamma_A \ = \text{ previous } \gamma_A;$$

$$\gamma_{2l+1} = \gamma_F; \tag{6.18a}$$

$$A = 1, 2. \cdots, 2l : \ \gamma_A' \ = \text{ previous } \gamma_A,$$

$$\gamma_{2l+1}' = -\gamma_F. \tag{6.18b}$$

That these are inequivalent follows from the relations

$$\gamma_{2l+1} = i^l \gamma_1 \gamma_2 \cdots \gamma_{2l}, \ \gamma_{2l+1}' = -i^l \gamma_1' \gamma_2' \cdots \gamma_{2l}'. \tag{6.19}$$

But the B_l generators built up from γ_A and from γ_A' are equivalent:

$$M_{AB} = \frac{i}{4}[\gamma_A, \gamma_B],$$

$$M_{AB}' = \frac{i}{4}[\gamma_A', \gamma_B'] = \gamma_F M_{AB} \gamma_F^{-1},$$

$$A, B = 1, 2, \cdots, 2l+1. \tag{6.20}$$

(As with $SO(2\ell)$, here the single UIR Δ is a double-valued UIR of $SO(2l+1)$, and a faithful UIR of the simply connected covering group $\overline{SO(2l+1)} = Spin(2l+1)$. These properties of the spinor irreps will be left implicit.)

We take the hermitian generators of the spinor UIR Δ of $SO(2l+1)$ to be M_{AB}. The commutation relations are as in Eq. (5.27) with A, B, C, D going over the range $1, 2, \cdots, 2l+1$. The space of the UIR is the same V of dimension 2^l used in Section 6.1. The weights present are, with reference to Eq. (6.8), the set $\{\frac{1}{2} \epsilon\}$, 2^l of them, the highest weight is $\Lambda^{(l)} = \frac{1}{2}(1, 1, \cdots, 1)$. All the weights are simple.

From this construction it follows that

$$\text{spinor UIR } \Delta \text{ of } SO(2l+1)\Big|_{\text{restricted to } SO(2l)} = \Delta^{(1)} \oplus \Delta^{(2)} \text{ of } SO(2l). \qquad (6.21)$$

For the convenience of the reader we give both an $SO(2l+1)$ summary analogous to (6.15), and a comparison of the two cases:

$$\gamma_A : \text{ dim } 2^l, \text{ irreducible, two inequivalent representations;}$$

$$M_{AB} : \text{ dim } 2^l, \text{ two irreducible equivalent representations;}$$

$$[M_{AB}, \gamma_C] = i(\delta_{BC}\gamma_A - \delta_{AC}\gamma_B);$$

$$W = \{\frac{1}{2}\,\epsilon | \epsilon_a = \pm 1, \ a = 1, 2, \cdots, l\}, \text{ all simple} \qquad (6.22)$$

Comparing B_l and D_l we have the results:

	$B_l = SO(2l+1)$	$D_l = SO(2l)$
γ_A	$A = 1, \cdots 2l+1$;	$A = 1, \cdots, 2l$;
	two inequivalent irreps,	one irrep up to equivalence,
	each of dim 2^l	of dim 2^l.
M_{AB}	$A, B = 1, \cdots, 2l+1$;	$A, B = 1, \cdots, 2l$; reducible,
	irreducible, one irrep	sum of two inequivalent
	up to equivalence, of dim 2^l	irreps, of dim 2^{l-1} each

(6.23)

Needless to stress, $M_{AB} = \frac{i}{4}[\gamma_A, \gamma_B]$ throughout.

6.3 Conjugation Properties of Spinor UIR's

There is a very intricate structure involved here, we work through the details. We need to construct the matrix C introduced in Eq. (1.88) when it exists, and examine its properties. We examine the two spinor UIR's of D_l first, namely $\Delta^{(1)}$ and $\Delta^{(2)}$; we take up Δ of B_l later.

The case of $D_l = SO(2l)$

The generators M_{AB} given in Eq. (6.6) lead upon exponentiation to the reducible UR $\Delta^{(1)} \oplus \Delta^{(2)}$ (which as has been mentioned is double-valued):

$$A \in D_l \ : \ \overline{U}(A) = \exp\left(-\frac{i}{2}\omega_{AB}M_{AB}\right)$$

$$= \text{ block diagonal in } V_1 \text{ and } V_2,$$

$$\omega_{AB} = -\omega_{BA} = \text{real.} \qquad (6.24)$$

We ask if this UR is self conjugate, i.e., whether we can find a matrix C such that

$$\overline{U}(A)^* = C\overline{U}(A)C^{-1},$$

$$\text{i.e., } -M_{AB}^T = CM_{AB}C^{-1}. \tag{6.25}$$

Such C can indeed be constructed by recursion in l. That is, we build it up in steps $l \to l+1, \cdots, D_l \to D_{l+1}, \cdots$. From the construction (6.4) of the γ's, and indicating l as a superscript, we see that

$$\gamma_A^{(l+1)} = \gamma_A^{(l)}\sigma_3^{(l+1)}, \quad A = 1, 2, \cdots, 2l;$$

$$\gamma_{2l+1 \text{ or } 2l+2}^{(l+1)} = \sigma_{1 \text{ or } 2}^{(l+1)}. \tag{6.26}$$

Therefore the D_{l+1} spinor generators and those for D_l are related as follows: with $A, B = 1, 2, \cdots, 2l$,

$$M_{AB}^{(l+1)} = M_{AB}^{(l)} \cdot \mathbb{1}^{(l+1)};$$

$$M_{A,2l+1}^{(l+1)} = \frac{i}{2}\gamma_A^{(l+1)}\gamma_{2l+1}^{(l+1)} = -\frac{1}{2}\gamma_A^{(l)}\sigma_2^{(l+1)};$$

$$M_{A,2l+2}^{(l+1)} = \frac{i}{2}\gamma_A^{(l+1)}\gamma_{2l+2}^{(l+1)} = \frac{1}{2}\gamma_A^{(l)}\sigma_1^{(l+1)};$$

$$M_{2l+1,2l+2}^{(l+1)} = H_{l+1} = -\frac{1}{2}\sigma_3^{(l+1)}. \tag{6.27}$$

Now suppose for $M_{AB}^{(l)}$ we have found C_l obeying Eq. (6.25):

$$M_{AB}^{(l)^T} = -C_l M_{AB}^{(l)} C_l^{-1}, \quad A, B = 1, 2, \cdots, 2l. \tag{6.28}$$

To construct C_{l+1} we try setting $C_{l+1} = C_l A_{l+1}$ when A_{l+1} is a suitable matrix in the two dimensional space of the $(l+1)^{\text{th}}$ Pauli triplet $\boldsymbol{\sigma}^{(l+1)}$. Then from (6.27) we find a set of conditions on A_{l+1}:

$$(i) \quad A \leq 2l, B = 2l+1:$$
$$\gamma_A^{(l)^T}\sigma_2^{(l+1)} = C_l\gamma_A^{(l)}C_l^{-1} \cdot A_{l+1}\sigma_2^{(l+1)}A_{l+1}^{-1};$$

$$(ii) \quad A \leq 2l, B = 2l+2:$$
$$\gamma_A^{(l)^T}\sigma_1^{(l+1)} = -C_l\gamma_A^{(l)}C_l^{-1} \cdot A_{l+1}\sigma_1^{(l+1)}A_{l+1}^{-1};$$

$$(iii) \quad A = 2l+1, \ B = 2l+2:$$
$$-\sigma_3^{(l+1)} = A_{l+1}\sigma_3^{(l+1)}A_{l+1}^{-1}. \tag{6.29}$$

Suppose now that at stage l we have been able to construct C_l such that, consistent with (6.28), we have

$$C_l \gamma_A^{(l)} C_l^{-1} = \epsilon_l \gamma_A^{(l)T}, \quad \epsilon_l = \pm 1, \tag{6.30}$$

where of course $A = 1, 2, \cdots, 2l$. Then depending on ϵ_l we can make a definite choice for A_{l+1} and get C_{l+1}:

$$\epsilon_l = +1 : \quad A_{l+1} = i\sigma_2^{(l+1)}, \quad C_{l+1} = iC_l\sigma_2^{(l+1)};$$

$$\epsilon_l = -1 : \quad A_{l+1} = \sigma_1^{(l+1)}, \quad C_{l+1} = C_l\sigma_1^{(l+1)}. \tag{6.31}$$

So C_l can be built up step by step, starting with say $A_1 = i\sigma_2^{(1)}$. The solution is

$$C_l = i\sigma_2^{(1)}\sigma_1^{(2)}i\sigma_2^{(3)}\sigma_1^{(4)}\cdots(i\sigma_2^{(l)} \text{ for odd } l, \sigma_1^{(l)} \text{ for even } l);$$

$$C_l\gamma_A^{(l)}C_l^{-1} = (-1)^l \gamma_A^{(l)T};$$

$$C_l M_{AB}^{(l)} C_l^{-1} = -M_{AB}^{(l)T}. \tag{6.32}$$

This means that ϵ_l of (6.30) is $(-1)^l$. Next we need to know the symmetry/antisymmetry properties of C_l and the behaviour with respect to γ_F. These are:

$$C_l^T = (-1)^{l(l+1)/2} C_l, \quad C_l\gamma_F^{(l)} = (-1)^l \gamma_F^{(l)} C_l. \tag{6.33}$$

This means that for even l, $\Delta^{(1)}$ and $\Delta^{(2)}$ are each self conjugate; while for odd l, they are conjugates of one another. In the even l case, we have further: $l(l+1)/2 = $ even or odd implies $\Delta^{(1)}$ and $\Delta^{(2)}$ are each potentially real or pseudoreal, respectively. All this leads to the following results for the spinor UIR's of D_l:

l	Symmetry of C_l	C_l and γ_F	$\Delta^{(1)}$	$\Delta^{(2)}$	
$4m$	symmetric	commute	pot. real	pot. real	
$4m+1$	antisymmetric	anticommute	$\sim \Delta^{(2)*}$	$\sim \Delta^{(1)*}$	(6.34)
$4m+2$	antisymmetric	commute	pseudoreal	pseudoreal	
$4m+3$	symmetric	anticommute	$\sim \Delta^{(2)*}$	$\sim \Delta^{(1)*}$	

The case of $B_l = SO(2l+1)$

The matrix C_l built up in Eq. (6.32) for D_l 'does the job' for B_l as well, as the γ_F relation (6.33) shows:

$$C_l \gamma_A^{(l)} C_l^{-1} = (-1)^l \gamma_A^{(l)T},$$

$$C_l M_{AB}^{(l)} C_l^{-1} = -M_{AB}^{(l)T}, \quad A, B = 1, 2, \cdots, 2l+1. \tag{6.35}$$

Therefore Δ is always a self conjugate UIR; this is natural, as there is only one spinor UIR for B_l. Further from the property of C_l in (6.33) we draw these conclusions:

$$l = 4m \text{ or } 4m+3: \quad C_l^T = C_l, \; \Delta \text{ is potentially real};$$

$$l = 4m+1 \text{ or } 4m+2: \quad C_l^T = -C_l, \; \Delta \text{ is pseudoreal}. \tag{6.36}$$

6.4 Combined Results for D_l and B_l

Combining the results for D_l and B_l we have a cycle of eight structure for the spinor UIR's of the real orthogonal groups:

Table 6.1   Cycle of eight structure for spinor UIR's

Group	Name	$\Delta^{(1)}, \Delta^{(2)}$	Δ
$SO(8m)$	D_{4m}	dim 2^{4m-1}, each real	—
$SO(8m+1)$	B_{4m}	—	dim 2^{4m}, real
$SO(8m+2)$	D_{4m+1}	dim 2^{4m}, conjugates	—
$SO(8m+3)$	B_{4m+1}	—	dim 2^{4m+1}, pseudoreal
$SO(8m+4)$	D_{4m+2}	dim 2^{4m+1}, each pseudoreal	—
$SO(8m+5)$	B_{4m+2}	—	dim 2^{4m+2}, pseudoreal
$SO(8m+6)$	D_{4m+3}	dim $\cdot 2^{4m+2}$, conjugates	—
$SO(8m+7)$	B_{4m+3}	—	dim 2^{4m+3}, real

An intriguing case is $SO(8)$ when $m = 1$: then the defining vector representation and the two spinor UIR's are all eight dimensional and real (in suitable bases).

Another interesting feature is the general pattern of the reduction of these UIR's under restriction to the 'next lower' canonical orthogonal subgroup. This can be conveyed in the form of a flow chart:

V_{2^l} : UIR Δ_{2l+1} of $SO(2l+1)$, dim 2^l, use $\gamma_1, \gamma_2, \cdots, \gamma_{2l+1}$

$$\downarrow \text{ restrict to } SO(2l)$$

$$\gamma_{2l+1} = -1: \qquad\qquad \gamma_{2l+1} = 1:$$

$V_{2^{l-1}}$: UIR $\Delta_{2l}^{(1)}$ of $SO(2l)$, dim 2^{l-1}, $V'_{2^{l-1}}$: UIR $\Delta_{2l}^{(2)}$ of $SO(2l)$, dim 2^{l-1}

use $\gamma_1, \gamma_2, \cdots, \gamma_{2l}$ use $\gamma_1, \gamma_2, \cdots, \gamma_{2l}$

$$\longleftarrow\text{-- restrict to } SO(2l-1) \text{ --}\longrightarrow$$

UIR Δ_{2l-1} of $SO(2l-1)$, dim 2^{l-1},   UIR Δ_{2l-1} of $SO(2l-1)$, dim 2^{l-1},

use $\gamma_1, \gamma_2, \cdots, \gamma_{2l-1}$ use $\gamma_1, \gamma_2, \cdots, \gamma_{2l-1}$

We see that the spinor UIR of $SO(2l+1)$ reduces to the direct sum of the two inequivalent spinor UIR's of $SO(2l)$; but each of the latter remains irreducible under $SO(2l-1)$.

6.5 Some Properties of Antisymmetric Tensors

We have seen in Chapter 5 that for both B_l and D_l, except for the spinor UIR's, all the other fundamental UIR's are provided by antisymmetric tensors of suitable ranks. This is seen in Eqs. (5.115, 5.118). However, certain UR's provided by antisymmetric tensors do not appear among the fundamental UIR's. We discuss these matters briefly in this last section of this chapter.

Let us begin with some preliminary notions relevant for all the proper real orthogonal groups $SO(n)$. With subscripts $A, B, \cdots$, ranging over $1, 2, \cdots, n$, a rank p antisymmetric tensor T with respect to $SO(n)$ has components $T_{A_1 A_2 \cdots A_p}$ which are completely antisymmetric in the subscripts (a property assumed in all that follows) and which transform by the rule

$$A \in SO(n): \quad T'_{A_1 A_2 \cdots A_p} = A_{A_1 B_1} A_{A_2 B_2} \cdots A_{A_p B_p} T_{B_1 B_2 \cdots B_p} \qquad (6.37)$$

The rank p obeys $0 \le p \le n$, and the corresponding UR of $SO(n)$ will be denoted by $A_p^{(n)}$. Now we can take advantage of the fact that there is a numerical invariant completely antisymmetric Levi–Civita tensor $\epsilon_{A_1 A_2 \cdots A_n}$ associated with $SO(n)$, which allows us to map the space of rank p tensors one-to-one invertibly onto the space of rank $n - p$ tensors:

$$T_{A_1 \cdots A_p} \to \tilde{T}_{A_1 \cdots A_{n-p}} = \frac{1}{p!} \epsilon_{A_1 \cdots A_{n-p} B_1 \cdots B_p} T_{B_1 \cdots B_p}. \tag{6.38}$$

This means that $A_p^{(n)}$ and $A_{n-p}^{(n)}$ are equivalent. So in listing tensors of distinct representation types we can limit ourselves as follows:

$$n = 2l, D_l : \quad p = 1, 2, \cdots, l - 1, l;$$
$$n = 2l + 1, B_l : \quad p = 1, 2, \cdots, l - 1, l. \tag{6.39}$$

In the former case, for $p = l$, the map (6.38) carries the space of tensors concerned onto itself.

Now when we compare the above with the lists of fundamental UIR's for D_l and B_l in Eqs. (5.115, 5.118), we see that in the D_l case the representations $A_{l-1}^{(2l)}$ and $A_l^{(2l)}$ do not occur (instead we have the two spinor UIR's $\Delta^{(1)}, \Delta^{(2)}$); and in the B_l case the representation $A_l^{(2l+1)}$ does not occur (instead we have the spinor UIR Δ). We now see how these 'missing' tensor representations can be obtained from direct products of the spinor UIR's. We take up B_l first as it is simpler, and only describe the basic facts.

Representation $A_l^{(2l+1)}$ of B_l

One can see quite easily that this is a UIR with highest weight (using the same notation as in Eq. (5.118))

$$A_l^{(2l+1)} : \quad \Lambda = e_1 + e_2 + \cdots + e_l = 2\Lambda^{(l)},$$
$$\Lambda^{(l)} = \text{highest weight of UIR } \Delta. \tag{6.40}$$

This immediately allows us to recognise that

$$A_l^{(2l+1)} = \text{leading UIR in direct product } \Delta \times \Delta. \tag{6.41}$$

Representations $\mathcal{A}_{l-1}^{(2l)}$, $\mathcal{A}_{l}^{(2l)}$ of D_l

Here the case $p = l - 1$ is easy to handle. The UR $\mathcal{A}_{l-1}^{(2l)}$ is irreducible and has the highest weight

$$\mathcal{A}_{l-1}^{(2l)} : \quad \Lambda = e_1 + e_2 + \cdots + e_{l-1} = \Lambda^{(l-1)} + \Lambda^{(l)}, \tag{6.42}$$

leading to the conclusion

$$\mathcal{A}_{l-1}^{(2l)} = \text{leading UIR in direct product } \Delta^{(1)} \times \Delta^{(2)}. \tag{6.43}$$

For $p = l$ the situation is more involved. Now the map (6.38) can be regarded as the action of an operator $\hat{\epsilon}$ on the space of D_l antisymmetric tensors of rank l:

$$\tilde{T} = \hat{\epsilon} T : \quad \tilde{T}_{A_1 \cdots A_l} = \frac{1}{l!} \epsilon_{A_1 \cdots A_l A_{l+1} \cdots A_{2l}} T_{A_{l+1} \cdots A_{2l}},$$

$$\hat{\epsilon}^2 = (-1)^l. \tag{6.44}$$

A detailed analysis shows that we can define two kinds of such tensors, eigentensors of $\hat{\epsilon}$, depending on the eigenvalues of $\hat{\epsilon}$:

$$\text{self dual tensors: } \hat{\epsilon} T = i^{l^2} T$$

$$= (1 \text{ or } i)T \text{ for } l \text{ even or odd;}$$

$$\text{antiself dual tensors: } \hat{\epsilon} T = -i^{l^2} T$$

$$= (-1 \text{ or } -i)T \text{ for } l \text{ even or odd.} \tag{6.45}$$

The corresponding UR's are in fact irreducible and the highest weights are:

$$\text{self dual case}: \quad \Lambda = e_1 + e_2 + \cdots + e_{l-1} + e_l = 2\Lambda^{(l)};$$

$$\text{antiself dual case}: \quad \Lambda = e_1 + e_2 + \cdots + e_{l-1} - e_l = 2\Lambda^{(l-1)}. \tag{6.46}$$

This leads to the conclusions

$$\text{self dual part of } \mathcal{A}_l^{(2l)} = \text{leading UIR in direct product } \Delta^{(2)} \times \Delta^{(2)},$$

$$\text{antiself dual part of } \mathcal{A}_l^{(2l)} = \text{leading UIR in direct product } \Delta^{(1)} \times \Delta^{(1)}. \tag{6.47}$$

Collecting these results we have a table covering four cases: As is evident from (6.45), for even l i.e., for $SO(8), SO(12), SO(16), \cdots$, the self dual-antiself dual separation can be made at the real level; while for odd l, i.e., for $SO(10), SO(14), SO(18), \cdots$, this separation requires working in the complex domain.

Table 6.2 The 'missing' tensor UIR's of B_l and D_l

Group	Antisymmetric tensor UIR's	Leading UIR in
$B_l = SO(2l+1)$	rank l	$\Delta \times \Delta$
$D_l = SO(2l)$	rank $l-1$	$\Delta^{(1)} \times \Delta^{(2)}$
$D_l = SO(2l)$	rank l, self dual	$\Delta^{(2)} \times \Delta^{(2)}$
$D_l = SO(2l)$	rank l, antiself dual	$\Delta^{(1)} \times \Delta^{(1)}$

In conclusion we mention that the nonunitary spinor representations of the real noncompact pseudo orthogonal groups $SO(p, q)$, $p, q \geq 1$, are important in some applications in physics. An account of these may be found in the book by Mukhi and Mukunda, as given below in the Biblography.

Problems

P6.1   The group $D_3 = SO(6)$ is locally isomorphic to its universal covering group $A_3 = SU(4)$ as seen in Eq. (5.84). Establish the connection of the two spinor UIR's $\Delta^{(1)}$, $\Delta^{(2)}$ of $\overline{SO(6)}$ to the defining UIR of $SU(4)$ and its complex conjugate.

P6.2   The defining real vector UIR of $SO(8)$, and its two spinor UIR's, are each of dimension eight; the latter two can be brought to real forms. Explore the kinds of connections – like homomorphisms and isomorphisms – that must exist among these three UIR's.

(This is a quite sophisticated situation, referred to as the Principle of Triality for the covering group $\overline{SO(8)}$. See for example [Cartan, 1966])

Bibliography

Brauer, R. and Weyl, H. (1935). Spinors in n dimensions. *Am. J. Math.* 57: 425

Cartan, E. (1966). *The Theory of Spinors.* Cambridge, Massachusetts: MIT Press.

Mukhi, S. and Mukunda, N. (2010). *Lectures on Advanced Mathematical Methods for Physicists.* New Delhi: Hindustan Book Agency.

notation in the space of the defining representation D. We use an orthonormal basis of vectors

$$|k\rangle, \quad k = 1, 2, \ldots, \quad \langle k | k' \rangle = \delta^{kk'}, \quad \sum_k |k\rangle\langle k| = \mathbb{1}. \tag{5.72}$$

7

Properties of Some Reducible Group Representations, and Systems of Generalised Coherent States

The basic ideas of group representation theory were presented in Chapter 1. For finite groups and for compact simple Lie groups, every representation can be assumed to be unitary. The basic building blocks, or units, of representation theory are then the unitary irreducible representations (UIR's), which are all finite dimensional. The UIR's of the permutation groups S_n, and of compact simple Lie groups, have been described in Chapters 2 and 5, respectively.

The two regular representations of any group G of interest have special properties which we will now draw upon: (i) The representation space is the space of all complex valued functions on G, made into a Hilbert space with a suitably defined inner product; (ii) there are two mutually commuting regular representations, the left and the right; (iii) in each of them, upon complete reduction, each UIR appears with multiplicity equal to its dimension. Thus for example for $G = SU(2)$, in each regular representation the spin j UIR $D^{(j)}, j = 0, 1/2, 1, \ldots$, appears $(2j + 1)$ times.

In the particular case when G is abelian, for instance $G = SO(2)$, each UIR is one dimensional, the two regular representations coincide, and each of them is a multiplicity free direct sum of all the UIR's.

Now we turn to general UR's of G. When a UR is fully reduced and expressed as a direct sum of UIR's, each distinct UIR occurs with some multiplicity. Thus the

UR as a whole is in principle completely determined up to unitary equivalence by these multiplicities. However, certain UR's may have special significance, reflecting the way they are constructed, and so merit special attention. We will study the following interesting UR's – one which we call the Schwinger representation of a group, and others obtained by an elegant 'process of induction' from UIR's of various Lie subgroups H in G. We conclude the chapter with a brief description of generalised coherent states based on UIR's of Lie groups. These are important in many physical contexts, and their analysis makes use of induced representations in an essential way.

7.1 The Schwinger Representation of a Group

The Schwinger construction for $SU(2)$ briefly described in Chapter 3 has very interesting features contrasting with the regular representations. Upon exponentiation of the generators in Eq. (3.103), we get an infinite dimensional UR of $SU(2)$ on the quantum mechanical Hilbert space of two simple harmonic oscillators, which is the direct sum of each UIR $D^{(j)}$ exactly once. This construction has been found useful in many physical problems. Its distinctive features are economy and completeness – unlike the regular representations, each UIR appears only once, and no UIR is absent. Thus it is much 'leaner' than either regular representation, while every UIR can be found in it.

Now we abstract these features and define what we shall call the Schwinger representation of any finite or compact simple Lie group. One may anticipate that this construction will turn out to be useful in many physics contexts.

7.1.1 Definition of Schwinger representation

For simplicity we adopt a notation suggested by the quantum theory of angular momentum. For any group G of interest, the UIR's will be labelled by the symbol j; and m', m will be row and column indices for the matrices within the UIR. (Thus, in the case of S_n dealt with in Chapter 2, j stands for a Young frame, while m', m are selected Young tableau. For the compact simple Lie groups dealt with in Chapter 5, j is the collection $\{N_a\}$ of l nonnegative integers labelling a general UIR; while m', m are the entries inside the vectors $|\mathbf{m}, \dots\rangle$ of an orthonormal basis (ONB), $\mathbf{m}$ being an l-component weight vector and the $\dots$ the eigenvalues of $\frac{1}{2}(r - 3l)$ additional mutually commuting Hermitian operators polynomial in the generators. Here r and l are order and rank of G.) Thus the UIR matrices $\mathcal{D}^{(j)}_{m'm}(g)$ act on vectors in a Hilbert space $\mathcal{H}^{(j)}$ of dimension N_j.

We now define the Schwinger representation of G. The Hilbert space for this representation is the direct sum of N_j dimensional subspaces,

$$\mathcal{H}_0 = \sum_j {}_\oplus \mathcal{H}^{(j)}, \qquad (7.1)$$

and the UR is also a direct sum of UIR's:

$$\mathcal{D}_0 = \sum_j {}_\oplus \mathcal{D}^{(j)}. \qquad (7.2)$$

Each UIR $\mathcal{D}^{(j)}$ is present exactly once, and acts on $\mathcal{H}^{(j)}$. Members of ONB's are written as $|j, m\rangle$:

$$\langle j', m' | j, m \rangle = \delta_{j'j} \delta_{m'm};$$
$$\mathcal{H}^{(j)} = \mathrm{Sp}\{|jm\rangle,\ j \text{ fixed}, m \text{ varying}\},$$
$$\mathcal{H}_0 = \mathrm{Sp}\{|j, m\rangle,\ j \text{ and } m \text{ varying}\}. \qquad (7.3)$$

Then for $g \in G$,

$$\langle j'm' | \mathcal{D}_0(g) | jm \rangle = \delta_{j'j} \mathcal{D}^{(j)}_{m'm}(g). \qquad (7.4)$$

In the case of a compact simple Lie group, the generators $J_{..}$ of $\mathcal{D}_0$ also are direct sums, $J^{(j)}$ belonging to $\mathcal{D}^{(j)}$ appearing in the jth place. Each of these obeys the commutation relations characteristic of G. In addition, for each j, the $J^{(j)}$ obey characteristic polynomial relations belonging to the jth UIR $\mathcal{D}^{(j)}$. However, the 'total' generators $J_{..}$ do not obey any such algebraic relations beyond the commutation relations of G.

The Schwinger construction (3.103) of the $SU(2)$ Lie algebra leads by definition to the Schwinger representation of $SU(2)$. We shall now describe a method to arrive at an equivalent form of this $SU(2)$ representation, starting from the regular representation. This method works also for $SO(3)$, $SU(3)$ and indeed $SU(n)$, $n \geq 4$. In the $SU(2)$ case we will also show how to pass from this version of the Schwinger representation to the one based on the Schwinger construction. We will deal with $SU(2)$, $SO(3)$, $SU(3)$ and $SU(n)$ in that sequence. For $SU(2)$ and $SO(3)$ we provide necessary details about the regular representations beyond those given in Chapter 3.

7.1.2 The $SU(2)$ case

The Euler angles description of elements $u \in SU(2)$ is given in Eq. (3.92) repeated here:

$$u(\theta, \psi, \phi) = e^{\frac{-i}{2}\theta\sigma_3}\, e^{\frac{-i}{2}\psi\sigma_2}\, e^{\frac{-i}{2}\phi\sigma_3},$$

$$0 \le \theta \le 4\pi, 0 \le \psi \le \pi, 0 \le \phi \le 2\pi. \tag{7.5}$$

The Hilbert space $\mathcal{H} = L^2(SU(2))$ of square integrable functions on $SU(2)$ is

$$\mathcal{H} = \{f(\theta, \psi, \phi) \mid \|f\|^2 = \frac{1}{16\pi^2}\int_0^{4\pi} d\theta \int_0^{2\pi} d\phi \int_0^{\pi} \sin\psi \; d\psi |f(\theta, \psi, \phi)|^2 < \infty\}. \tag{7.6}$$

We will write $f(u)$ or $f(\theta, \psi, \phi)$ interchangeably.

The left and right regular representations of $SU(2)$ act on $\mathcal{H}$ by unitary operators $U(u)$, $\tilde{U}(u)$, respectively:

$$(U(u')f)(u) = f(u'^{-1}u), \quad (\tilde{U}(u')f)(u) = f(uu'), \tag{7.7}$$

and obey

$$U(u')U(u) = U(u'u), \quad \tilde{U}(u')\tilde{U}(u) = \tilde{U}(u'u), \quad \tilde{U}(u')U(u) = U(u)\tilde{U}(u'). \tag{7.8}$$

Their hermitian generators are

$$J_1 = i\left(\cos\theta \cot\psi \frac{\partial}{\partial\theta} + \sin\theta \frac{\partial}{\partial\psi} - \frac{\cos\theta}{\sin\psi}\frac{\partial}{\partial\phi}\right),$$

$$J_2 = i\left(\sin\theta \cot\psi \frac{\partial}{\partial\theta} - \cos\theta \frac{\partial}{\partial\psi} - \frac{\sin\theta}{\sin\psi}\frac{\partial}{\partial\phi}\right),$$

$$J_3 = -i\frac{\partial}{\partial\theta}; \tag{a}$$

$$\tilde{J}_1 = i\left(-\frac{\cos\phi}{\sin\psi}\frac{\partial}{\partial\theta} + \sin\phi\frac{\partial}{\partial\psi} + \cos\phi \cot\psi \frac{\partial}{\partial\phi}\right),$$

$$\tilde{J}_2 = i\left(\frac{\sin\phi}{\sin\psi}\frac{\partial}{\partial\theta} + \cos\phi\frac{\partial}{\partial\psi} - \sin\phi \cot\psi \frac{\partial}{\partial\phi}\right),$$

$$\tilde{J}_3 = i\frac{\partial}{\partial\phi}; \tag{b} \tag{7.9}$$

and obey

$$[J_r, J_s] = i\epsilon_{rst}J_t, \quad [\tilde{J}_r, \quad \tilde{J}_s] = i\epsilon_{rst}\tilde{J}_t, \quad [J_r, \tilde{J}_s] = 0. \tag{7.10}$$

The simultaneous complete reduction of both regular representations into UIR's is achieved by using as an orthonormal basis for $\mathcal{H}$ the (normalised) unitary representation matrices from all the UIR's of $SU(2)$. These are given in Eq. (3.98a) along with the properties given in Eqs. (3.98b, 3.98c):

$$D_{mn}^{(j)}(\theta, \psi, \phi) = e^{-im\theta - in\phi} d_{mn}^{(j)}(\psi),$$

$$d_{mn}^{(j)}(\psi) = \langle jm|e^{-i\psi J_2}|jn\rangle = \text{real orthogonal},$$

$$j = 0, 1/2, 1, \dots ; m, n = j, j-1, \dots, -j. \tag{7.11}$$

With respect to the inner product (7.6) in $\mathcal{H}$, the functions $(2j+1)^{1/2} D_{mn}^{(j)}(u)$ form an orthonormal basis. Since, as is easily checked,

$$(U(u')D_{mn}^{(j)})(u) = \sum_{m'} D_{m'm}^{(j)}(u')^* D_{m'n}^{(j)}(u), \qquad (a)$$

$$(\tilde{U}(u')D_{mn}^{(j)})(u) = \sum_{n'} D_{n'n}^{(j)}(u')D_{mn'}^{(j)}(u), \qquad (b) \tag{7.12}$$

the complete reduction of both regular representations is achieved in this basis. For the reduction of $U(u)$ into UIR's the second index n in $D_{mn}^{(j)}(u)$ counts the multiplicity, for $\tilde{U}(u)$ the first index m plays this role.

The two regular representations share the same Casimir invariant:

$$J_r J_r = \tilde{J}_r \tilde{J}_r = J^2, \tag{7.13}$$

and the D-matrices obey

$$(J^2, J_3, \tilde{J}_3)\, D_{mn}^{(j)}(\theta, \psi, \phi) = (j(j+1), -m, n)\, D_{mn}^{(j)}(\theta, \psi, \phi). \tag{7.14}$$

In passing we mention that the UIR $D^{(j)}$ of $SU(2)$ is equivalent to its complex conjugate (see later).

Now we identify a subspace $\mathcal{H}_0$ in $\mathcal{H}$ which carries essentially the Schwinger representation of $SU(2)$ as a part of the left regular representation $U(u)$. Since now the eigenvalue n of $\tilde{J}_3$ is the multiplicity index, by an operator condition we pick up the maximum possible value j for n and so define $\mathcal{H}_0$:

$$\mathcal{H}_0 = \{f(u) \in \mathcal{H} | (\tilde{J}_1 + i\tilde{J}_2)f(u) = 0\}$$
$$= Sp\{(2j+1)^{1/2} D_{mj}^{(j)}(\theta, \psi, \phi), j = 0, 1/2, 1, \dots, m = j, j-1, \dots, -j\} \subset \mathcal{H}.$$

$$(7.15)$$

We see that $U(u)$ restricted to $\mathcal{H}_0$ gives a UR of $SU(2)$ which contains each UIR $D^{(j)}$ exactly once, which is just the Schwinger representation of $SU(2)$.

To explicitly relate this to what is obtained from the Schwinger construction, we start from the known forms of the functions $D_{mj}^{(j)}(\theta, \psi, \phi)$:

$$D_{mj}^{(j)}(\theta, \psi, \phi) = \sqrt{\frac{(2j)!}{(j+m)!(j-m)!}} \xi^{j-m} \eta^{j+m},$$

$$\xi = e^{\frac{i}{2}(\theta-\phi)} \sin\frac{\psi}{2}, \quad \eta = e^{-\frac{i}{2}(\theta+\phi)} \cos\frac{\psi}{2}.$$

$$(7.16)$$

We know from Eq. (7.14) that these are J_3 eigen functions with eigenvalue $-m$, and it is easily checked that

$$(J_1 + iJ_2)D_{mj}^{(j)}(\theta, \psi, \phi) = -\sqrt{(j+m)(j-m+1)}D_{m-1,j}^{(j)}(\theta, \psi, \phi).$$

$$(7.17)$$

To recover the standard forms for the basis functions of SU(2) UIR's, i.e., to 'undo' the complex conjugation in Eq. (7.12(a)), some relabelling is required. We therefore define as an orthonormal basis for $\mathcal{H}_0$ the family of functions

$$y_{jm}(\theta, \psi, \phi) = (-1)^{j-m}(2j+1)^{1/2} D_{-m,j}^{(j)}(\theta, \psi, \phi)$$

$$= \sqrt{\frac{(2j+1)!}{(j+m)!(j-m)!}} \xi^{j+m}(-\eta)^{j-m}.$$

$$(7.18)$$

As square integrable functions on $SU(2)$ they do obey

$$\frac{1}{16\pi^2} \int_0^{4\pi} d\theta \int_0^{2\pi} d\phi \int_0^\pi \sin\psi \, d\psi \; y_{j'm'}(\theta, \psi, \phi)^* y_{jm}(\theta, \psi, \phi) = \delta_{j'j}\delta_{m'm}, \quad (7.19)$$

and under action by $U(u)$ they transform in the standard way via the UIR $D^{(j)}$ of $SU(2)$.

The orthonormality relation (7.19) can be expressed in an alternative way. Introduce two independent complex variables z_1, z_2 by

$$z_1 = \rho\xi, z_2 = -\rho\eta,$$
$$|z_1|^2 + |z_2|^2 = \rho^2, \quad 0 \leq \rho < \infty. \tag{7.20}$$

The uniform integration measure over the two complex planes contains as a factor the $SU(2)$ volume element:

$$\mathrm{d}^2 z_1 \, \mathrm{d}^2 z_2 = \pi^2 \rho^2 \mathrm{d}\rho^2 \, \mathrm{d}u,$$
$$\mathrm{d}u = \frac{1}{16\pi^2} \mathrm{d}\theta \mathrm{d}\phi \sin\psi \, \mathrm{d}\psi. \tag{7.21}$$

Then Eq. (7.19) takes the form

$$(2j+1)! \iint \frac{\mathrm{d}^2 z_1 \, \mathrm{d}^2 z_2}{\pi \quad \pi} \delta(\rho^2 - 1) u_{j'm'}\left(\frac{z_1}{\rho}, \frac{z_2}{\rho}\right)^* u_{jm}\left(\frac{z_1}{\rho}, \frac{z_2}{\rho}\right) = \delta_{j'j}\delta_{m'm},$$
$$u_{jm}(\xi, \eta) = \xi^{j+m}\eta^{j-m}/\sqrt{(j+m)!(j-m)!} \tag{7.22}$$

Now the last two factors in the integrand here are actually independent of ρ, and the Kronecker deltas on the right hand side arise from the integration over $SU(2)$ with measure $\mathrm{d}u$. This allows us to replace $\delta(\rho^2 - 1)$ by any real positive function $f_j(\rho^2)$ provided

$$\int_0^\infty \mathrm{d}\rho^2 \, \rho^2 f_j(\rho^2) = 1, \tag{7.23}$$

and then Eq. (7.22) will remain valid in the form

$$(2j+1)! \iint \frac{\mathrm{d}^2 z_1 \, \mathrm{d}^2 z_2}{\pi \quad \pi} f_j(\rho^2)(\rho^2)^{-2j} u_{j'm'}(z_1, z_2)^* u_{jm}(z_1, z_2) = \delta_{jj'}\delta_{mm'}. \tag{7.24}$$

A suggestive choice consistent with condition (7.23) is

$$f_j(\rho^2) = (\rho^2)^{2j} e^{-\rho^2}/(2j+1)!, \tag{7.25}$$

which then leads to

$$\iint \frac{\mathrm{d}^2 z_1 \, \mathrm{d}^2 z_2}{\pi \quad \pi} e^{-|z_1|^2 - |z_2|^2} u_{j'm'}(z_1, z_2)^* u_{jm}(z_1, z_2) = \delta_{j'j}\delta_{m'm}. \tag{7.26}$$

This allows us to make contact with the Schwinger $SU(2)$ construction. The Bargmann representation of the canonical commutation relations (3.102) for two quantum mechanical simple harmonic oscillators uses a Hilbert space $\mathcal{H}_0$ made up of entire functions of two independent complex variables. The relevant equations are:

$$\mathcal{H}_0 = \Big\{ f(z_1, z_2) = \text{entire function} \mid \|f\|^2$$

$$= \iint \frac{d^2 z_1}{\pi} \frac{d^2 z_2}{\pi} e^{-|z_1|^2 - |z_2|^2} |f(z_1, z_2)|^2 < \infty \Big\},$$

$$(a_\alpha^\dagger f)(z_1, z_2) = z_\alpha f(z_1, z_2), \quad (a_\alpha f)(z_1, z_2) = \frac{\partial f(z_1, z_2)}{\partial z_\alpha}, \alpha = 1, 2. \quad (7.27)$$

The orthonormal basis in which the Schwinger representation $\mathcal{D}_0$ of $SU(2)$ generated by the Schwinger construction is fully reduced is just the set of functions

$$u_{jm}(z_1, z_2) = \frac{z_1^{j+m} z_2^{j-m}}{\sqrt{(j+m)!(j-m)!}}, \quad (7.28)$$

so the connection is complete.

7.1.3 The $SO(3)$ case

This case involves some changes compared to $SU(2)$. In the defining real orthogonal UIR of $SO(3)$, Eq. (3.13), the Euler angles description of a general element $A(\theta, \psi, \phi)$ is given in Eq. (3.24) repeated here:

$$A(\theta, \psi, \phi) = \begin{pmatrix} \cos\theta & -\sin\theta & 0 \\ \sin\theta & \cos\theta & 0 \\ 0 & 0 & 1 \end{pmatrix} \begin{pmatrix} \cos\psi & 0 & \sin\psi \\ 0 & 1 & 0 \\ -\sin\psi & 0 & \cos\psi \end{pmatrix}$$

$$\times \begin{pmatrix} \cos\phi & -\sin\phi & 0 \\ \sin\phi & \cos\phi & 0 \\ 0 & 0 & 1 \end{pmatrix}, \quad 0 \leq \theta, \phi \leq 2\pi, 0 \leq \psi \leq \pi. \quad (7.29)$$

The normalised volume element on $SO(3)$ is

$$dA = \frac{1}{8\pi^2} d\theta \, d\phi \sin\psi \, d\psi. \quad (7.30)$$

The regular representations of $SO(3)$ act on $\mathcal{H} = L^2(SO(3))$:

$$\mathcal{H} = \{ f(\theta, \psi, \phi) \mid \|f\|^2 = \int_{SO(3)} dA \, |f(\theta, \psi, \phi)|^2 < \infty \}. \quad (7.31)$$

The mutually commuting left and right regular representation actions are:

$$(U(A')f)(A) = f(A'^{-1}A), \quad (\tilde{U}(A')f)(A) = f(AA'). \quad (7.32)$$

Their generators $L_r, \tilde{L}_r$ have the same expressions as in Eq. (7.9), and their commutation relations follow the pattern of Eq. (7.10), so need not be repeated. The orthonormal basis functions realising the complete reductions of both regular representations are $(2l+1)^{1/2} D^{(l)}_{mn}(\theta, \psi, \phi)$ for $l = 0, 1, 2, \ldots$; $m, n = l, l-1, \ldots, -l$; and $-m, n$ are eigenvalues of $L_3, \tilde{L}_3$, respectively.

Following the same steps as with $SU(2)$, we isolate a subspace $\mathcal{H}_0$ in $\mathcal{H}$ supporting the Schwinger representation $\mathcal{D}_0$ of $SO(3)$:

$$\mathcal{H}_0 = \{f(\theta, \psi, \phi) \in \mathcal{H} \mid (\tilde{L}_1 + i\tilde{L}_2) f(\theta, \psi, \phi) = 0\}$$

$$= Sp\{(2l+1)^{1/2} D^{(l)}_{m,l}(\theta, \psi, \phi), l = 0, 1, 2, \ldots ; m = l, l-1, \ldots, -l\} \subset \mathcal{H}.$$

$$(7.33)$$

The particular D-functions here again have a simple single-term structure:

$$D^{(l)}_{m,l}(\theta, \psi, \phi) = \sqrt{\frac{(2l)!}{(l+m)!(l-m)!}} e^{-im\theta - il\phi} \left(\cos \frac{\psi}{2}\right)^{l+m} \left(\sin \frac{\psi}{2}\right)^{l-m}. \quad (7.34)$$

The identification of basis functions transforming in the standard way under $U(A)$ action again needs some relabelling:

$$\mathcal{Y}_{lm}(\theta, \psi, \phi) = (-1)^{l-m}(2l+1)^{1/2} D^{(l)}_{-m,l}(\theta, \psi, \phi)$$

$$= (-1)^{l-m} \sqrt{\frac{(2l+1)!}{(l+m)!(l-m)!}} e^{im\theta - il\phi} \left(\cos \frac{\psi}{2}\right)^{l-m} \left(\sin \frac{\psi}{2}\right)^{l+m}.$$

$$(7.35)$$

A familiar way of realising the Schwinger representation of $SO(3)$ is by using the geometrical action of rotations on functions $f(\theta, \psi) \in L^2(\mathbb{S}^2)$ (θ, ψ are the azimuthal and polar angles usually written as ϕ, θ), with spherical harmonics $Y_{lm}(\psi, \theta)$ as basis functions. This is equivalent as a UR to the realisation developed above, though the carrier spaces and basis functions are different. In particular, the $\mathcal{Y}_{lm}(\theta, \psi, \phi)$ are all single-term expressions, while the $Y_{lm}(\psi, \theta)$ are generally sums of many terms.

7.1.4 The $SU(3)$ case

The $SU(2)$ and $SO(3)$ Schwinger representations are naturally very similar, moreover both are rank one Lie groups with a common Lie algebra. As a different example we now look at the $SU(3)$ case, a rank two Lie group. The structure and description of the $SU(3)$ UIR's are a special case of the general theory given in Chapter 5, and will be recalled below.

The defining representation of $SU(3)$ is the set of all three dimensional unitary unimodular matrices, and it is an eight parameter Lie group. In the tensor notation the generators and commutation relations for any UIR or UR are as in Eq. (5.70) for the case $l = 2$, and are recalled here: with $\lambda, \mu, \ldots = 1, 2, 3$,

$$A^{\lambda\dagger}{}_{\mu} = A^{\mu}{}_{\lambda}, A^{\lambda}{}_{\lambda} = 0;$$

$$[A^{\lambda}{}_{\mu}, A^{\rho}{}_{\sigma}] = \delta^{\lambda}{}_{\sigma}A^{\rho}{}_{\mu} - \delta^{\rho}{}_{\mu}A^{\lambda}{}_{\sigma}. \tag{7.36}$$

The UIR's of $SU(3)$ are best described using the canonical subgroup chain. The $SU(2) \times U(1)$ subgroup is taken to be generated by $A^{1}{}_{2}, A^{2}{}_{1}, A^{1}{}_{1} - A^{2}{}_{2}$ and $-A^{3}{}_{3}$. In the standard isospin or angular momentum notation we identify the $SU(2)$ and $U(1)$ generators

$$\hat{I}_3 = A^{1}{}_{1} - A^{2}{}_{2}, \hat{I}_+ = \sqrt{2}A^{1}{}_{2}, \hat{I}_- = \sqrt{2}A^{2}{}_{1}; \hat{Y} = -A^{3}{}_{3}. \tag{7.37}$$

A general UIR of $SU(3)$ is labelled by a pair (p, q) with $p, q = 0, 1, 2, \ldots$ independently. (The defining '3' representation is $(1, 0)$, its complex conjugate 3* is $(0, 1)$.) The dimension of (p, q) is

$$N_{p,q} = \frac{1}{2}(p + 1)(q + 1)(p + q + 2). \tag{7.38}$$

An orthonormal basis within (p, q) is written as $|p, q; I, I_3, Y\rangle$. Here I, I_3 are the $SU(2)$ quantum numbers – $I = 0, 1/2, 1, \ldots$; $I_3 = I, I-1, \ldots, -I$ – and Y is the eigenvalue of the U(1) generator $\hat{Y}$. The (multiplicity free) '$I - Y$ multiplets' or spectrum within (p, q) is given by a simple counting procedure:

$$r = 0, 1, 2, \ldots, p; s = 0, 1, 2, \ldots, q:$$

$$I = \frac{1}{2}(r + s), Y = r - s + \frac{2}{3}(q - p); I_3 = I, I - 1, \ldots, -I. \tag{7.39}$$

The nonhermitian generators $A^{1}{}_{2}, A^{1}{}_{3}, A^{2}{}_{3}$ cause changes in the labels inside the basis vectors:

$$A^{1}{}_{2} : I, I_3, Y \to I, I_3 + 1, Y;$$

$$A^{1}{}_{3} : I, I_3, Y \to I \pm 1/2, I_3 + 1/2, Y + 1;$$

$$A^{2}{}_{3} : I, I_3, Y \to I \pm 1/2, I_3 - 1/2, Y + 1. \tag{7.40}$$

Thus either Υ increases by unity or Υ stays unchanged while I_3 increases by unity. The unique basis state in (p, q) annihilated by all $A^\lambda{}_\mu, \lambda < \mu$, is the one for $r = p, s = 0$, namely

$$|p, q; \frac{p}{2}, \frac{p}{2}, \frac{1}{3}(p + 2q)\rangle, \tag{7.41}$$

and this is the highest weight vector in the UIR.

With this information about the UIR's, we can analyse the regular representations. The Hilbert space is $\mathcal{H} = L^2(SU(3))$:

$$\mathcal{H} = \{f(g) \mid \|f\|^2 = \int_{SU(3)} dg \, |f(g)|^2 < \infty\}, \tag{7.42}$$

with dg the normalised invariant volume element on $SU(3)$. The left and right regular representations $U(g)$, $\tilde{U}(g)$ act as in Eqs. (7.7, 7.32), with generators $A^\lambda{}_\mu, \tilde{A}^\lambda{}_\mu$. The set of all normalised unitary representation matrices

$$(N_{p,q})^{1/2} D^{(p,q)}_{I\,I_3\,\Upsilon;\tilde{I}\,\tilde{I}_3\,\tilde{\Upsilon}}(g) \tag{7.43}$$

gives an orthonormal basis in $\mathcal{H}$ in which both $U(g)$ and $\tilde{U}(g)$ are fully reduced. The subspace $\mathcal{H}_0$ supporting a Schwinger representation $\mathcal{D}_0$ of $SU(3)$ is identified by

$$\mathcal{H}_0 = \{f(g) \in \mathcal{H} | \tilde{A}^\lambda{}_\mu f = 0, \lambda < \mu\}$$
$$= \mathrm{Sp}\{(N_{p,q})^{1/2} D^{(p,q)}_{I\,I_3\,\Upsilon;\frac{1}{2}p,\frac{1}{2}p,\frac{1}{3}(p+2q)}(g), \; p\,q\,I\,I_3\,\Upsilon \text{ varying}\}. \tag{7.44}$$

With respect to the action by $U(g)$ we see that each UIR occurs exactly once (here we use $(p, q)^* = (q, p)$) in $\mathcal{H}_0$.

This UR of $SU(3)$ was constructed long ago by Beg and Ruegg. It is clear that the method works for $SU(n)$ for all $n \geq 4$: in the space of the regular representations, one picks out $\mathcal{H}_0$ as the subspace annihilated by $\tilde{A}^\lambda{}_\mu$ for all $\lambda < \mu$.

7.2 Induced Representations on Coset Spaces, the Reciprocity Theorem

We now turn to a brief account of UR's of a group G obtained by the method of induction starting from UIR's of a subgroup $H \subset G$. As we will see, on account of the fundamental Reciprocity Theorem, with respect to the reduction into UIR's of G, these UR's are typically 'leaner' than either regular representation though possibly not as much as the Schwinger representation.

7.2.1 The inducing construction

As in Section 7.1.1, let the complete set of UIR's of G be written in matrix form as $\left(\mathcal{D}^{(j)}_{m'm}(g)\right)$, acting on Hilbert spaces $\mathcal{H}^{(j)}$ of dimensions N_j. Let $H \subset G$ be some Lie subgroup. For the induction method we need to make use of some UIR $D^{(0)}(\,\cdot\,)$ of H in some Hilbert space $\mathcal{H}_0$ of some dimension; there is, however, no need to set up a notation able to handle all possible UIR's of H in Hilbert spaces of various dimensions. The present $\mathcal{H}_0$ is to be distinguished from the space appearing in Eq. (7.1).

The UR of G to be constructed will be denoted by $\mathcal{D}^{(\mathrm{ind},D^{(0)})}_H$: it is the induced representation arising from the UIR $D^{(0)} = \left(D^{(0)}_{ab}(\,\cdot\,)\right)$ of the subgroup $H \subset G$ on the Hilbert space $\mathcal{H}_0$ (which is not explicitly indicated).

At this point we describe the Peter–Weyl theorem for the matrix elements of UIR's of G, as these results will be useful later. As with the $SO(3)$, $SU(2)$ cases dealt with in Chapter 3, for each G there is an invariant normalised integration process, i.e., an invariant 'volume element' dg on G, having the properties

$$\int_G dg\, f(g) = \int_G dg\, \left(f(g_0 g)\ \text{or}\ f(g g_0)\ \text{or}\ f(g^{-1})\right), \ \text{any fixed } g_0 \in G;$$

$$\int_G dg\, 1 = 1, \tag{7.45}$$

for any function $f(g)$ on G. The Hilbert space of square integrable functions on G is

$$L^2(G,\mathbb{C}) = \left\{ f(g) \in \mathbb{C} \ \middle|\ \|f\|^2 = \int_G dg\, |f(g)|^2 < \infty \right\}. \tag{7.46}$$

Then, with respect to the invariant measure, the orthogonality and completeness properties of the UIR matrix elements are:

$$\int_G dg\, \mathcal{D}^{(j')}_{m'_1 m'_2}(g)^* \mathcal{D}^{(j)}_{m_1 m_2}(g) = \frac{1}{N_j} \delta_{j'j} \delta_{m'_1 m_1} \delta_{m'_2 m_2};$$

$$\sum_{jmm'} N_j \mathcal{D}^{(j)}_{mm'}(g) \mathcal{D}^{(j)}_{mm'}(g')^* = \delta(g',g), \tag{7.47}$$

with $\delta(g',g)$ the invariant Dirac delta function on G with respect to the volume element dg.

Now back to the inducing construction.

Consider vector valued functions $\phi : G \rightarrow \mathcal{H}_0$ satisfying a (right) covariance condition with respect to H:

$$g \in G \longrightarrow \phi(g) \in \mathcal{H}_0;$$

$$h \in H : \phi(gh) = D^{(0)}(h^{-1})\phi(g),$$

$$i.e., \quad \phi_a(gh) = \sum_b D^{(0)}_{ba}(h)^* \phi_b(g). \tag{7.48}$$

We now define an (left) action by G on such ϕ by operators $\mathcal{U}(\cdot)$:

$$g \in G: \quad \left(\mathcal{U}(g)\phi\right)(g') = \phi(g^{-1}g'). \tag{7.49}$$

The representation property is obvious, as is the compatibility of the covariance condition (7.48) with the action (7.49), i.e., the latter respects the former. We now proceed to complete the construction and simplify it as much as possible.

Let $Q = G/H$ be the space of right cosets in G with respect to H. We may think of it as similar to a classical mechanical configuration space; its dimension is the difference of dimensions of G and H. (For example, $SU(2)/U(1) \simeq SO(3)/SO(2) \simeq \mathbb{S}^2$.) A point $q \in Q$ defines an H-coset in G:

$$q \in Q : q = g_0 H = \{g_0 h \mid \text{all } h \in H\} \subset G, \text{ some } g_0 \in G. \tag{7.50}$$

A natural 'origin' in Q is the coset containing the identity, i.e., H itself:

$$q_0 = \{eh \mid \text{all } h \in H\} = \{H\} \in Q. \tag{7.51}$$

The (left) action of G on Q is given as follows:

$$q = g_0 H \in Q, g \in G \rightarrow q' = gq = gg_0 H - \{gg_0 h \mid g_0 \text{ fixed, all } h \in II\} \in Q. \tag{7.52}$$

A choice of (local) coset representatives $Q \rightarrow G$ is a choice of $\ell(q) \in G$ for each $q \in Q$ such that

$$q = \ell(q)q_0. \tag{7.53}$$

It is now clear that the 'independent information' contained in ϕ obeying condition (7.48) is given by its 'values' at coset representatives:

$$q \in Q: \phi_0(q) = \phi(\ell(q));$$

$$h \in H, g = \ell(q)h \in G : \phi(g) = D^{(0)}(h^{-1})\phi_0(q) ;$$

$$(\phi(g), \phi(g))_{\mathcal{H}_0} = (\phi_0(q), \phi_0(q))_{\mathcal{H}_0}. \tag{7.54}$$

On these the action by G is easily computed:

$$\phi' = \mathcal{U}(g)\phi : \phi_0'(q) = \phi'(\ell(q)) = \phi(g^{-1}\ell(q))$$
$$= \phi(\ell(g^{-1}q)\ell(g^{-1}q)^{-1}g^{-1}\ell(q))$$
$$= D^{(0)}(\ell(q)^{-1}g\ell(g^{-1}q))\phi_0(g^{-1}q). \tag{7.55}$$

Here we used the fact that for any $q \in Q, g \in G$,

$$\ell(q)^{-1}g\ell(g^{-1}q)q_0 = q_0, \ i.e., \ \ell(q)^{-1}g\ell(g^{-1}q) \in H, \tag{7.56}$$

since under the G action (7.52) on Q, the stability group or the 'little group' of the origin q_0 is H.

Thus from each $\phi : G \to \mathcal{H}_0$ we obtain the corresponding $\phi_0 : Q \to \mathcal{H}_0$; and the G-action (7.49) on ϕ translates into the G-action (7.55) on ϕ_0.

We finally define the Hilbert space inner product of these 'wave functions' in such a way that the operators $\mathcal{U}(g)$ are unitary. We use the following notation:

$$L^2(Q, \mathcal{H}_0) = \left\{ \phi_0(q) \in \mathcal{H}_0 \mid q \in Q, \ ||\phi_0||^2 = \int d\mu(q)(\phi_0(q), \phi_0(q))_{\mathcal{H}_0} < \infty \right\}. \tag{7.57}$$

Here $d\mu(q)$ is the G-invariant normalised volume element on Q arising as a factor in

$$dg = d\mu(q)dh \tag{7.58}$$

where dg, dh are the invariant normalised volume elements on G and H, respectively. It is now obvious that unitarity of $D^{(0)}$ on $\mathcal{H}_0$ leads to unitarity of $\mathcal{U}(g)$ on $L^2(Q, \mathcal{H}_0)$. This $\text{UR}\,\mathcal{U}(\cdot)$ of G acting on the Hilbert space $L^2(Q, \mathcal{H}_0)$ is the induced representation $\mathcal{D}_H^{(\text{ind}, D^{(0)})}$.

Combining Eqs. (7.55, 7.57) we see that we can introduce an (ideal) orthonormal basis $|q, a\rangle$ for $L^2(Q, \mathcal{H}_0)$ with these properties:

$$\phi_{0,a}(q) = \langle q, a | \phi_0 \rangle,$$
$$\langle q', a' | q, a \rangle = \delta(q', q)\delta_{a',a};$$
$$\mathcal{U}(g)|q, a\rangle = \sum_{a'} D_{a'a}^{(0)}(\ell(gq)^{-1}g\ell(q))|gq, a'\rangle. \tag{7.59}$$

This can be regarded as a standard (Wigner) form of $\mathcal{D}_H^{(\text{ind}, D^{(0)})}$. Here $\delta(q, q')$ is the G-invariant Dirac delta function on Q.

7.2.2 The reciprocity theorem

Now we come to the important question: how often does a general UIR $\mathcal{D}^{(j)}$ of G occur in the induced representation $\mathcal{D}_H^{(\text{ind},D^{(0)})}$ of G upon complete reduction into irreducibles? The answer is given by the Frobenius–Mackey Reciprocity Theorem:

Theorem: The UR $\mathcal{D}_H^{(\text{ind},D^{(0)})}$ of G contains the UIR $\mathcal{D}^{(j)}$ of G as many times as the latter contains the UIR $D^{(0)}$ of H upon restriction from G to H.

Obviously, this multiplicity of occurrence of $\mathcal{D}^{(j)}$ cannot exceed N_j, which is why any induced representation is 'leaner' than the regular representation of G (except when $H = \{e\}$, $\mathcal{H}_0 = \{\mathbb{C}\}$, $D^{(0)} = $ trivial). A physics motivated proof of this theorem, and a way to choose orthonormal state labels to handle the multiple occurrence of each UIR $\mathcal{D}^{(j)}$ within $\mathcal{D}_H^{(\text{ind},D^{(0)})}$, can be constructed.

7.2.3 Some Schwinger representations as induced representations

For some low dimensional groups G the interested reader can verify quite easily the following instructive relations:

$$\text{Schwinger Representation of } SU(2) = \mathcal{D}_{U(1)}^{(\text{ind},0)} \oplus \mathcal{D}_{U(1)}^{(\text{ind},1/2)}; \quad (a)$$

$$\text{Schwinger Representation of } SO(3) = \mathcal{D}_{SO(2)}^{(\text{ind},0)}; \quad (b)$$

$$\text{Schwinger Representation of } SU(3) = \mathcal{D}_{SU(2)}^{(\text{ind},0)}. \quad (c) \quad (7.60)$$

In the $SU(2)$ case, the $U(1)$ subgroup is, say, that generated by $\frac{1}{2}\sigma_3$ in the defining representation, and the superscripts 0 and $1/2$ denote possible eigenvalues of this generator. In the $SO(3)$ case, the superscript 0 denotes the trivial representation of $SO(2)$. For $SU(3)$, the subgroup $SU(2)$ is, say, generated by $\hat{I}_3, \hat{I}_\pm$ in Eq. (7.37), and again the superscript 0 denotes the trivial UIR of $SU(2)$. As Eq. (7.39) shows, each UIR (p, q) of $SU(3)$ contains exactly one state with vanishing isospin: $I = I_3 = 0, Y = \frac{2}{3}(q - p)$.

However, the result above for $SU(3)$ does not continue for $SU(n), n \geq 4$. More precisely, the Schwinger representation of $SU(n), n \geq 4$, is not an induced representation for any choice of UIR of the canonical $SU(n-1)$ subgroup of $SU(n)$.

7.3 Generalised Coherent State Systems

This is the third topic of this chapter, to some extent connected with induced representation theory. First some useful background from quantum mechanics.

7.3.1 Coherent states for the quantum mechanical harmonic oscillator

The standard harmonic oscillator coherent states in quantum mechanics were discovered by Schrödinger in 1926 as a way to saturate the Heisenberg Uncertainty Principle. Many years later, from the early 1960s onwards, as their properties began to be better understood, they came to play a critical role in the field of Quantum Optics. (However, as early as in 1932, von Neumann had already explored many of their unusual properties.) We recall the basic essentials in the context of a single mode quantised radiation field. This involves any numbers of photons all in the same single photon state.

The photon annihilation and creation operators $\hat{a}, \hat{a}^\dagger$ are related to hermitian quadrature operators $\hat{q}, \hat{p}$ (like position and momentum operators for a material oscillator) by

$$\hat{a} = \frac{1}{\sqrt{2}}(\hat{q} + i\hat{p}), \quad \hat{a}^\dagger = \frac{1}{\sqrt{2}}(\hat{q} - i\hat{p}), \tag{7.61}$$

and obey the Boson commutation relation

$$[\hat{q}, \hat{p}] = i\mathbb{1} \rightarrow [\hat{a}, \hat{a}^\dagger] = \mathbb{1}. \tag{7.62}$$

The Fock states $|n\rangle, n = 0, 1, 2, \cdots$, form an orthonormal basis for the Hilbert space of the single mode field, with the definition and properties

$$|n\rangle = \frac{(\hat{a}^\dagger)^n}{\sqrt{n!}}|0\rangle, \quad \hat{a}^\dagger\hat{a}|n\rangle = n|n\rangle;$$

$$\langle m|n\rangle = \delta_{m,n}, \quad \sum_{n=0}^{\infty}|n\rangle\langle n| = \mathbb{1}. \tag{7.63}$$

With $\hat{a}^\dagger\hat{a}$ being the photon number operator, the state $|n\rangle$ is an n-photon pure state.

The coherent states $|z\rangle$ are labelled by a complex number $z \in \mathbb{C}$. Their definitions and some essential properties are:

$$|z\rangle = e^{-\frac{1}{2}|z|^2} \sum_{n=0}^{\infty} \frac{z^n}{\sqrt{n!}} |n\rangle; \qquad (a)$$

$$\hat{a}|z\rangle = z|z\rangle; \qquad (b)$$

$$\langle z'|z\rangle = \exp\left(-\frac{1}{2}|z' - z|^2 + i\mathrm{Im}(z'^*z)\right); \qquad (c)$$

$$\int \frac{d^2z}{\pi} |z\rangle\langle z| = \mathbb{1} \qquad (d) \qquad (7.64)$$

The last property here is called the 'Resolution of Unity'. It shows that any state vector $|\psi\rangle$ can be expanded in terms of the coherent states as

$$|\psi\rangle = \int \frac{d^2z}{\pi} |z\rangle\langle z|\psi\rangle. \qquad (7.65)$$

However, due to the nonorthogonality of any two coherent states, Eq. (7.64c), these states are *overcomplete*, and there is no unique way to expand a general $|\psi\rangle$ in terms of them. The particular expansion (7.65) is characterised by the property

$$\langle z^*|\psi\rangle = e^{-\frac{1}{2}|z|^2} f(z),$$

$$f(z) = \sum_{n=0}^{\infty} \langle n|\psi\rangle \frac{z^n}{\sqrt{n!}} = \text{entire analytic function of } z. \qquad (7.66)$$

This is what underlies the Bargmann representation of the canonical commutation relations (3.102), used in Eq. (7.27) for two degrees of freedom.

The expansion (7.65) expresses the (over)completeness of the coherent states at the Hilbert space vector level. Exploiting the overcompleteness, it was discovered that for any operator $\hat{A}$ (of a large class, namely of Hilbert–Schmidt type) there is a unique Diagonal Representation using projections onto the coherent states:

$$\hat{A} = \int \frac{d^2z}{\pi} \phi(z) |z\rangle\langle z|. \qquad (7.67)$$

For hermitian $\hat{A}$, $\phi(z)$ is real, but a distribution in general of a very singular kind.

We see that with coherent states the concept of (over)completeness comes up at two levels – at the state vector level, and at the operator level using projections onto these states.

In the generalisation of the coherent state idea proposed by Klauder, the emphasis is on the existence of the Resolution of Unity, but no other significant operator based conditions or properties are demanded. In particular a Diagonal Representation for operators generalising Eq. (7.67) is not demanded; and in examples one finds that there may be no such representation for a general operator.

The generalisation due to Perelomov is within the framework of UIR's of Lie groups in Hilbert spaces, and to start with neither a Resolution of Unity nor a Diagonal Representation for operators is demanded. Whether one or the other of them is valid can vary from case to case. We now present a sketch of this generalisation.

7.3.2 Coherent states within UIR's of Lie groups

We use some of the ideas and notations from Section 7.2 in developing the second important notion of Generalised Coherent States. The ingredients needed are: a Lie group G; a UIR $\mathcal{D}^{(j_0)}(\,\cdot\,)$ of G on a Hilbert space $\mathcal{H}^{(j_0)}$ of dimension N_{j_0}; and a chosen normalised fiducial vector $\psi_0 \in \mathcal{H}^{(j_0)}$. For definiteness we may assume G to be compact, and then N_{j_0} is finite; but with suitable changes we can extend the treatment to include the Heisenberg–Weyl group, the group $SU(1,1)$, etc.

The orbit $\vartheta(\psi_0)$ of ψ_0 under action by G is a collection of normalised vectors in $\mathcal{H}^{(j_0)}$:

$$\vartheta(\psi_0) = \{\psi(g) = \mathcal{D}^{(j_0)}(g)\psi_0 \mid g \in G\} \subset \mathcal{H}^{(j_0)}. \qquad (7.68)$$

Such a collection $\{\psi(g)\}$ is by definition a system of Generalised Coherent States (GCS).

The orbit $\vartheta(\psi_0)$ has an immediate identification as a coset space. The stability group H_0 of ψ_0 is defined to be the subgroup

$$H_0 = \{g \in G \mid \mathcal{D}^{(j_0)}(g)\psi_0 = \psi_0\} \subset G. \qquad (7.69)$$

It is then clear that $\vartheta(\psi_0)$ is essentially the space G/H_0 of right cosets: using the notation of Eq. (7.50),

$$\vartheta(\psi_0) \simeq Q_0 = G/H_0 \qquad (7.70)$$

A general element $g \in G$ is the product of a coset representative $\ell(q) \in G$ and an element of H_0 (as in Eq. (7.54)):

$$g \in G \colon g = \ell(q)h, \quad q \in Q_0, h \in H_0. \qquad (7.71)$$

So using Eq. (7.69),

$$\psi(g) = \psi(\ell(q)) \equiv \psi_0(q). \tag{7.72}$$

To see whether a Resolution of Unity exists in this framework, we can check that by Schur's Lemma

$$\int_G dg \; \psi(g)\psi(g)^\dagger \equiv \int_{Q_0} d\mu(q) \; \psi_0(q)\psi_0(q)^\dagger = c\,\mathbb{1}, \tag{7.73}$$

for some real positive constant c. In the compact case, with the total volumes of G, $\mathcal{H}_0$ and Q_0 normalised to unity and N_{j_0} finite, we find $c = 1/N_{j_0}$, so a Resolution of Unity does exist. Any $\psi \in \mathcal{H}^{(j_0)}$ can be expanded as an integral over the GCS $\{\psi(g)\} \equiv \{\psi_0(q)\}$. If G is noncompact, however, the finiteness of c is not assured, and it can depend on the UIR of G concerned.

At this point, we apply the methods of harmonic analysis, using an appropriate UR of G acting on square integrable functions on Q_0, to extract the G-representation contents of the GCS $\{\psi(g)\}$. The key fact is that the vectors $\psi_0(q)$ belong to the linear space $\mathcal{H}^{(j_0)}$. Therefore, a natural complete orthonormal set of functions on Q_0, a system of 'spherical harmonics', can be used to project out the 'Fourier' components of $\psi_0(q)$, with well defined irreducible behaviours under G. Later $\psi_0(q)$ can be synthesised from these components. It is here that we run into induced representations.

A comparison of the steps involved in building induced representations on the one hand, and GCS on the other, will guide us. With induced representations, the goal is to construct a Hilbert space and then a UR of G on it. The steps are:

(i) Choose a group G, a subgroup H, UIR $D^{(0)}$ of H on $\mathcal{H}_0$;

(ii) Introduce functions $\phi : G \to \mathcal{H}_0$ obeying the covariance condition involving $D^{(0)}$ under right H-action;

(iii) Introduce the coset space $Q = G/H$ and functions $\phi_0 : Q \to \mathcal{H}_0$;

(iv) Define the Hilbert space $L^2(Q, \mathcal{H}_0)$ carrying the UR $\mathcal{D}_H^{(\mathrm{ind}, D^{(0)})}$ of G.

For GCS, in comparison:

(i) Choose G, and its UIR $\mathcal{D}^{(j_0)}$ on $\mathcal{H}^{(j_0)}$;

(ii) Choose $\psi_0 \in \mathcal{H}^{(j_0)}$, with stability group $H_0 \subset G$;

(iii) The GCS consist of vectors $\psi(g) \equiv \psi_0(q)$ in the orbit $\vartheta(\psi_0)$ of ψ_0, with $q \in Q_0 = G/H_0$.

(iv) The 'independent information' contained in $\psi(g)$, and G-action thereon, are expressed by (cf. Eq. (7.72)):

$$\psi(g) = \psi(\ell(q)) = \psi_0(q) = \mathcal{D}^{(j_0)}(\ell(q))\,\psi_0, \quad \psi_0(q_0) = \psi_0;$$

$$\mathcal{D}^{(j_0)}(g)\,\psi_0(q) = \psi(g\ell(q)) = \psi_0(gq). \tag{7.74}$$

From this comparison we see that for the harmonic analysis of $\psi_0(q)$ we need the induced representation $\mathcal{D}^{(\mathrm{ind},0)}_{H_0}$ of G, built up from the trivial UIR '0' of the stability group H_0 of ψ_0, acting on $L^2(Q_0,\mathbb{C})$:

$$L^2(Q_0,\mathbb{C}) = \left\{ f(q) \in \mathbb{C} \mid q \in Q_0, \ \|f\|^2 = \int_{Q_0} d\mu(q)|f(q)|^2 < \infty \right\};$$

$$g \in G : (\mathcal{U}(g)f)(q) = f(g^{-1}q). \tag{7.75}$$

By the Reciprocity Theorem, this induced UR of G contains a general UIR $\mathcal{D}^{(j)}$ of G as many times as the latter contains the trivial UIR '0' of H_0 upon restriction to H_0. This is the number of independent vectors in $\mathcal{H}^{(j)}$ invariant under the action by $\mathcal{D}^{(j)}(h)$ for all $h \in H_0$. Let us denote a maximal orthonormal choice of such vectors by

$$\phi^{(j,\lambda)} = (\phi^{(j,\lambda)}_m) \in \mathcal{H}^{(j)}, \lambda = 1,2,\cdots ; \quad (a)$$

$$\phi^{(j,\lambda')\dagger}\phi^{(j,\lambda)} = \delta_{\lambda'\lambda}; \quad (b)$$

$$h \in H_0 : \mathcal{D}^{(j)}(h)\,\phi^{(j,\lambda)} = \phi^{(j,\lambda)}. \quad (c) \tag{7.76}$$

(Of course we omit those j for which there are no such vectors in $\mathcal{H}^{(j)}$.) Then an orthonormal basis for $L^2(Q_0,\mathbb{C})$ is given by

$$\Upsilon^{(j,\lambda)}_m(q) = \sum_{m'} N^{1/2}_j \mathcal{D}^{(j)}_{mm'}(\ell(q))\phi^{(j,\lambda)}_{m'},$$

$$\Upsilon^{(j,\lambda)}_m(q_0) = N^{1/2}_j \phi^{(j,\lambda)}_m. \tag{7.77}$$

We see that there are as many independent 'spherical harmonics' on Q_0 of representation type j as $\mathcal{H}^{(j)}$ contains independent H_0-invariant vectors, and λ counts this multiplicity. The basic properties of these functions are:

$$\Upsilon_m^{(j,\lambda)}(gq) = \sum_{m'} \mathcal{D}_{mm'}^{(j)}(g)\Upsilon_{m'}^{(j,\lambda)}(q); \qquad (a)$$

$$\int_{Q_0} d\mu(q)\Upsilon_{m'}^{(j',\lambda')}(q)^*\Upsilon_m^{(j,\lambda)}(q) = \delta_{\lambda'\lambda}\delta_{j'j}\delta_{m'm}; \qquad (b)$$

$$\sum_{j,\lambda,m} \Upsilon_m^{(j,\lambda)}(q)\Upsilon_m^{(j,\lambda)}(q')^* = \delta(q',q). \qquad (c) \qquad (7.78)$$

In the last completeness relation we have the G-invariant Dirac delta function on Q_0 with respect to the volume element $d\mu(q)$.

Now we use Eqs. (7.78) to carry out the harmonic analysis of $\psi_0(q)$. The results, as may be expected, will be very simple. Since for $j = j_0$, $\psi_0 \in \mathcal{H}^{(j_0)}$ is (by definition of H_0) invariant under H_0, in this case $\lambda \geq 1$. With no loss of generality we assume that

$$\psi_0 = \phi^{(j_0,1)} \in \mathcal{H}^{(j_0)}. \qquad (7.79)$$

Then the GCS, the vectors contained in $\vartheta(\psi_0)$, are:

$$\psi(g) = \mathcal{D}^{(j_0)}(g)\phi^{(j_0,1)}$$

$$= \mathcal{D}^{(j_0)}(\ell(q)h)\phi^{(j_0,1)},$$

$$\text{i.e., } \psi_0(q) = \mathcal{D}^{(j_0)}(\ell(q))\phi^{(j_0,1)},$$

$$\text{i.e., } \psi_{0,m_0}(q) = \frac{1}{\sqrt{N_{j_0}}}\Upsilon_{m_0}^{(j_0,1)}(q). \qquad (7.80)$$

As functions on Q_0 with values in $\mathcal{H}^{(j_0)}$, they involve only the $(j_0,1)$ 'spherical harmonics':

$$\int_{Q_0} d\mu(q)\Upsilon_m^{(j,\lambda)}(q)^*\psi_{0,m_0}(q) = \frac{1}{\sqrt{N_{j_0}}}\delta_{jj_0}\delta_{\lambda,1}\delta_{mm_0}. \qquad (7.81)$$

Using this along with the completeness relation (7.78c) just gives back Eq. (7.80).

Examples

We present two simple examples to bring out the concepts and see the formalism 'in action'.

(i) $G = SU(2), \mathcal{D}^{(j_0)} = $ defining two dimensional representation, $j_0 = 1/2$, $\dim \mathcal{H}^{(j_0)} = 2$

The matrices of this representation are given in Eq. (7.5). The choice of Ψ_0 and its consequences are:

$$\Psi_0 = \begin{pmatrix} 1 \\ 0 \end{pmatrix}: \quad H_0 = \{e\}, \text{ trivial subgroup}; \quad Q_0 = SU(2) \simeq \mathbb{S}^3. \tag{7.82}$$

The GCS are all $SU(2)$ transforms of Ψ_0:

$$\Psi(\theta, \psi, \phi) = u(\theta, \psi, \phi)\Psi_0 = e^{-i\phi/2} \begin{pmatrix} e^{-i\theta/2} \cos \psi/2 \\ e^{i\theta/2} \sin \psi/2 \end{pmatrix},$$

$$0 \le \theta \le 4\pi, \ 0 \le \phi \le 2\pi, \ 0 \le \psi \le \pi. \tag{7.83}$$

The Resolution of Unity is easily checked:

$$\frac{1}{16\pi^2} \int_0^{4\pi} d\theta \int_0^{2\pi} d\phi \int_0^\pi \sin \psi \, d\psi \, \Psi(\theta, \psi, \phi)\Psi(\theta, \psi, \phi)^\dagger = \frac{1}{2} \mathbb{1}_{2\times 2}. \tag{7.84}$$

Furthermore, $\Psi(\theta, \psi, \phi)$ is just the first column of $u(\theta, \psi, \phi)$ illustrating Eqs. (7.81)

(ii) $G = SO(3), \mathcal{D}^{(j_0)} = $ defining (real) three dimensional representation, $j_0 = 1$, $\dim \mathcal{H}^{(j_0)} = 3$

The matrices of this representation are given in Eq. (7.29). We choose Ψ_0 so as to get a nontrivial H_0:

$$\Psi_0 = \begin{pmatrix} 0 \\ 0 \\ 1 \end{pmatrix}: \quad H_0 = SO(2) = \{A(0, 0, \phi), 0 \le \phi \le 2\pi\};$$

$$Q_0 = SO(3)/SO(2) \simeq \mathbb{S}^2. \tag{7.85}$$

The GCS vectors in $\vartheta(\psi_0)$, are:

$$\Psi(\theta, \psi, \phi) = A(\theta, \psi, \phi)\Psi_0 = \begin{pmatrix} \sin \psi \cos \theta \\ \sin \psi \sin \theta \\ \cos \psi \end{pmatrix}, \quad 0 \le \psi \le \pi, \ 0 \le \theta \le 2\pi. \tag{7.86}$$

These constitute the sphere $\mathbb{S}^2$ with polar angle ψ and azimuthal angle θ. As a simple case of Eq. (7.81), $\Psi(\theta, \psi, \phi)$ is the third column of $A(\theta, \psi, \phi)$.

The Resolution of Unity is easily checked:

$$\frac{1}{8\pi^2} \int_0^{2\pi} d\theta \int_0^{2\pi} d\phi \int_0^\pi \sin\psi \, d\psi \, \Psi(\theta,\psi,\phi)\Psi(\theta,\psi,\phi)^\dagger$$

$$= \frac{1}{4\pi} \int_0^{2\pi} d\theta \int_0^\pi \sin\psi \, d\psi \begin{pmatrix} \sin\psi\cos\theta \\ \sin\psi\sin\theta \\ \cos\psi \end{pmatrix} (\sin\psi\cos\theta \quad \sin\psi\sin\theta \quad \cos\psi)$$

$$= \frac{1}{3}\mathbb{1}_{3\times 3}. \tag{7.87}$$

Many more examples with varying levels of complexity may be found in the literature.

7.3.3 Existence of diagonal representation for operators

The question of the existence of a Diagonal Representation for all operators in the space of a GCS system needs a more elaborate analysis. This is at the operator, not vector, level. We sketch some features of it.

For a chosen fiducial vector $\psi_0 \in \mathcal{H}^{(j_0)}$, the space of the UIR $\mathcal{D}^{(j_0)}$ of G, the pure state projection operator is

$$\psi_0 \longrightarrow \hat\rho_0 = \psi_0\psi_0^\dagger. \tag{7.88}$$

Its orbit under G-action is the family of rank one projection operators

$$\vartheta(\hat\rho_0) = \{\psi(g)\psi(g)^\dagger = \hat\rho(g) = \mathcal{D}^{(j_0)}(g)\hat\rho_0\mathcal{D}^{(j_0)}(g)^\dagger \mid g \in G\} \tag{7.89}$$

arising from the vectors $\psi(g)$. The stability group of $\hat\rho_0$ is a subgroup of G defined by

$$H = \{g \in G \mid \mathcal{D}^{(j_0)}(g)\hat\rho_0\,\mathcal{D}^{(j_0)}(g)^\dagger = \hat\rho_0\} \subset G. \tag{7.90}$$

Clearly, H_0 of Eq. (7.69) is an invariant subgroup of H. The possible relationships between H_0 and H are determined by the nature of the quotient group H/H_0:

$$\text{Case A: } H/H_0 = \text{trivial or discrete;}$$

$$\text{Case B: } H/H_0 = U(1). \tag{7.91}$$

In Case A, the phase of ψ_0 can be altered by G-action only in discrete fashion. Case B means that there are elements in G whose effect on ψ_0 is to alter its phase by any specified amount.

For the analysis of $\vartheta(\hat{\rho}_0)$ the relevant coset space is $Q = G/H$. In Case A, Q_0 is a discrete cover of Q; while in Case B, Q has dimension one less than Q_0, so locally $Q_0 \simeq Q \times U(1)$. The harmonic analysis of $\hat{\rho}(g)$ requires use of the induced representation $\mathcal{D}_H^{(\text{ind},0)}$ of G on $L^2(Q, \mathbb{C})$, similar in spirit to the treatment of Section 7.3.2 for $\psi(g)$.

To answer the question whether any operator $\hat{A}$ acting on $\mathcal{H}^{(j_0)}$ can be expressed in the form

$$\hat{A} = \int_G dg \ a(g) \ \hat{\rho}(g) \tag{7.92}$$

for suitable $a(g)$, and if so whether this is unique, one has to carry out the harmonic analysis of $\hat{\rho}(g)$ and compare the results with the Clebsch–Gordan analysis of the direct product of UIR's $\mathcal{D}^{(j_0)} \times \mathcal{D}^{(j_0)*}$. The former determines the set of irreducible tensor operators under G which can be extracted from $\hat{\rho}(g)$, while the latter determines the set of irreducible unit tensors on $\mathcal{H}^{(j_0)}$ which always form a complete operator basis. The necessary and sufficient conditions for existence of the representation (7.92) for all $\hat{A}$ are that the latter tensor operators be expressible in terms of the former, the contents of $\hat{\rho}(g)$.

Through simple examples, provided the values of certain Clebsch–Gordan coefficients of G are known, one sees that there are cases where the Diagonal Representation (7.92) does hold for all $\hat{A}$, and others where this is not so. For example,

- $G = SU(2)$, spin j_0 UIR $\mathcal{D}^{(j_0)}$ for any $j_0 \geq 1/2$, $\psi_0 = |j_0, m_0\rangle$, $m_0 \neq 0$: Diagonal Representation exists;

- $G = SU(3)$, adjoint octet UIR $\mathcal{D}^{(1,1)}$, $\psi_0 = |1, 0, 0\rangle$ in the I, I_3, Y state labelling scheme in Section 7.1.4 : Diagonal Representation not possible in general.

Of course, in both cases we do have the Resolution of Unity.

Problems

P7.1   Derive the expressions (7.9) for the $SU(2)$ regular representation generators, and check Eq. (7.13).

P7.2   Verify the statements made in Eq. (7.60).

P7.3   As seen in Eq. (7.64), the coherent states $|z\rangle$ (respectively $\langle z'|$) are right eigenstates of $\hat{a}$ (respectively left eigenstates of $\hat{a}^\dagger$), and $\langle z'|z\rangle$ never vanishes. From this show formally that any operator $\hat{A}$ can be expressed as a function of $\hat{a}^\dagger$ and $\hat{a}$ in normal ordered form.

Bibliography

Beg, M. A. B. and Ruegg, H. (1965). A Set of Harmonic Functions for the Group SU(3). *J. Math. Phys.* 6: 677.

Chaturvedi, S., Marmo, G., Mukunda, N., et al. (2006). The Schwinger Representation of a Group {Concept and Applications}. *Rev. Math. Phys.* 18: 887.

Edmonds, A. R. (1996). *Angular Momentum in Quantum Mechanics.* Princeton, New Jersey: Princeton University Press.

Klauder, J. R. (1960). The Action Option and a Feynman Quantization of Spinor Fields in Terms of Ordinary c-Numbers. *Ann. Phys.* 11: 123.

Klauder, J. R. (1963). Continuous Representation Theory-1 {Postulates of Continuous Representation Theory}. *J. Math. Phys.* 4: 1058.

Klauder, J. R. and Sudarshan, E. C. G. (1968). *Fundamentals of Quantum Optics.* New York: W. A. Benjamin.

Klauder, J. R. and Skagerstam, B-S. (1985). *Coherent States: Applications in Physics and Mathematical Physics.* Singapore: World Scientific Publishing Company.

Mackey, G. W. (1963). *Group Representations in Hilbert Space.* Providence, Rhode Island: American Mathematical Society.

Mukunda, N., Arvind. Chaturvedi, S. and Simon, R. (2003). Generalised Coherent States and the Diagonal Representation for Operators. *J. Math. Phys.* 44: 2479

Perelomov, A. (1986). *Generalized Coherent States and Their Applications.* Springer-Verlag.

Schwinger, J. (1952). On Angular Momentum, USAEC Report NYO-3071. Reprint (1965) *Quantum Theory of Angular Momentum.* Beidenharn, L. C., and van Dam, H., eds. New York: Academic Press.

Sudarshan, E. C. G. (1963). Equivalence of Semi Classical and Quantum Mechanical Descriptions of Statistical Light Beams. *Physical Review Letters.* 10: 277.

8

Structure and Some Properties and Applications of the Groups $Sp(2n, \mathbb{R})$

Classical mechanics in phase space involves the concepts of Poisson brackets and canonical transformations. In (nonrelativistic) quantum mechanics, based on fundamental Cartesian variables, the kinematics is expressed via the Heisenberg canonical commutation relations (CCR's). One now has unitary transformations on wave functions in Hilbert space. In both cases, Hamiltonians quadratic in the canonical variables lead to linear equations of motion whose solutions are linear canonical transformations. In all this, the key role is played by the family of real symplectic groups $Sp(2n, \mathbb{R})$. They are the 'noncompact' versions of the classical compact Lie groups $C_l = USp(2l)$, and their defining representations are in real even dimensional vector spaces. In addition to their significance in the canonical formulation of classical mechanics and its counterpart in quantum mechanics, these groups are important in several specific quantum mechanical problems as well as in classical and quantum optics.

We look at some aspects of these groups in this chapter. While we describe the main features, we will not present complete proofs of all statements made, but a motivated reader should be able to supply the details and proceed to make practical uses of these groups.

In both classical and quantum mechanics we have in mind only Cartesian fundamental variables. So for n degrees of freedom, or n canonical pairs, the classical phase space is $\mathbb{R}^{2n}$, while in quantum mechanics we have the Hilbert space $\mathcal{H} = L^2(\mathbb{R}^n)$.

Since the symplectic groups are somewhat unfamiliar to most students of physics, we study the case $n = 1$ to begin with, and later go on to general n.

8.1 The Group $Sp(2, \mathbb{R})$

For one degree of freedom with Cartesian variables, in classical mechanics we have real canonical variables q and p, varying from $-\infty$ to ∞, and the Poisson brackets (PB)

$$\{q, p\} = 1, \quad \{q, q\} = \{p, p\} = 0. \tag{8.1}$$

The phase space is the two dimensional plane $\mathbb{R}^2$. For the corresponding system in quantum mechanics we have an irreducible pair of hermitian operators $\hat{q}, \hat{p}$ obeying the Heisenberg CCR's

$$[\hat{q}, \hat{p}] = i, \quad [\hat{q}, \hat{q}] = [\hat{p}, \hat{p}] = 0 \quad (\hbar = 1). \tag{8.2}$$

In the usual Schrödinger solution to these relations we have a Hilbert space $\mathcal{H} = L^2(\mathbb{R})$ with $\hat{q}, \hat{p}$ acting as follows:

$$\mathcal{H} = \left\{ \psi(q) \in \mathbb{C} \mid \|\psi\|^2 = \int_{-\infty}^{\infty} dq |\psi(q)|^2 < \infty \right\};$$

$$(\hat{q}\psi)(q) = q\psi(q), \quad (\hat{p}\psi)(q) = -i \frac{d\psi(q)}{dq}. \tag{8.3}$$

(Strictly speaking, since $\hat{q}$ and $\hat{p}$ are unbounded operators, their domains should be carefully specified; we assume such precautions are implicitly observed.)

Now consider linear homogeneous transformations on q, p which in the classical case preserve the PB's (8.1) and in the quantum case the CCR's (8.2). In either case, if we write

$$\begin{pmatrix} q' \\ p' \end{pmatrix} = \begin{pmatrix} a & b \\ c & d \end{pmatrix} \begin{pmatrix} q \\ p \end{pmatrix}, \quad a, b, c, d \text{ real}, \tag{8.4}$$

then the only nontrivial requirement is

$$\{q', p'\} = \{aq + bp, cq + dp\} = ad - bc = 1. \tag{8.5}$$

This leads to the defining matrix representation of the two dimensional real symplectic group $Sp(2, \mathbb{R})$:

$$Sp(2, \mathbb{R}) = \left\{ S = \begin{pmatrix} a & b \\ c & d \end{pmatrix} \text{ real } 2 \times 2 \text{ matrix} \mid \det S = ad - bc = 1 \right\}. \tag{8.6}$$

(It is because of the low dimensionality that the condition that the PB's or CCR's be preserved amounts simply to a determinant condition; for $n \geq 2$ degrees of freedom, the conditions are more demanding.) This is a three parameter or three dimensional Lie group, as we have four real parameters obeying one condition. The group composition law is given by matrix multiplication. The identity or unit element corresponds to the unit matrix, while inverses are given as

$$\begin{pmatrix} a & b \\ c & d \end{pmatrix}^{-1} = \begin{pmatrix} d & -b \\ -c & a \end{pmatrix}. \tag{8.7}$$

The Iwasawa decomposition and the group manifold

Some useful subgroups which are also helpful in determining the overall structure of $Sp(2, \mathbb{R})$ as a manifold are these:

$K-$phase space rotations:

$$\begin{pmatrix} q' \\ p' \end{pmatrix} = \begin{pmatrix} \cos \frac{\phi}{2} & -\sin \frac{\phi}{2} \\ \sin \frac{\phi}{2} & \cos \frac{\phi}{2} \end{pmatrix} \begin{pmatrix} q \\ p \end{pmatrix}, \quad 0 \leq \phi \leq 4\pi;$$

$A-$reciprocal scale changes:

$$\begin{pmatrix} q' \\ p' \end{pmatrix} = \begin{pmatrix} e^{\frac{\eta}{2}} & 0 \\ 0 & e^{-\frac{\eta}{2}} \end{pmatrix} \begin{pmatrix} q \\ p \end{pmatrix}, \quad -\infty < \eta < \infty;$$

$N-$'lens' subgroup:

$$\begin{pmatrix} q' \\ p' \end{pmatrix} = \begin{pmatrix} 1 & 0 \\ \xi & 1 \end{pmatrix} \begin{pmatrix} q \\ p \end{pmatrix}, \quad -\infty < \xi < \infty. \tag{8.8}$$

The 'names' K, A, N come from the general theory of noncompact simple Lie groups. K is basically $SO(2)$, a maximal compact subgroup of $Sp(2, \mathbb{R})$; A is an abelian subgroup, and N a nilpotent one. As a simple instance of a general theorem, the Iwasawa decomposition theorem, one can globally express each element $S \in Sp(2, \mathbb{R})$ uniquely as the product of three factors, one each from these subgroups K, A and N. (Always K is maximal compact, A is abelian and N is nilpotent). Then one finds

$$S = \begin{pmatrix} a & b \\ c & d \end{pmatrix} = \begin{pmatrix} 1 & 0 \\ \xi & 1 \end{pmatrix} \begin{pmatrix} e^{\eta/2} & 0 \\ 0 & e^{-\eta/2} \end{pmatrix} \begin{pmatrix} \cos \frac{\phi}{2} & -\sin \frac{\phi}{2} \\ \sin \frac{\phi}{2} & \cos \frac{\phi}{2} \end{pmatrix},$$

$$\xi = (ac + bd)/(a^2 + b^2) \in (-\infty, \infty),$$

$$\eta = \ln\left(a^2 + b^2\right) \in (-\infty, \infty),$$

$$\phi = 2\arg\left(a - ib\right) \in (-2\pi, 2\pi] \text{ or } [0, 4\pi). \tag{8.9}$$

It is clear that the three factors are globally well defined and unambiguous, and vary continuously with S. Therefore the *manifold structure* is seen to be the Cartesian product

$$Sp(2, \mathbb{R}) \sim \mathbb{S}^1 \times \mathbb{R}^2, \tag{8.10}$$

which means it is connected but infinitely multiply connected due to the $\mathbb{S}^1$ factor.

8.1.1 Generators and commutation relations in defining representation

If we write elements near the identity in exponential form and retain only linear terms, we find:

$$S \simeq 1 - i\epsilon J, \quad |\epsilon| \ll 1:$$

$$S \in Sp(2, \mathbb{R}) \Leftrightarrow J = \text{pure imaginary traceless.} \tag{8.11}$$

Thus J is a real linear combination of σ_2, $i\sigma_3$ and $i\sigma_1$, We choose the generators in the defining representation (8.6), in agreement with common convention, as:

$$J_0 = \frac{1}{2}\sigma_2, \quad K_1 = \frac{i}{2}\sigma_3, \quad K_2 = \frac{i}{2}\sigma_1. \tag{8.12}$$

These lead to the commutation relations

$$[J_0, K_1] = iK_2, \quad [J_0, K_2] = -iK_1, \quad [K_1, K_2] = -iJ_0. \tag{8.13}$$

Notice that in the defining nonunitary representation, J_0 is hermitian while K_1 and K_2 are antihermitian. The special elements in Eq. (8.8) belonging to important subgroups are in exponential form:

$$K = \begin{pmatrix} \cos\frac{\phi}{2} & -\sin\frac{\phi}{2} \\ \sin\frac{\phi}{2} & \cos\frac{\phi}{2} \end{pmatrix} = e^{-i\phi J_0};$$

$$A = \begin{pmatrix} e^{\frac{\eta}{2}} & 0 \\ 0 & e^{-\frac{\eta}{2}} \end{pmatrix} = e^{-i\eta K_1};$$

$$N = \begin{pmatrix} 1 & 0 \\ \xi & 1 \end{pmatrix} = e^{-i\xi(J_0 + K_2)}, \tag{8.14}$$

so as a product of exponentials the most general element is

$$S(\phi,\eta,\xi) = e^{-i\xi(J_0+K_2)}e^{-i\eta K_1}e^{-i\phi J_0}. \tag{8.15}$$

As was the case with compact simple Lie groups, here too because of the low dimension there are some coincidental isomorphisms:

$$Sp(2,\mathbb{R}) \sim SL(2,\mathbb{R}) = \text{group of real two dimensional}$$
$$\text{unimodular matrices}$$
$$\sim SU(1,1) = \text{group of complex unimodular}$$
$$\text{two dimensional pseudounitary matrices.} \tag{8.16}$$

But in higher dimensions these groups branch out into different families. There is also a connection to the proper homogeneous Lorentz group in two space and one time dimension:

$$Sp(2,\mathbb{R}), SL(2,\mathbb{R}), SU(1,1) = \text{two-fold covers of } SO(2,1). \tag{8.17}$$

That is, there are two-to-one homomorphisms from $Sp(2,\mathbb{R})$, $SL(2,\mathbb{R})$, $SU(1,1)$ to $SO(2,1)$. This is similar to the relation between $SU(2)$ and $SO(3)$.

Nontrivial unitary representations of $Sp(2,\mathbb{R})$ are necessarily infinite dimensional. Their generators obeying the commutation relations (8.13) are all hermitian. It is *impossible* to find finite dimensional hermitian solutions to (8.13), other than the trivial one. All finite dimensional (irreducible) representations of $Sp(2,\mathbb{R})$ are necessarily nonunitary – in them we can always assume that J_0 is hermitian while K_1 and K_2 are antihermitian. These representations of $Sp(2,\mathbb{R})$ are obtained very easily by an 'analytic continuation' of the finite dimensional UIR's of $SU(2)$ by the so-called Weyl unitary trick:

$$J_1, J_2, J_3 \text{ of } SU(2) \rightarrow J_0 = J_3, \ K_1 = iJ_1, \ K_2 = iJ_2 \text{ for } Sp(2,\mathbb{R}). \tag{8.18}$$

In addition to all these, $Sp(2,\mathbb{R})$ also possesses infinite dimensional nonunitary irreducible representations, indeed infinitely many of them!

8.1.2 Quantum mechanics and the metaplectic group $Mp(2)$

Now we consider the consequences of the Heisenberg CCR's (8.2) being preserved by $Sp(2,\mathbb{R})$ transformations. Here a particular *double-valued* reducible UR of $Sp(2,\mathbb{R})$

is involved; actually it is a true reducible UR of a double covering group of $Sp(2, \mathbb{R})$ called the metaplectic group $Mp(2)$. We have the relationships

$$Mp(2) \xrightarrow{2 \text{ to } 1} Sp(2, \mathbb{R}) \xrightarrow{2 \text{ to } 1} SO(2, 1): \text{homomorphisms.} \qquad (8.19)$$

How does this come about? According to a celebrated theorem of Stone and von Neumann, apart from unitary equivalence there is a unique irreducible hermitian solution to the CCR's (8.2), given in the Schrödinger form by Eq. (8.3). Now apply an $Sp(2, \mathbb{R})$ matrix S to $\hat{q}, \hat{p}$ and define,

$$S \in Sp(2, \mathbb{R}): \hat{q}' = a\hat{q} + b\hat{p}, \; \hat{p}' = c\hat{q} + d\hat{p}. \qquad (8.20)$$

Since $\hat{q}'$ and $\hat{p}'$ are also a hermitian irreducible solution to the CCR's (8.2), they must be related to $\hat{q}$ and $\hat{p}$ by some unitary operator $\bar{U}(S)$ dependent on S:

$$S \in Sp(2, \mathbb{R}): \hat{q}' = a\hat{q} + b\hat{p} = \bar{U}(S)^{-1}\hat{q}\bar{U}(S),$$

$$\hat{p}' = c\hat{q} + d\hat{p} = \bar{U}(S)^{-1}\hat{p}\bar{U}(S),$$

$$\bar{U}(S)^{\dagger}\bar{U}(S) = \bar{U}(S)\bar{U}(S)^{\dagger} = \mathbb{1}. \qquad (8.21)$$

This $\bar{U}(S)$ is clearly determined by S up to an arbitrary S-dependent phase factor; irreducibility of $\hat{q}$ and $\hat{p}$ ensures that this is the only freedom in the choice of $\bar{U}(S)$. Now we carry out a second such transformation S', appeal to the Stone–von Neumann theorem, and reach the conclusion:

$$S', S \in Sp(2, \mathcal{R}): \bar{U}(S')\bar{U}(S) = (\text{phase factor dependent}$$

$$\text{on } S' \text{ and } S) \times \bar{U}(S'S). \qquad (8.22)$$

Thus we have here a UR of $Sp(2, \mathbb{R})$ up to phases. Now we ask if the freedom of an S-dependent phase in each $\bar{U}(S)$ can be used to simplify the right hand side in Eq. (8.22), possibly removing the S' and S dependent phase completely and leading to a true UR of $Sp(2, \mathbb{R})$. The answer is – not quite! We can adjust the phases of individual $\bar{U}(S)$'s so as to reduce the phase factor on the right in Eq. (8.22) at best to ± 1, so a sign ambiguity remains. Thus we can achieve:

$$S', S \in Sp(2, \mathbb{R}): \bar{U}(S')\bar{U}(S) = \pm\bar{U}(S'S), \qquad (8.23)$$

but cannot go any further. This is why the operators $\bar{U}(S)$ are said to give a 'double-valued' UR of $Sp(2, \mathbb{R})$. Stated otherwise, they give a true UR of another group, the metaplectic group $Mp(2)$ which is a two-fold covering of $Sp(2, \mathbb{R})$. Strictly speaking, then, we should change the argument S in $\bar{U}(S)$ to an element of $Mp(2)$. However, we continue with the above notation and simply say that with $\bar{U}(S)$ we have a double-valued UR of $Sp(2, \mathbb{R})$.

The generators of this metaplectic UR are found quite easily. We go to the infinitesimal forms of Eqs. (8.8, 8.21) and then 'synthesise' the hermitian generators $\hat{J}_0, \hat{K}_1, \hat{K}_2$ of $\bar{U}(S)$:

$$\begin{pmatrix} \hat{q}' \\ \hat{p}' \end{pmatrix} \simeq \begin{pmatrix} 1 & -\frac{\phi}{2} \\ \frac{\phi}{2} & 1 \end{pmatrix} \begin{pmatrix} \hat{q} \\ \hat{p} \end{pmatrix} \simeq (\mathbb{1} + i\phi\hat{J}_0) \begin{pmatrix} \hat{q} \\ \hat{p} \end{pmatrix} (\mathbb{1} - i\phi\hat{J}_0):$$

$$\hat{J}_0 = -\frac{1}{4}(\hat{q}^2 + \hat{p}^2);$$

$$\begin{pmatrix} \hat{q}' \\ \hat{p}' \end{pmatrix} \simeq \begin{pmatrix} 1 + \frac{\eta}{2} & 0 \\ 0 & 1 - \frac{\eta}{2} \end{pmatrix} \begin{pmatrix} \hat{q} \\ \hat{p} \end{pmatrix} \simeq (\mathbb{1} + i\eta\hat{K}_1) \begin{pmatrix} \hat{q} \\ \hat{p} \end{pmatrix} (\mathbb{1} - i\eta\hat{K}_1):$$

$$\hat{K}_1 = \frac{1}{4}(\hat{q}\hat{p} + \hat{p}\hat{q});$$

$$\begin{pmatrix} \hat{q}' \\ \hat{p}' \end{pmatrix} \simeq \begin{pmatrix} 1 & 0 \\ \xi & 1 \end{pmatrix} \begin{pmatrix} \hat{q} \\ \hat{p} \end{pmatrix} \simeq (\mathbb{1} + i\xi(\hat{J}_0 + \hat{K}_2)) \begin{pmatrix} \hat{q} \\ \hat{p} \end{pmatrix} (\mathbb{1} - i\xi(\hat{J}_0 + \hat{K}_2)):$$

$$\hat{J}_0 + \hat{K}_2 = -\frac{1}{2}\hat{q}^2. \tag{8.24}$$

Of course each generator is determined up to an additive real constant, which have been chosen so that the commutation relations (8.13) are obeyed:

$$\hat{J}_0 = -\frac{1}{4}(\hat{q}^2 + \hat{p}^2), \ \hat{K}_1 = \frac{1}{4}(\hat{q}\hat{p} + \hat{p}\hat{q}), \ \hat{K}_2 = \frac{1}{4}(\hat{p}^2 - \hat{q}^2);$$

$$[\hat{J}_0, \hat{K}_1] = i\hat{K}_2, \ [\hat{J}_0, \hat{K}_2] = -i\hat{K}_1, \ [\hat{K}_1, \hat{K}_2] = -i\hat{J}_0. \tag{8.25}$$

In terms of the nonhermitian oscillator operators

$$\hat{a} = \frac{1}{\sqrt{2}}(\hat{q} + i\hat{p}), \ \hat{a}^\dagger = \frac{1}{\sqrt{2}}(\hat{q} - i\hat{p}), \ [\hat{a}, \hat{a}^\dagger] = 1, \tag{8.26}$$

the $Mp(2)$ generators are

$$\hat{J}_0 = -\frac{1}{2}\left(\hat{a}^\dagger\hat{a} + \frac{1}{2}\right), \ \hat{K}_1 = \frac{i}{4}(\hat{a}^{\dagger 2} - \hat{a}^2), \ \hat{K}_2 = -\frac{1}{4}(\hat{a}^{\dagger 2} + \hat{a}^2). \tag{8.27}$$

All these generators are hermitian quadratic in $\hat{q}$ and $\hat{p}$ naturally, so that upon commutation with $\hat{q}$ and $\hat{p}$ we get expressions linear in $\hat{q}$ and $\hat{p}$. As a result they are all even parity operators, that is, they commute with the parity operator $\hat{P}$. In the notation of Eq. (8.3),

$$\psi(q) \in \mathcal{H} \colon (\hat{P}\psi)(q) = \psi(-q) \in \mathcal{H},$$

$$\hat{P}(\hat{q} \text{ or } \hat{p})\hat{P}^{-1} = -\hat{q} \text{ or } -\hat{p};$$

$$\hat{P}(\hat{J}_0 \text{ or } \hat{K}_1 \text{ or } \hat{K}_2)\hat{P}^{-1} = \hat{J}_0 \text{ or } \hat{K}_1 \text{ or } \hat{K}_2. \tag{8.28}$$

(We recognise that the matrix $-\mathbb{1}$ belongs to $Sp(2, \mathcal{R})$, and in fact lies in the centre; in the metaplectic representation it is realised as the parity operation $\hat{P}$.) So if we define two mutually orthogonal subspaces $\mathcal{H}^{(\pm)}$ of $\mathcal{H}$ as made up of even/odd parity states, eigenspaces of $\hat{P}$:

$$\mathcal{H}^{(\pm)} = \{\psi(q) \in \mathcal{H} | \, \psi(-q) = \pm\psi(q)\} \subset \mathcal{H}, \tag{8.29}$$

they are invariant under action by the metaplectic unitary operators $\bar{U}(S)$ and their generators:

$$\bar{U}(S)\mathcal{H}^{(+)} = \mathcal{H}^{(+)}, \quad \bar{U}(S)\mathcal{H}^{(-)} = \mathcal{H}^{(-)}. \tag{8.30}$$

In each of these subspaces $\bar{U}(S)$ is irreducible. Thus the UR $\bar{U}(S)$ of $Mp(2)$ acting on $\mathcal{H}$ is the direct sum of two UIR's, one each on $\mathcal{H}^{(+)}$ and on $\mathcal{H}^{(-)}$, and these are inequivalent. This is seen from the fact that the spectra of $\hat{J}_0$ differ in the two cases:

$$\mathcal{H}^{(+)}\text{: eigenvalues of } \hat{J}_0 = -\frac{1}{2} \times \left(\frac{1}{2}, \frac{5}{2}, \frac{9}{2}, \cdots\right);$$

$$\mathcal{H}^{(-)}\text{: eigenvalues of } \hat{J}_0 = -\frac{1}{2} \times \left(\frac{3}{2}, \frac{7}{2}, \frac{11}{2}, \cdots\right); \tag{8.31}$$

These two UIR's of $Mp(2)$ are usually referred to as $D_{1/4}^{(-)}$ and $D_{3/4}^{(-)}$, respectively, going back to a notation introduced by V. Bargmann, so we have:

$$\text{UR } \{\bar{U}(S)\} \text{ of } Mp(2) \text{ on } \mathcal{H} = \text{ UIR } D_{-1/4}^{(-)} \text{ on } \mathcal{H}^{(+)} \oplus \text{ UIR } D_{-3/4}^{(-)} \text{ on } \mathcal{H}^{(-)}. \tag{8.32}$$

At this point, we can see a hint of the $Mp(2)$-$Sp(2, \mathbb{R})$ relationship. In the defining representation (8.6, 8.14) of $Sp(2, \mathbb{R})$, we see that the 'compact generator' J_0 obeys $e^{-4\pi i J_0} = \mathbb{1}$. On the other hand, we see from Eq. (8.31) that in the metaplectic UR this generator obeys

$$e^{-4\pi i \hat{J}_0} = -\mathbb{1}, \quad e^{-8\pi i \hat{J}_0} = +\mathbb{1} \quad \text{on } \mathcal{H}. \tag{8.33}$$

Essentially this is why $Mp(2)$ is a two-fold cover of $Sp(2, \mathbb{R})$.

The generalised Huyghens kernel

As has been mentioned, the notation $\bar{U}(S)$ is somewhat deficient in the sense that $S \in Sp(2,\mathbb{R})$, but we know that we are dealing with a UR of $Mp(2)$. Subject to this, in other words if S is sufficiently close to the identity in $Mp(2)$, we can fairly easily obtain the result:

$$S = \begin{pmatrix} a & b \\ c & d \end{pmatrix} \in Sp(2,\mathbb{R}):$$

$$\langle q' | \bar{U}(S) | q \rangle = \frac{e^{-i\pi/4}}{\sqrt{|b|}} \exp\{i(dq'^2 - 2q'q + aq^2)/2b\} \text{ if } b \neq 0,$$

$$\exp(iacq^2/2)\delta\left(\frac{q'}{a} - q\right) \text{ if } b = 0. \tag{8.34}$$

This is called the generalised Huyghens kernel for one degree of freedom, and it is useful in classical wave optics as well as in quantum mechanics. The solution to the Schrödinger equation, in the form of the Feynman kernel, for free particles and the harmonic oscillator, are particular cases of (8.34) as their Hamiltonians are quadratic.

8.2 The Group $Sp(2n,\mathbb{R})$

Now we consider the general case of classical and quantum systems with n Cartesian degrees of freedom. We assemble the basic positions and momenta into $2n$-component column vectors as follows:

$$\text{Classical: } \xi = (\xi_a) = (q_1, \cdots, q_n, p_1, \cdots, p_n)^T, \ a = 1, 2, \cdots, 2n;$$

$$\text{Quantum: } \hat{\xi} = (\hat{\xi}_a) = (\hat{q}_1, \cdots, \hat{q}_n, \hat{p}_1, \cdots, \hat{p}_n)^T. \tag{8.35}$$

We have adopted a convention: list all the coordinates first, then all the momenta. Another sometimes more convenient convention is to list each canonical pair in sequence: $q_1, p_1, q_2, p_2, \cdots, q_n, p_n$; however, we will use (8.35). The classical PB relations and the quantum CCR's can be written in concise matrix forms:

$$\{\xi_a, \xi_b\} = \beta_{ab},$$

$$[\hat{\xi}_a, \hat{\xi}_b] = i\beta_{ab}, \ (\hbar = 1),$$

$$\beta = (\beta_{ab}) = \begin{pmatrix} 0 & \mathbb{1}_{n\times n} \\ -\mathbb{1}_{n\times n} & 0 \end{pmatrix}, \ \beta^2 = -\mathbb{1}_{2n\times 2n}. \tag{8.36}$$

This real antisymmetric $2n \times 2n$ matrix β is called the 'symplectic metric matrix'.

In quantum mechanics we have to generalise Eq. (8.3) to n dimensions. The Hilbert space $\mathcal{H}$ on which the $\hat{\xi}_a$ act irreducibly can be described in many ways (as is, indeed, the case for $n = 1$ as well). The most familiar is the Schrödinger description using complex-valued wave functions on $\mathbb{R}^n$, that is, elements of $L^2(\mathbb{R}^n)$. The $\hat{q}_r$ act multiplicatively while the $\hat{p}_r$ are differential operators:

$$\mathcal{H} = \{\psi(\boldsymbol{q}) \in \mathbb{C} \mid \|\psi\|^2 = \int_{\mathbb{R}^n} d^n q |\psi(\boldsymbol{q})|^2 < \infty\};$$

$$(\hat{q}_r \psi)(\boldsymbol{q}) = q_r \psi(\boldsymbol{q}), \ (\hat{p}_r \psi)(\boldsymbol{q}) = -i \frac{\partial}{\partial q_r} \psi(\boldsymbol{q}),$$

$$\boldsymbol{q} = (q_1, \cdots, q_n) \in \mathbb{R}^n. \tag{8.37}$$

We now consider linear homogeneous transformations $\xi \to \xi', \hat{\xi} \to \hat{\xi}'$ of the form

$$\xi'_a = S_{ab} \xi_b, \ \hat{\xi}'_a = S_{ab} \hat{\xi}_b, \tag{8.38}$$

maintaining reality or hermiticity, respectively. If we require that the relations (8.36) be preserved, we find that the real $2n \times 2n$ matrix S must obey

$$S \beta S^T = \beta. \tag{8.39}$$

This leads to the defining representation of the group $Sp(2n, \mathbb{R})$:

$$Sp(2n, \mathbb{R}) = \{S = \text{real } 2n \times 2n \text{ matrix} \mid S \beta S^T = \beta\}. \tag{8.40}$$

If in place of β we had the unit matrix, this would have been the orthogonal group $O(2n)$. Thus in a sense $Sp(2n, \mathbb{R})$ is 'diametrically opposite' to the group of rotations in Euclidean $2n$-dimensional space. Indeed, symplectic geometry, based on the matrix β viewed as metric, is strikingly different from Euclidean geometry and exists only in even dimensional (real) spaces. Yet it is basic to the study of Cartesian canonical variables, their PB's and CCR's.

Here are some important properties derivable from Eqs. (8.39, 8.40):

(i) $Sp(2n, \mathbb{R})$ is of dimension $n(2n + 1)$, as we confirm later.

(ii) $\beta \in Sp(2n, \mathbb{R})$: it basically interchanges positions and momenta, so in quantum mechanics it corresponds to Fourier transformation. Then $\beta^2 = -\mathbb{1} \in Sp(2n, \mathbb{R})$ corresponds to the parity operation $\hat{P}$.

(iii) $S \in Sp(2n, \mathbb{R}) \Rightarrow S^T, \ S^{-1} = \beta S^T \beta, \ -S \in Sp(2n, \mathbb{R})$.

(iv) $\det S = +1$: This is particularly interesting, as (8.39) leads in the first instance to $\det S = \pm 1$. The proof that $\det S = +1$ is quite subtle, and we indicate it later.

(v) $S \in Sp(2n, \mathbb{R}) \Rightarrow$ eigenvalue spectrum of S is invariant under reflection about the real axis, and through the unit circle

$$\left(re^{i\theta} \rightarrow \frac{1}{r} e^{i\theta} \right); \tag{8.41}$$

the eigenvalues ± 1 have even multiplicities.

Property (i) will emerge easily at the Lie algebra level. While (ii), (iii) are easy, (v) is a consequence of S and $(S^{-1})^T$ being real and related by a similarity transformation.

For some purposes it is useful to write S in terms of four block matrices each of dimension n:

$$S = \begin{pmatrix} A & B \\ C & D \end{pmatrix} \in Sp(2n, \mathbb{R}) \Leftrightarrow$$

$$AB^T, \ CD^T \ \text{symmetric}, \ AD^T - BC^T = \mathbb{1}_{n \times n} \Leftrightarrow$$

$$\text{Since } \ S^T \in Sp(2n, \mathbb{R}) \ \text{too,}$$

$$A^T C, \ B^T D \ \text{symmetric}, \ A^T D - C^T B = \mathbb{1}_{n \times n}. \tag{8.42}$$

(These are the generalisations of the simple determinant condition in Eq. (8.6) for $n = 1$.) While it is easy to check (as mentioned above) that $S \in Sp(2n, \mathbb{R})$ implies $S^T \in Sp(2n, \mathbb{R})$, it is not so easy to pass directly from the first set of conditions above to the second set, aside from reconstituting A, B, C, D into S and then passing to S^T.

Complex form of $Sp(2n, \mathbb{R})$

The β matrix reflects the precise way in which the real q's and p's have been put together in (8.35) to form the $2n$-component ξ with real entries. Sometimes it is convenient, for instance in dealing with modes of the radiation field, to work with the n-dimensional nonhermitian version of (8.26):

$$\hat{a}_j = \frac{1}{\sqrt{2}}(\hat{q}_j + i\hat{p}_j), \ \hat{a}_j^\dagger = \frac{1}{\sqrt{2}}(\hat{q}_j - i\hat{p}_j), \ j = 1, 2, \cdots, n. \tag{8.43}$$

We arrange these into a new column vector $\hat{\xi}^{(c)}$ with nonhermitian entries:

$$\hat{\xi}^{(c)} = (\hat{\xi}_a^{(c)}) = (\hat{a}_1, \cdots, \hat{a}_n, \hat{a}_1^\dagger, \cdots, \hat{a}_n^\dagger)^T = \Omega\hat{\xi},$$

$$\hat{\xi} = \Omega^\dagger \hat{\xi}^{(c)},$$

$$\Omega = \frac{1}{\sqrt{2}}\begin{pmatrix} \mathbb{1} & i\mathbb{1} \\ \mathbb{1} & -i\mathbb{1} \end{pmatrix}, \quad \Omega^{-1} = \Omega^\dagger = \frac{1}{\sqrt{2}}\begin{pmatrix} \mathbb{1} & \mathbb{1} \\ -i\mathbb{1} & i\mathbb{1} \end{pmatrix}. \tag{8.44}$$

Then the CCR's in (8.36) can be expressed in two equivalent ways:

$$[\hat{\xi}_a^{(c)}, \hat{\xi}_b^{(c)}] = \beta_{ab},$$

$$[\hat{\xi}_a^{(c)}, \hat{\xi}_b^{(c)\dagger}] = (\Sigma_3)_{ab},$$

$$\Sigma_3 = \begin{pmatrix} \mathbb{1} & 0 \\ 0 & -\mathbb{1} \end{pmatrix}. \tag{8.45}$$

Now when we subject $\hat{\xi}$ to the real transformation $S \in Sp(2n, \mathbb{R})$, $\hat{\xi}^{(c)}$ experiences an equivalent complex transformation:

$$\hat{\xi}' = S\hat{\xi} \Leftrightarrow \hat{\xi}'^{(c)} = S^{(c)}\hat{\xi}^{(c)},$$

$$S^{(c)} = \Omega S \Omega^{-1} = \frac{1}{2}\begin{pmatrix} A + D + i(C - B) & A - D + i(B + C) \\ A - D - i(B + C) & A + D + i(B - C) \end{pmatrix} \tag{8.46}$$

So $S^{(c)}$ is just a convenient complex form of the real transformation S, much like the passage from Cartesian to spherical components of spherical tensors.

8.2.1 Useful subgroups of $Sp(2n, \mathbb{R})$

We now identify several interesting and useful subgroups. (Remember that unlike $O(2n, \mathbb{R})$, $Sp(2n, \mathbb{R})$ is noncompact.) We describe them in terms of the block matrices A, B, C, D. Some of them have not appeared in earlier chapters.

(a) $GL(n, \mathbb{R})$: This is the group of all $n \times n$ real nonsingular matrices, 'embedded' in $Sp(2n, \mathbb{R})$ in this way:

$$A \in GL(n, \mathbb{R}), \quad B = C = 0, \quad D = (A^{-1})^T, \tag{8.47}$$

so the q's transform linearly among themselves, the p's transform contragrediently, and the PB's and CCR's are preserved.

(b) $O(n, \mathbb{R})$: This is that part of $GL(n, \mathbb{R})$ in which q's and p's change in the same manner,

$$A = D \in O(n, \mathbb{R}), \quad B = C = 0. \tag{8.48}$$

These are thus real orthogonal rotations in configuration space.

(c) $U(n)$: This is the maximal compact subgroup in $Sp(2n, \mathbb{R})$. We display it fully:

$$U = X - i\Upsilon \in U(n), \quad U^\dagger U = UU^\dagger = \mathbb{1}_{n \times n}:$$

$$S(X, \Upsilon) = \begin{pmatrix} X & \Upsilon \\ -\Upsilon & X \end{pmatrix} \in Sp(2n, \mathbb{R}),$$

$$A = D = X, \quad B = -C = \Upsilon. \tag{8.49}$$

Note that while the determinant of $U = X - i\Upsilon$ need not be real, that of $S(X, \Upsilon)$ always assumes the value $+1$. We see from Eq. (8.46) that in the complex form

$$S^{(c)}(X, \Upsilon) = \begin{pmatrix} U & 0 \\ 0 & U^* \end{pmatrix}, \tag{8.50}$$

corresponding to transformations of the $\hat{a}_j$ among themselves, and of the $\hat{a}_j^\dagger$ by the complex conjugate transformations.

It is important to realise that $O(2n, \mathbb{R})$ is not contained in $Sp(2n, \mathbb{R})$. What is true is that

$$O(2n, \mathbb{R}) \cap Sp(2n, \mathbb{R}) = U(n) = \{S(X, \Upsilon)\}. \tag{8.51}$$

Thus, $U(n)$ is that subgroup of phase space rotations which are also canonical, i.e., preserve classical PB's and quantum mechanical CCR's.

Among the three subgroups so far described, we have the expected relation

$$O(n, \mathbb{R}) = GL(n, \mathbb{R}) \cap U(n). \tag{8.52}$$

(d) $T^{(f)}$: This is the subgroup in which the p's are unchanged, so we call it the 'free propagation' subgroup:

$$A = D = \mathbb{1}, \quad B = B^T, \quad C = 0:$$

$$\hat{q}' = \hat{q} + B\hat{p}, \quad \hat{p}' = \hat{p}. \tag{8.53}$$

On account of (8.42), B has to be symmetric.

(e) $T^{(l)}$: This is the conjugate of $T^{(f)}$, now the q's are unchanged:

$$A = D = \mathbb{1}, \; B = 0, \; C = C^T:$$

$$\hat{q}' = \hat{q}, \hat{p}' = \hat{p} + C\hat{q}. \tag{8.54}$$

From applications in optics, this is called the 'lens' subgroup.

(f) $\mathcal{A}$: This is the diagonal abelian subgroup of independent reciprocal scale changes in each canonical pair q_j, p_j:

$$A = \mathrm{diag}(\kappa_1, \kappa_2, \cdots, \kappa_n), \; \kappa_j > 0; \; B = C = 0;$$

$$D = \mathrm{diag}(\kappa_1^{-1}, \kappa_2^{-1}, \cdots, \kappa_n^{-1}),$$

$$S = \mathcal{A}(\boldsymbol{\kappa}) = \mathrm{diag}(\kappa_1, \cdots, \kappa_n, \kappa_1^{-1}, \cdots, \kappa_n^{-1}). \tag{8.55}$$

(g) $\mathcal{N}$: this is the nilpotent subgroup important for the global Iwasawa decomposition. Here the choices are:

$$A = \begin{pmatrix} 1 & & & \\ \vdots & 1 & & \mathbf{0} \\ & & \ddots & \\ \cdots\cdots & & & 1 \end{pmatrix}, \; B = 0, \; D = (A^{-1})^T, \; A^T C \text{ symmetric};$$

$$S = \mathcal{N}(A, C) = \begin{pmatrix} A & 0 \\ C & (A^{-1})^T \end{pmatrix}. \tag{8.56}$$

The dimensionalities of these subgroups are:

$$GL(n, \mathbb{R}), \; U(n), \; \mathcal{N} \;\text{———}\; n^2;$$

$$T^{(f)}, \; T^{(l)} \;\text{———}\; \frac{1}{2}n(n+1);$$

$$O(n, \mathbb{R}) \;\text{———}\; \frac{1}{2}n(n-1);$$

$$\mathcal{A} \;\text{———}\; n. \tag{8.57}$$

8.2.2 Global decompositions for $Sp(2n, \mathbb{R})$

We present four useful ways of expressing any $S \in Sp(2n, \mathbb{R})$ as a product of specially chosen factors, either two or three in number.

(a) **Polar decomposition:** Here any $S \in Sp(2n, \mathbb{R})$ can be expressed uniquely as the product of two factors, one from the maximal compact subgroup $K = U(n)$, and the other belonging to an important subset (not a subgroup!) $\pi(n)$ defined in this way:

$$\pi(n) = \{S \in Sp(2n, \mathbb{R}) \mid S^T = S, \ S \text{ positive definite}\} \subset Sp(2n, \mathbb{R}). \quad (8.58)$$

The decomposition reads:

$$S \in Sp(2n, \mathbb{R}): \ S = S(X, \Upsilon)P \text{ uniquely,}$$

$$S(X, \Upsilon) \in K, \ \ P \in \pi(n). \quad (8.59)$$

By conjugating P with $S(X, \Upsilon)$ one could write the two factors in the opposite sequence. This is a globally valid decomposition, and it helps us see that of the two possibilities $\det S = \pm 1$ allowed by (8.39), the choice $\det S = 1$ is selected.

(b) **Euler decomposition:** Here we have three factors, each taken from a subgroup of $Sp(2n, \mathbb{R})$, but there is some nonuniqueness in the choice of factors. Two factors belong to K, the third from those elements of $\pi(n)$ which are diagonal and do form a subgroup. These last are just the reciprocal scale changes denoted by $\mathcal{A}(\kappa)$ in Eq. (8.55). The decomposition is:

$$S \in Sp(2n, \mathbb{R}): S = S(X_1, \Upsilon_1)\mathcal{A}(\kappa)S(X_2, \Upsilon_2). \quad (8.60)$$

If we add the numbers of free parameters in the factors, one gets the dimension $n(2n+1)$ of $Sp(2n, \mathbb{R})$. This means that the nonuniqueness in this decomposition is of a discrete, not a continuous, nature. It arises essentially from the freedom to order the first n diagonal elements in $\mathcal{A}(\kappa)$ in various ways.

(c) **Pre-Iwasawa decomposition:** In Eq. (8.9) we described the Iwasawa decomposition theorem for $Sp(2, \mathbb{R})$. An analogous theorem holds for $Sp(2n, \mathbb{R})$ as well, but as a first step towards it we find another useful result. This is a unique three factor decomposition, which arises from the attempt to reduce the off-diagonal block B in a general $S \in Sp(2n, \mathbb{R})$ to zero by using an element of $K = U(n)$ on the right. The result reads:

$$S = \begin{pmatrix} A & B \\ C & D \end{pmatrix} = \begin{pmatrix} \mathbb{1} & 0 \\ C_0 A_0^{-1} & \mathbb{1} \end{pmatrix} \begin{pmatrix} A_0 & 0 \\ 0 & A_0^{-1} \end{pmatrix} \begin{pmatrix} X & \Upsilon \\ -\Upsilon & X \end{pmatrix},$$

$$A_0 = (AA^T + BB^T)^{1/2},$$

$$X - i\Upsilon = A_0^{-1}(A - iB), \ C_0 = (CA^T + DB^T)A_0^{-1}. \quad (8.61)$$

Here A_0 is to be chosen symmetric positive definite, and all factors are then unique. It can be checked that $C_0 A_0^{-1}$ is symmetric, so the first factor belongs to the 'lens' subgroup $T^{(l)}$ of Eq. (8.54). The second factor lies in the intersection $GL(n, \mathbb{R}) \cap \pi(n)$, which is not a subgroup. This particular decomposition is of importance in obtaining the n-dimensional generalised Huyghens kernel, which we describe later.

(d) **Iwasawa decomposition:** The polar and the pre-Iwasawa decompositions are similar since they involve unique factors, but each factor may not be taken from a subgroup. The Euler decomposition (8.60) solves this problem, but in the process uniqueness is lost. The Iwasawa decomposition possesses both virtues: it is global and involves unique factors each from a characteristic subgroup. The n-dimensional generalisation of Eq. (8.9) says that any $S \in Sp(2n, \mathbb{R})$ can be expressed uniquely as a product of three factors

$$S = \begin{pmatrix} A & 0 \\ C & (A^{-1})^T \end{pmatrix} \mathcal{A}(\boldsymbol{\kappa}) S(X, \Upsilon), \tag{8.62}$$

belonging to the subgroups $\mathcal{N}$ of Eq. (8.56), $\mathcal{A}$ of Eq. (8.55) and $K = U(n)$ of Eq. (8.49), respectively. The dimensionalities of these subgroups, n^2, n and n^2, add up properly to $n(2n + 1)$.

8.2.3 $Sp(2n, \mathbb{R})$ Lie algebra in the defining representation and in general

Let us take a matrix S close to the identity, and examine Eq. (8.39):

$$S \simeq \mathbb{1} - i\epsilon J, \; S\beta S^T = \beta \Rightarrow J\beta = -\beta J^T,$$

$$\text{i.e., } (\beta J)^T = \beta J \text{ or } (J\beta)^T = J\beta. \tag{8.63}$$

In addition, since we have extracted the quantum mechanical i, J is pure imaginary. Thus in the defining representation we get all generators by taking all real symmetric $2n \times 2n$ matrices and multiplying them on the left or on the right by $i\beta$. Let us take the former route and use the natural choice of independent real symmetric matrices. Then the simplest basis for $Sp(2n, \mathbb{R})$ in the defining representation is

$$X_{ab}^{(0)} = X_{ba}^{(0)}, \; a, b = 1, 2, \cdots, 2n:$$

$$(X_{ab}^{(0)})_{cd} = i(\delta_{ad}\beta_{cb} + \delta_{bd}\beta_{ca}). \tag{8.64}$$

We see that the number of generators, hence the order of $Sp(2n, \mathbb{R})$, is $n(2n+1)$. The commutation relations are easily obtained:

$$[X_{ab}^{(0)}, X_{cd}^{(0)}] = i(\beta_{ac} X_{bd}^{(0)} + \beta_{bc} X_{ad}^{(0)} + \beta_{ad} X_{cb}^{(0)} + \beta_{bd} X_{ca}^{(0)}). \tag{8.65}$$

These are the $Sp(2n, \mathbb{R})$ Lie algebra relations, and a general generator matrix J is a real linear combination of $X_{ab}^{(0)}$.

It follows that in any representation of $Sp(2n, \mathbb{R})$, finite or infinite dimensional, unitary or not, the generators $X_{ab} = X_{ba}$ obey

$$[X_{ab}, X_{cd}] = i(\beta_{ac} X_{bd} + \beta_{bc} X_{ad} + \beta_{ad} X_{cb} + \beta_{bd} X_{ca}). \tag{8.66}$$

The adjoint transformation law valid in any representation is

$$S \in Sp(2n, \mathbb{R}): D(S) X_{ab} D(S)^{-1} = S_{ac}^{-1} S_{bd}^{-1} X_{cd}. \tag{8.67}$$

To go further it is useful to bring in split index notation:

$$a, b, \cdots = 1, 2, \cdots, 2n: a \to r\alpha, \ b \to s\beta, \cdots, \ r, s = 1, \cdots, n; \ \alpha, \beta = 1, 2;$$

$$a = 1, 2, \cdots, n: r = 1, 2, \cdots, n; \ \alpha = 1;$$

$$a = n+1, n+2, \cdots, 2n: r = 1, 2, \cdots, n; \ \alpha = 2; \tag{8.68}$$

Thus $r, s, \cdots$ label the canonical pairs, while $\alpha, \beta = 1, 2$ correspond to q and p within a pair. In this notation β is

$$\beta_{ab} \equiv \beta_{r\alpha, s\beta} = \delta_{rs} \epsilon_{\alpha\beta}, \ \epsilon = i\sigma_2 = \begin{pmatrix} 0 & 1 \\ -1 & 0 \end{pmatrix}. \tag{8.69}$$

We then split up X_{ab} in this way:

$$X_{ab} \equiv X_{r\alpha, s\beta}: X_{r1, s1} = V_{rs} = V_{sr};$$

$$X_{r1, s2} = W_{rs};$$

$$X_{r2, s2} = Z_{rs} = Z_{sr}. \tag{8.70}$$

There are $\frac{1}{2}n(n+1)$ independent V's, n^2 independent W's, and $\frac{1}{2}n(n+1)$ independent Z's, adding up to $n(2n+1)$. For $Sp(2, \mathbb{R})$, we have $r = s = 1$, and then

$$V = -2(J_0 + K_2), \ W = 2K_1, \ Z = 2(K_2 - J_0). \tag{8.71}$$

In the notations (8.70), the commutation relations (8.66) break up as follows:

$$[W_{rs}, W_{uv}] = i(\delta_{rv} W_{us} - \delta_{us} W_{rv});$$

$$[W_{rs}, V_{uv}] = -i(\delta_{us} V_{rv} + \delta_{vs} V_{ru});$$

$$[W_{rs}, Z_{uv}] = i(\delta_{ru} Z_{sv} + \delta_{rv} Z_{su});$$

$$[V_{rs}, Z_{uv}] = i(\delta_{ru} W_{sv} + \delta_{su} W_{rv} + \delta_{rv} W_{su} + \delta_{sv} W_{ru});$$

$$[V, V] = [Z, Z] = 0. \tag{8.72}$$

For the subgroups listed earlier the generators can be identified:

Table 8.1   Some important subalgebras of $Sp(2n, \mathbb{R})$

Subgroup	Generators
$GL(n, \mathbb{R})$	W_{rs}
$SO(n, \mathbb{R})$	$J_{rs} = W_{sr} - W_{rs}$
$U(n)$	$J_{rs}, Q_{rs} = V_{rs} + Z_{rs}$ or $A_{rs} = \frac{1}{2}(Q_{rs} - iJ_{rs})$
$T^{(f)}$	Z_{rs}
$T^{(l)}$	V_{rs}
$\mathcal{A}$	$W_{rr}, \ r = 1, 2, \cdots, n$
$\mathcal{N}$	W_{rs} for $r < s$, and all V_{rs}

Since $Sp(2n, \mathbb{R})$ is noncompact (and simple), any nontrivial finite dimensional representation is necessarily nonunitary. On the other hand, $U(n)$ is the maximal compact subgroup, all of whose representations are unitary. We can analyse $Sp(2n, \mathbb{R})$ from this point of view, that is, in a $U(n)$-adapted form. The generators outside of $U(n)$ can be arranged into complex combinations with definite tensor behaviours under $U(n)$. These combinations are:

$$T_{rs} = T_{sr} = V_{rs} - Z_{rs} - i(W_{rs} + W_{sr}),$$

$$\bar{T}_{rs} = \bar{T}_{sr} = V_{rs} - Z_{rs} + i(W_{rs} + W_{sr}). \tag{8.73}$$

Then the $U(n)$-adapted form of the Lie algebra relations (8.72) is:

$$[A_{rs}, A_{uv}] = \delta_{su} A_{rv} - \delta_{rv} A_{us};$$

$$[A_{rs}, T_{uv}] = \delta_{su} T_{rv} + \delta_{sv} T_{ru};$$

$$[A_{rs}, \bar{T}_{uv}] = -\delta_{ru} \bar{T}_{sv} - \delta_{rv} \bar{T}_{su};$$

$$[T_{rs}, \bar{T}_{uv}] = -4(\delta_{ru} A_{sv} + \delta_{rv} A_{su} + \delta_{su} A_{rv} + \delta_{sv} A_{ru});$$

$$[T, T] = [\bar{T}, \bar{T}] = 0. \tag{8.74}$$

These forms mean that T_{rs} and $\bar{T}_{rs}$ are second rank symmetric tensors under $U(n)$, of contravariant and covariant types, respectively. The overall picture of the hermiticity properties of the generators is as follows:

Table 8.2 Hermiticity properties of $Sp(2n, \mathbb{R})$ generators

	$W_{rs} - W_{sr}$	$W_{rs} + W_{sr}$	$V_{rs} + Z_{rs}$	$V_{rs} - Z_{rs}$	$A_{rs}^{\dagger}$	$T_{rs}^{\dagger}$
Finite dimensional representations	hermitian	anti hermitian	hermitian	anti hermitian	A_{sr}	$-\bar{T}_{rs}$
Infinite dimensional UR	hermitian	hermitian	hermitian	hermitian	A_{sr}	$\bar{T}_{rs}$

In addition, there are infinite dimensional nonunitary representations.

8.2.4 The metaplectic group $Mp(2n)$, actions on $\hat{q}$'s and $\hat{p}$'s

Now we go back to Eq. (8.38) in the quantum case. By the Stone–von Neumann theorem, the irreducible $\hat{q}'$'s and $\hat{p}'$'s are unitarily related to the original $\hat{q}$'s and $\hat{p}$'s. For any $S, S' \in Sp(2n, \mathbb{R})$:

$$\hat{\xi}_a' = S_{ab}\hat{\xi}_b = \bar{U}(S)^{-1}\hat{\xi}_a \bar{U}(S),$$

$$\bar{U}(S)^{\dagger}\bar{U}(S) = \bar{U}(S)\bar{U}(S)^{\dagger} = \mathbb{1};$$

$$\bar{U}(S')\bar{U}(S) = \pm\bar{U}(S'S). \tag{8.75}$$

After using the freedom in the choice of a phase factor in each $\bar{U}(S)$ to simplify the composition law as much as possible, a $\pm$ sign ambiguity remains. Thus these operators $\bar{U}(S)$ (inappropriately written!) actually provide a true faithful UR of a group $Mp(2n)$ which is a double covering of $Sp(2n, \mathbb{R})$. The hermitian generators of this UR are

$$\widehat{W}_{rs} = \frac{1}{2}\{\hat{q}_r, \hat{p}_s\}, \quad \widehat{V}_{rs} = \hat{q}_r\hat{q}_s, \quad \widehat{Z}_{rs} = \hat{p}_r\hat{p}_s, \tag{8.76}$$

so they make up all hermitian quadratic expressions in $\hat{q}$'s and $\hat{p}$'s.

As with $Sp(2, \mathbb{R})$ for one degree of freedom, the $Mp(2n)$ generators (8.76) commute with the parity operator $\hat{P}$ which reverses the signs of all $\hat{\xi}_a$. Therefore we can decompose $\mathcal{H} = L^2(\mathbb{R}^n)$ of Eq. (8.37) into the two orthogonal subspaces $\mathcal{H}^{(\pm)}$ of even/odd parity wave functions, and they then carry two inequivalent UIR's of $Mp(2n)$. (As was mentioned in the $Sp(2, \mathbb{R})$ case, here too the matrix $-\mathbb{1}$ belongs to $Sp(2n, \mathbb{R})$, lies in its centre, and is realised as parity $\hat{P}$ in the metaplectic representation.)

Among the $Mp(2n)$ generators (8.76), there are some 'compact' and the remaining 'noncompact' combinations. Their characteristic differences show up if we express them in terms of the nonhermitian $\hat{a}$'s and $\hat{a}^\dagger$'s: referring to Table 8.2,

$$\text{Compact generators}: \ \widehat{W}_{rs} - \widehat{W}_{sr} = i(\hat{a}_s^\dagger \hat{a}_r - \hat{a}_r^\dagger \hat{a}_s),$$

$$\widehat{V}_{rs} + \widehat{Z}_{rs} = (\hat{a}_r^\dagger \hat{a}_s + \hat{a}_s^\dagger \hat{a}_r + \delta_{rs}),$$

$$\text{noncompact generators}: \ \widehat{W}_{rs} + \widehat{W}_{sr} = i(\hat{a}_r^\dagger \hat{a}_s^\dagger - \hat{a}_r \hat{a}_s),$$

$$\widehat{V}_{rs} - \widehat{Z}_{rs} = (\hat{a}_r^\dagger \hat{a}_s^\dagger + \hat{a}_r \hat{a}_s). \tag{8.77}$$

From Table 8.1, we can see that the compact generators belong to $U(n)$ and conserve 'total photon number', meaning that they commute with the operator $\widehat{N} = \hat{a}_r^\dagger \hat{a}_r$. For this reason, $U(n)$ transformations are said to be 'passive'. On the other hand, the noncompact generators do not conserve $\widehat{N}$, are used to construct the tensor operators T_{rs}, $\bar{T}_{rs}$ in Eq. (8.73), and are antihermitian in finite dimensional representations.

Going further, we can see that in the relevant UR of $Mp(2n)$, single exponentials of i times real linear combinations of the compact generators give us operators of the form $\bar{U}(S(X, Y))$; while single exponentials of i times real linear combinations of the noncompact generators give us operators of the form $\bar{U}(P)$, $P \in \pi(n)$ of Eq. (8.58). For this reason the latter are called 'squeezing transformations' in the quantum optical context. Then the global polar decomposition (8.59) says that any metaplectic unitary transformation is uniquely the product of a compact passive factor and a noncompact 'active' factor.

8.2.5 The generalised Huyghens kernel in n dimensions

The generalised Huyghens kernel for $n = 1$ has been presented in Eq. (8.34). To obtain the corresponding result for any n, we may use the pre-Iwasawa decomposition (8.61). In the metaplectic representation, certain elements of $Sp(2n, \mathbb{R})$ act very simply in the Schrödinger representation (8.37) of the $\hat{q}$'s and $\hat{p}$'s. We first list these results:

$$A \in GL(n, \mathbb{R}), \ S(A) = \begin{pmatrix} A & 0 \\ 0 & (A^{-1})^T \end{pmatrix}:$$

$$\bar{U}(S(A))|\boldsymbol{q}\rangle = |\det A|^{1/2}|A\boldsymbol{q}\rangle,$$

$$\langle \boldsymbol{q}|\bar{U}(S(A)) = |\det A|^{-1/2}\langle A^{-1}\boldsymbol{q}|;$$

$$\mathcal{A}(\boldsymbol{\kappa}) \in \mathcal{A}: \ \bar{U}(\mathcal{A}(\boldsymbol{\kappa})) = \exp\left(-i\sum_{r=1}^{n} \ln \kappa_r \cdot \widehat{W}_{rr}\right),$$

$$\bar{U}(\mathcal{A}(\boldsymbol{\kappa}))|\boldsymbol{q}\rangle = \left(\prod_{r=1}^{n} \kappa_r\right)^{1/2} |\kappa_1 q_1, \cdots, \kappa_n q_n\rangle,$$

$$\langle\boldsymbol{q}|\bar{U}(\mathcal{A}(\boldsymbol{\kappa})) = \left(\prod_{r=1}^{n} \kappa_r\right)^{-1/2} \langle\kappa_1^{-1} q_1, \cdots, \kappa_n^{-1} q_n|;$$

$$L(C) = \begin{pmatrix} \mathbb{1} & 0 \\ C & \mathbb{1} \end{pmatrix} \in T^{(l)}, \quad C^T = C:$$

$$\bar{U}(L(C)) = \exp\left(\frac{i}{2} C_{rs} \widehat{V}_{rs}\right),$$

$$\bar{U}(L(C))|\boldsymbol{q}\rangle = \exp\left(\frac{i}{2}\boldsymbol{q}^T C \boldsymbol{q}\right)|\boldsymbol{q}\rangle. \tag{8.78}$$

Using these special cases of $\bar{U}(S)$ and Eq. (8.61), the generalised n-dimensional Huyghens kernel can be computed fairly easily. We present the result only for the generic case where the submatrix B is nonsingular:

$$S = \begin{pmatrix} A & B \\ C & D \end{pmatrix} \in Sp(2n, \mathbb{R}), \ \det B \neq 0:$$

$$\langle\boldsymbol{q}'|\bar{U}(S)|\boldsymbol{q}\rangle = \frac{e^{-i\pi n/4}}{|\det B|^{1/2}} \exp\left\{\frac{i}{2}\left(\boldsymbol{q}'^T DB^{-1}\boldsymbol{q}' - 2\boldsymbol{q}^T B^{-1}\boldsymbol{q}' + \boldsymbol{q}^T B^{-1}A\boldsymbol{q}\right)\right\}. \tag{8.79}$$

When $\det B = 0$, the kernel collapses to a lower dimensional expression with several delta function factors.

8.3 Quantum Variance Matrices, $Sp(2n, \mathbb{R})$ Invariant Uncertainty Principles

Recall the usual statement of the uncertainty principle for quantum systems based on one degree of freedom. If $\hat{\rho}$ is the density matrix of a general state, the means and variances – or uncertainties – of coordinate and momentum are

$$\langle\hat{q}\rangle = \mathrm{Tr}(\hat{\rho}\hat{q}), \quad \langle\hat{p}\rangle = \mathrm{Tr}(\hat{\rho}\hat{p}),$$

$$(\Delta q)^2 = \mathrm{Tr}(\hat{\rho}(\hat{q} - \langle\hat{q}\rangle)^2), \ (\Delta p)^2 = \mathrm{Tr}(\hat{\rho}(\hat{p} - \langle\hat{p}\rangle)^2). \tag{8.80}$$

Then the familiar Heisenberg form of the uncertainty principle is

$$(\Delta q)^2 (\Delta p)^2 \geq \frac{1}{4}. \tag{8.81}$$

It is easy to see that of all the transformations of $Sp(2,\mathbb{R})$ that can act on $\hat{q}$ and $\hat{p}$, this statement is invariant under reciprocal scale transformations alone apart from Fourier transformation which essentially exchanges q and p.

A more stringent uncertainty principle, often called the Schrödinger form, is also known. It involves one more quantity of the nature of a cross term:

$$\Delta(q,p) = \frac{1}{2}\mathrm{Tr}(\hat{\rho}\{\hat{q} - \langle \hat{q} \rangle, \, \hat{p} - \langle \hat{p} \rangle\})$$

$$= \frac{1}{2}\mathrm{Tr}(\hat{\rho}\{\hat{q}, \hat{p}\}) - \langle \hat{q} \rangle \langle \hat{p} \rangle, \tag{8.82}$$

and the statement is best given in terms of the 'variance matrix' for the state $\hat{\rho}$:

$$V(\hat{\rho}) = \begin{pmatrix} (\Delta q)^2 & \Delta(q,p) \\ \Delta(q,p) & (\Delta p)^2 \end{pmatrix},$$

$$\det V(\hat{\rho}) = (\Delta q)^2 (\Delta p)^2 - (\Delta(q,p))^2 \geq \frac{1}{4}. \tag{8.83}$$

This form of the uncertainty principle implies the more familiar (8.81); more importantly, it is $Sp(2,\mathbb{R})$ invariant, as we will soon realise.

Let us now sketch the generalisation to n degrees of freedom. We will allow elements of $Sp(2n,\mathbb{R})$ to act (via the metaplectic representation) on density matrices $\hat{\rho}$ according to the rule

$$S \in Sp(2n,\mathbb{R}): \hat{\rho}' = \bar{U}(S)\hat{\rho}\,\bar{U}(S)^{-1}. \tag{8.84}$$

The means and variances of the canonical operators in the state $\hat{\rho}$ are:

$$\langle \hat{\xi}_a \rangle = \mathrm{Tr}(\hat{\rho}\hat{\xi}_a);$$

$$V_{ab}(\hat{\rho}) = \frac{1}{2}\mathrm{Tr}(\hat{\rho}\{\hat{\xi}_a - \langle \hat{\xi}_a \rangle, \hat{\xi}_b - \langle \hat{\xi}_b \rangle\})$$

$$= \frac{1}{2}\mathrm{Tr}(\hat{\rho}\{\hat{\xi}_a, \hat{\xi}_b\}) - \langle \hat{\xi}_a \rangle \langle \hat{\xi}_b \rangle. \tag{8.85}$$

Taken together, these variances make up the $2n \times 2n$ variance matrix $V(\hat{\rho})$ associated with the state $\hat{\rho}$. It is clear that $V(\hat{\rho})$ is real symmetric positive semidefinite.

This $2n \times 2n$ matrix can be broken up into four $n \times n$ blocks, to show the appearance of coordinates and momenta:

$$V(\hat{\rho}) = \begin{pmatrix} V_1(\hat{\rho}) & V_2(\hat{\rho}) \\ V_2(\hat{\rho})^T & V_3(\hat{\rho}) \end{pmatrix}$$

$$(V_1(\hat{\rho}))_{rs} = \text{Tr}(\hat{\rho}\hat{q}_r\hat{q}_s) - \langle\hat{q}_r\rangle\langle\hat{q}_s\rangle,$$

$$(V_2(\hat{\rho}))_{rs} = \frac{1}{2}\text{Tr}(\hat{\rho}\{\hat{q}_r, \hat{p}_s\}) - \langle\hat{q}_r\rangle\langle\hat{p}_s\rangle,$$

$$(V_3(\hat{\rho}))_{rs} = \text{Tr}(\hat{\rho}\hat{p}_r\hat{p}_s) - \langle\hat{p}_r\rangle\langle\hat{p}_s\rangle. \tag{8.86}$$

This is clearly a multidimensional generalisation of the expressions in Eqs. (8.80, 8.82).

The effect of the transformation (8.84) on the variance matrix is immediate. From Eq. (8.75) we find for any $S \in Sp(2n, \mathbb{R})$:

$$\hat{\rho}' = \bar{U}(S)\hat{\rho}\bar{U}(S)^{-1}: \langle\hat{\xi}_a\rangle' = S_{ab}\langle\hat{\xi}_b\rangle,$$

$$V(\hat{\rho}') = SV(\hat{\rho})S^T. \tag{8.87}$$

Note that for general $S \in Sp(2n, \mathbb{R})$, this is *not* a similarity transformation but what is called a *congruence* transformation. It is a similarity only if $S = S(X, Y) \in U(n)$, the maximal compact subgroup of $Sp(2n, \mathbb{R})$.

With the help of the CCR's in the form (8.36) we see that

$$V_{ab}(\hat{\rho}) = \text{Tr}(\hat{\rho}\hat{\xi}_a\hat{\xi}_b) - \langle\hat{\xi}_a\rangle\langle\hat{\xi}_b\rangle - \frac{i}{2}\beta_{ab},$$

$$\text{i.e., } \text{Tr}(\hat{\rho}(\hat{\xi}_a - \langle\hat{\xi}_a\rangle)(\hat{\xi}_b - \langle\hat{\xi}_b\rangle)) = (V(\hat{\rho}) + \frac{i}{2}\beta)_{ab}. \tag{8.88}$$

From here it immediately follows that for any quantum mechanical state $\hat{\rho}$, since $\hat{\rho}$ is hermitian positive semidefinite,

$$V(\hat{\rho}) + \frac{i}{2}\beta \geq 0. \tag{8.89}$$

The matrix occurring here is hermitian, and we claim that this is an $Sp(2n, \mathbb{R})$-covariant statement of the uncertainty principles for n degrees of freedom. It is easy to check that for $n = 1$ this reduces precisely to the statement (8.83).

The proof of our claim depends on the fundamental theorem of Williamson, originally proved in 1936. This theorem studies the maximum simplification that can be achieved in a real symmetric $2n \times 2n$ matrix V if we are allowed to transform it to $V' = SVS^T$ for all $S \in Sp(2n, \mathbb{R})$. For general V, we may not be able to choose an S such that V' is diagonal; however, the theorem assures us that if V is positive semidefinite we can find S such that V' is diagonal. A particularly elegant and economical proof of Williamson's theorem may be found in [Simon et al. 1999].

Assume now that V' has been rendered diagonal. By suitable further rescaling and ordering of degrees of freedom we can reach the following conclusion: for V real symmetric positive semidefinite, for suitable $S \in Sp(2n, \mathbb{R})$

$$V' = SVS^T = \text{diag}(\kappa_1, \kappa_2, \cdots, \kappa_n, \ \kappa_1, \kappa_2, \cdots, \kappa_n),$$

$$0 < \kappa_1 \leq \kappa_2 \leq \cdots \leq \kappa_n. \qquad (8.90)$$

We may call this the *Williamson normal form* of V. Note that in general the κ_r are *not* the eigenvalues of V.

In the normal form (8.90), we have equal uncertainties $\Delta q_r = \Delta p_r = \kappa_r^{1/2}$ for the operators in each canonical pair, *and all off-diagonal variances vanish*. Therefore the complete statement of the uncertainty principles for all degrees of freedom reads

$$\kappa_r \geq \frac{1}{2}, \quad r = 1, 2, \cdots, n. \qquad (8.91)$$

Moreover, if these conditions are obeyed, the variance matrix V' can certainly be physically realised, for instance by choosing $\hat{\rho}'$ to be a product of suitable Gaussian pure states.

Putting all these results together, one can easily convince oneself that the condition (8.91) can be expressed directly in terms of the original variance matrix $V(\hat{\rho})$ without having to pass to its Williamson normal form, and the required statement is precisely the matrix inequality (8.89)!

8.4 $SO(2l)$ **Spinor UIR's and Metaplectic UR of** $Sp(2n, \mathbb{R})$ – **A Comparison**

We conclude this chapter with a somewhat intriguing comparison of the spinor UIR's of the groups $SO(2l)$ dealt with in Chapter 6, and the double-valued metaplectic UR of $Sp(2n, \mathbb{R})$ which has been described in this one. This is most conveniently done by compiling a list of important features characterising the spinor UIR's of $SO(2l)$, then a corresponding list involving $Sp(2n, \mathbb{R})$ and $Mp(2n)$. (We could have written $SO(2n)$ but use the notation already set up in this case.) This will reveal important similarities and of course differences.

$SO(2l)$ **Spinors**

(a) Order of $SO(2l)$ is $l(2l-1)$.

(b) Dirac algebra: $\{\gamma_A, \gamma_B\} = 2\delta_{AB}$,
$$\gamma_A^\dagger = \gamma_A, \text{ irreducible}, A, B = 1, 2, \cdots, 2l.$$

(c) Fact: up to unitary equivalence, there is only one irreducible representation, of dimension 2^l.

(d) Given γ_A, $\gamma_A' = -\gamma_A$ is necessarily unitarily implementable,
$$-\gamma_A = \gamma_F \gamma_A \gamma_F^{-1}, \quad \gamma_F = \gamma_F^{-1} = \gamma_F^\dagger.$$

(e) Any $S \in SO(2l)$: by (c),
$$\gamma_A' = S_{AB}\gamma_B = \text{ necessarily } \bar{U}(S)^{-1}\gamma_A \bar{U}(S),$$

for some unitary $\bar{U}(S)$ determined up to S-dependent phase, which can be adjusted so that
$$\bar{U}(S')\bar{U}(S) = \pm \bar{U}(S'S).$$

(f) Generators of $\bar{U}(S)$ are quadratic in γ's:
$$M_{AB} = \frac{i}{4}[\gamma_A, \gamma_B]$$

(g) By (d), both M_{AB} and $\bar{U}(S)$ commute with γ_F, resulting in the reduction into UIR's on the eigenspaces of γ_F:
$$\bar{U}(S) = \Delta^{(1)}(S) \oplus \Delta^{(2)}(S)$$

(h) The UIR's $\Delta^{(1)}(S)$, $\Delta^{(2)}(S)$ are inequivalent double-valued representations of $SO(2l)$, each of dimension 2^{l-1}.

We remark in passing that while in Chapter 6 the spinor UIR's were arrived at by constructing the generators M_{AB} obeying the **$SO(2l)$** Lie algebra relations, here we have used the fact (c) to conclude that the UR $\bar{U}(S)$ must exist.

Now let us make up a corresponding list for $Sp(2n, \mathbb{R})$.

Metaplectic UIR's of $Sp(2n, \mathbb{R})$

(a) Order of $Sp(2n, \mathbb{R})$ is $n(2n+1)$.

(b) Heisenberg algebra: $[\hat{\xi}_a, \hat{\xi}_b] = i\beta_{ab}$, $\hat{\xi}_a^\dagger = \hat{\xi}_a$, irreducible; $a, b = 1, 2, \cdots, 2n$.

(c) Fact: up to unitary equivalence (Stone–von Neumann) there is only one irreducible representation, of infinite dimension.

(d) Given $\hat{\xi}_a$, $\hat{\xi}'_a = -\hat{\xi}_a$ is necessarily unitarily implementable, $-\hat{\xi}_a = \hat{P}\hat{\xi}_a\hat{P}^{-1}$, $\hat{P} = \hat{P}^{-1} = \hat{P}^\dagger =$ parity operator.

(e) Any $S \in Sp(2n, \mathbb{R})$: by (c),

$$\hat{\xi}'_a = S_{ab}\hat{\xi}_b = \text{necessarily } \bar{U}(S)^{-1}\hat{\xi}_a\bar{U}(S),$$

for some unitary $\bar{U}(S)$ determined up to S-dependent phase, which can be adjusted so that

$$\bar{U}(S')\bar{U}(S) = \pm\bar{U}(S'S).$$

(f) Generators of $\bar{U}(S)$ are quadratic in $\hat{\xi}$'s:

$$\widehat{X}_{ab} = \frac{1}{2}\{\hat{\xi}_a, \hat{\xi}_b\},$$

as is clear from Eqs. (8.70, 8.76).

(g) By (d), both $\widehat{X}_{ab}$ and $\bar{U}(S)$ commute with $\hat{P}$, resulting in the reduction of $\bar{U}(S)$ into two inequivalent UIR's on the eigenspaces of $\hat{P}$ made up of even/odd parity wave functions $\psi(\mathbf{q})$.

(h) The two UIR's contained in $\bar{U}(S)$ are double valued representations of $Sp(2n, \mathbb{R})$, each of infinite dimension.

Problems

P8.1   For the matrices S of the group $Sp(2, \mathbb{R})$ show that

$$S\beta S^T = \beta \Leftrightarrow \det S = 1$$

P8.2   Verify the isomorphisms stated in Eq. (8.16) and the homomorphisms in Eq. (8.17).

P8.3   Show that the Lie algebra $Sp(4, \mathbb{R})$ is isomorphic to the Lie algebra $SO(3,2)$ of the real pseudo-orthogonal de Sitter group $SO(3,2)$. This is a noncompact analogue of the local isomorphism of $C_2 = USp(4)$ and $B_2 = SO(5)$ noted in Eq. (5.84), and has been used in Dirac (1963) and Simon et al. (1985).

P8.4   Use Williamson's theorem to show that any given real symmetric $2n \times 2n$ matrix V obeying Eq. (8.89) is the variance matrix of (in general many) quantum states $\hat{\rho}$, $V = V(\hat{\rho})$, of a quantum system involving n Cartesian canonical pairs of operators.

Bibliography

Arvind, Dutta, B., Mukunda, N. and Simon, R. (1995). The Real Symplectic Groups in Quantum Mechanics and Optics. *Pramana – Journal of Physics*. 45: 471.

Dirac, P. A. M. (1963). A Remarkable Representation of the 3 + 2 de Sitter Group. *J. Math. Phys.* 4: 901.

Simon, R., Sudarshan E. C. G. and Mukunda, N. (1985). Anisotropic Gaussian Schell-Model Beams: Passage through Optical Systems and Associated Invariants. *Phys. Rev. A.* 31: 2419.

Simon, R. and Mukunda, N. (1993). The Two-dimensional Symplectic and Meta-plectic Groups and Their Universal Cover, in Symmetries in Science–VI. *The Rotation Group to Quantum Algebras.* Gruber, B., ed. New York: Plenum Press. 659.

Simon, R., Chaturvedi, S. and Srinivasan, V. (1999). Congruences and Canonical Forms for Positive Matrices: Applications to the Schweinler–Wigner Extremum Principle. *J. Math. Phys.* 40: 3632

Weil, A. (1964). Sur certains groupes d'operateurs unitaires. *Acta Mathematica.* 111: 143.

9

Wigner's Theorem, Ray Representations and Neutral Elements

In this chapter we look at some aspects of group representation theory specifically pertaining to the 'needs' of quantum mechanics, and involving quite subtle features. We begin by recalling the basic mathematical – or perhaps better, the kinematical framework of quantum mechanics. With this preparation, we define the concept of a symmetry operation in quantum mechanics in the manner of Wigner, followed by a description of his celebrated unitary–antiunitary theorem. A proof of this theorem, well known for its elegance, was given by V. Bargmann in 1964. We indicate the structure of this proof, and then present two other recent proofs which afford considerable insight into the conceptualisation of symmetry in quantum mechanics.

Wigner's theorem leads us to examine ray representations, or representations up to phases, for Lie groups. Such representations have important consequences for the generator commutation relations, bringing in the concept of neutral elements and a certain degree of freedom or flexibility in the choice of generators. We study the extent to which this flexibility can be used to simplify, or possibly completely eliminate, neutral elements in the commutation relations associated with a given Lie group.

Neutral elements appear also in the context of realisations of Lie groups via canonical transformations in the phase space formalism of classical mechanics. Comparison of the situations in the two cases, classical and quantum mechanics helps us appreciate that whereas the origins of neutral elements are different, their algebraic properties are common.

9.1 Hilbert and Ray Space Descriptions of Pure Quantum States

Quantum mechanics uses vectors in a suitable complex Hilbert space $\mathcal{H}$ to describe the pure states – states of maximum possible information – of a quantum system. Depending on the system, the dimension of $\mathcal{H}$ may be finite or infinite. Each unit vector $\psi \in \mathcal{H}$ determines uniquely a certain pure physical state. However, this is a many-to-one rather than a one-to-one relationship, since for any phase α, the vector $e^{i\alpha}\psi$ determines the same pure state as ψ. Overall phases are unobservable and unphysical. In spite of this, the use of $\mathcal{H}$ is very convenient as one can express the *Superposition Principle* very simply. If $\psi_1, \psi_2, \cdots$, are any vectors in $\mathcal{H}$, each determining a corresponding pure state, and $c_1, c_2, \cdots$ are any complex numbers such that

$$\psi = c_1 \psi_1 + c_2 \psi_2 + \cdots \in \mathcal{H} \tag{9.1}$$

is nonzero, then ψ also determines, in general after normalisation, a certain pure state. Here we use the vector space nature of $\mathcal{H}$, and assume there are no superselection rules.

The inner product (ϕ, ψ) between two unit vectors in $\mathcal{H}$ thus cannot have an unambiguous physical meaning. However, the modulus squared of this scalar product has a physical meaning as a probability:

$$(\phi, \psi) = \text{unobservable probability amplitude},$$

$$|(\phi, \psi)|^2 = \text{observable 'transition' probability.} \tag{9.2}$$

There are then two options: either use the machinery of Hilbert space, permit the presence of unobservable overall phases, but express the Superposition Principle very simply; or eliminate these overall phases completely, construct mathematical objects which correspond to pure states in a one-to-one manner and yield the above probabilities more directly, but give the Superposition Principle a somewhat hidden quality. The best is to be able to work at both levels, and to pass between them whenever necessary. Here, we set up the necessary mathematical apparatus.

Starting from $\mathcal{H}$, we define the 'unit sphere' in $\mathcal{H}$ as the set of all unit vectors:

$$\mathcal{B} = \{\psi \in \mathcal{H} | (\psi, \psi) = 1\} \subset \mathcal{H}. \tag{9.3}$$

This is a subset, not a subspace, in $\mathcal{H}$. It is the map $\mathcal{B} \to$ pure states that is many-to-one. To convert this into a one-to-one map, we define the space of *unit rays* to be equivalence classes of phase related vectors $\{e^{i\alpha}\psi, \psi \in \mathcal{B}\}$. Thus a unit ray is the entire collection $\{e^{i\alpha}\psi\}$ where we keep $\psi \in \mathcal{B}$ fixed and let α run from 0 to 2π. Each such collection can be represented very simply and easily by the corresponding pure state density matrix

$$\rho(\psi) = |\psi\rangle\langle\psi| \text{ or } \psi\psi^{\dagger}, \quad \rho(e^{i\alpha}\psi) = \rho(\psi), \tag{9.4}$$

so we define

$$\mathcal{R} = \text{space of unit rays} = \{\rho(\psi) | \psi \in \mathcal{B}\}. \tag{9.5}$$

(We can, if needed, also define non-normalised rays starting from any nonzero ψ.) The map $\mathcal{R} \to$ pure states is clearly one-to-one, and in Bargmann's words,

Any significant statement in quantum theory is (therefore) a statement about unit rays.

The probabilities (9.2) are directly expressible as

$$|(\phi, \psi)|^2 = \text{Tr}(\rho(\phi)\rho(\psi)), \tag{9.6}$$

and the relation between $\mathcal{B}$ and $\mathcal{R}$ is that there is a projection map π from the former to the latter:

$$\pi : \mathcal{B} \to \mathcal{R} : \psi \in \mathcal{B} \to \pi(\psi) = \rho(\psi) = |\psi\rangle\langle\psi| \in \mathcal{R}. \tag{9.7}$$

Here are some simple and interesting facts. If $\mathcal{H}$ is of finite (complex) dimension N, then $\mathcal{B}$ is of real dimension $(2N - 1)$, while $\mathcal{R}$ is of real dimension $2(N - 1)$ which is always even:

$$\dim \mathcal{H} = N \Rightarrow \mathcal{B} \simeq \mathbb{S}^{2N-1}, \quad \mathcal{R} = CP^{N-1}. \tag{9.8}$$

(In the case $N = 2$, $\mathcal{B} \simeq \mathbb{S}^3$ and $\mathcal{R} \simeq \mathbb{S}^2$ is called the Poincaré or Bloch sphere.) Ray space $\mathcal{R}$ is a 'phase space' in the sense of classical mechanics, and carries a natural 'symplectic structure' leading to Poisson brackets. In addition, it is a Riemannian space, with a natural Riemannian metric called in this context the Fubini–Study metric.

However, the ray space $\mathcal{R}$ is somewhat difficult to visualise geometrically, as it is not a vector space. In particular the problem with expression of the Superposition Principle arises from the fact that in any nontrivial case it is impossible to express $\rho(c_1\psi_1 + c_2\psi_2 + \cdots)$ (after normalisation) in terms of $\rho(\psi_1), \rho(\psi_2), \cdots$. In spite of this being so, it is interesting that the existence or nonexistence of a vector level superposition (9.1) can be seen directly at the ray space level. We present the details at this point, bringing in an important class of expressions known as the 'Bargmann invariants' originally introduced by him in the course of his proof of Wigner's theorem referred to earlier.

For some n, let a set of distinct pure state density matrices $\rho_1, \rho_2, \cdots, \rho_n \in \mathcal{R}$ be given. Then choose vectors $\psi_1, \psi_2, \cdots, \psi_n \in \mathcal{B}$ such that, in the notation of Eq. (9.7),

$$\psi_j \in \pi^{-1}(\rho_j), \quad \rho_j = \rho(\psi_j), \quad j = 1, 2, \cdots, n. \tag{9.9}$$

Each of $\psi_1, \psi_2, \cdots, \psi_n$ is fixed up to a phase. Clearly, if $\mathcal{H}$ is of finite dimension N, if $n > N$, the vectors $\psi_1, \psi_2, \cdots, \psi_n$ are definitely linearly dependent. Then some one of them is expressible in terms of the others in the sense of Eq. (9.1). If on the other hand $n \leq N$, the ψ_j may be linearly independent or dependent. But which of these two possibilities occurs in a given situation is determined by the ρ_j alone at the ray space level! In other words, in case the ψ_j in Eq. (9.9) are linearly dependent, this property can be detected given the ρ_j alone.

The proof of this statement is quite instructive as it directly leads to the important and useful Bargmann invariants (BI). Having chosen some ψ_j in accordance with Eq. (9.9), form the $n \times n$ matrix of scalar products among them:

$$M(\psi_1, \psi_2, \cdots, \psi_n) = ((\psi_j, \psi_k)). \tag{9.10}$$

This is clearly hermitian positive semidefinite. If we change the ψ_j to ψ'_j by multiplication by independent phase factors, leaving the ρ_j unchanged:

$$\psi'_j = e^{i\alpha_j}\psi_j, \quad j = 1, 2, \cdots, n \tag{9.11}$$

then

$$M(\psi'_1, \psi'_2, \cdots, \psi'_n) = \mathcal{D}(\boldsymbol{\alpha})^\dagger M(\psi_1, \psi_2, \cdots, \psi_n)\mathcal{D}(\boldsymbol{\alpha}),$$

$$\mathcal{D}(\boldsymbol{\alpha}) = \text{diag}(e^{i\alpha_1}, e^{i\alpha_2}, \cdots, e^{i\alpha_n}). \tag{9.12}$$

This preserves the determinant of $M(\psi_1, \psi_2, \cdots \psi_n)$, so it must be expressible in terms of the ρ_j alone. Then its vanishing is the unambiguous sign of the existence of a linear relationship among the ψ_j, and therefore also among the ψ'_j for any choice of the α_j. We can see this explicitly starting from low values of n.

For $n = 2$ we have a 2×2 matrix involving two unit vectors $\psi_1, \psi_2 \in \mathcal{H}$:

$$\det M(\psi_1, \psi_2) = \begin{vmatrix} 1 & (\psi_1, \psi_2) \\ (\psi_2, \psi_1) & 1 \end{vmatrix} = 1 - |(\psi_1, \psi_2)|^2$$

$$= 1 - \text{Tr}(\rho_1 \rho_2). \tag{9.13}$$

This is real nonnegative. For $n = 3$ we have a 3×3 matrix and something new appears:

$$\det M(\psi_1, \psi_2, \psi_3) = \begin{vmatrix} 1 & (\psi_1, \psi_2) & (\psi_1, \psi_3) \\ (\psi_2, \psi_1) & 1 & (\psi_2, \psi_3) \\ (\psi_3, \psi_1) & (\psi_3, \psi_2) & 1 \end{vmatrix}$$

$$= 1 - \text{Tr}(\rho_2 \rho_3) - \text{Tr}(\rho_3 \rho_1) - \text{Tr}(\rho_1 \rho_2)$$

$$+ \Delta_3(\psi_1, \psi_2, \psi_3) + \Delta_3(\psi_1, \psi_3, \psi_2),$$

$$\Delta_3(\psi_1, \psi_2, \psi_3) = (\psi_1, \psi_2)(\psi_2, \psi_3)(\psi_3, \psi_1) = \text{Tr}(\rho_1 \rho_2 \rho_3). \tag{9.14}$$

The expression $\Delta_3(\psi_1, \psi_2, \psi_3)$ is the *third order* BI, and if the dimension of $\mathcal{H}$ is at least two it is in general complex. As shown, it is expressible in terms of ray space quantities. In these cases we conclude: if $n = 2$ and $\det M(\psi_1, \psi_2)$ vanishes, ψ_2 is a phase times ψ_1 and $\rho_2 = \rho_1$; if $n = 3$ and $\det M(\psi_1, \psi_2, \psi_3)$ vanishes, the vectors ψ_1, ψ_2, ψ_3 are linearly dependent and yet, in general, none of ρ_1, ρ_2, ρ_3 is expressible in terms of the other two.

The case of general n leads to higher order BI's: for any set of unit vectors $\phi_1, \phi_2, \cdots, \phi_m$:

$$\Delta_m(\phi_1, \phi_2, \cdots, \phi_m) = (\phi_1, \phi_2)(\phi_2, \phi_3) \cdots (\phi_{m-1}, \phi_m)(\phi_m, \phi_1)$$

$$= \text{Tr}(\rho(\phi_1)\rho(\phi_2) \cdots \rho(\phi_m)), \quad m = 2, 3, \cdots. \tag{9.15}$$

Then, going back to the $n \times n$ matrix in Eq. (9.10):

$$\det M(\psi_1, \psi_2, \cdots, \psi_n) = \begin{vmatrix} 1 & (\psi_1, \psi_2) & \cdots & (\psi_1, \psi_n) \\ (\psi_2, \psi_1) & 1 & \cdots & (\psi_2, \psi_n) \\ \cdots & \cdots & \cdots & \cdots \\ (\psi_n, \psi_1) & (\psi_n, \psi_2) & \cdots & 1 \end{vmatrix}$$

$$= \epsilon_{k_1 k_2 \cdots k_n} (\psi_1, \psi_{k_1})(\psi_2, \psi_{k_2}) \cdots (\psi_n, \psi_{k_n}). \tag{9.16}$$

This is a series of $n!$ terms carrying $\pm$ signs, each term a product of some BI's of various orders m, as each ψ_j appears once as a left factor and once as a right factor in scalar products. The BI's that appear are of orders $m = 2, 3, \cdots, n$.

We can sum up this discussion of the Superposition Principle of quantum mechanics in these statements:

Vectors $\psi_1, \psi_2, \cdots \psi_n$ linearly independent $\Leftrightarrow \det M(\psi_1, \psi_2, \cdots, \psi_n) > 0$

$\quad \Leftrightarrow$ pure states $\rho_1, \rho_2, \cdots, \rho_n$ are 'physically independent', and

$\qquad$ necessarily $n \leq N$; $\hfill$ (9.17a)

Vectors $\psi_1, \psi_2, \cdots, \psi_n$ linearly dependent $\Leftrightarrow \det M(\psi_1, \psi_2, \cdots, \psi_n) =$

$\quad$ (an expression in BI's of orders $m \leq n$ among $\rho_1, \rho_2, \cdots \rho_n) = 0$

$\quad \Leftrightarrow$ pure states $\rho_1, \rho_2, \cdots, \rho_n$ are 'physically dependent', superposition

$\quad$ principle is at work, even though no one of the ρ_j may be expressible

$\quad$ in terms of the others. $\hfill$ (9.17b)

This discussion of the Superposition Principle deepens our understanding of the relationship between Hilbert and ray spaces, and has led us to the BI's.

9.2 Wigner Symmetry and Unitary–Antiunitary Theorem

What follows is a mainly kinematical analysis, with no particular Hamiltonian in mind. The *Wigner definition of a symmetry operation* – in short a Wigner symmetry or W-S – is that it is a one-to-one onto invertible map $\Omega : \mathcal{R} \to \mathcal{R}$, a mapping of pure states onto pure states, which preserves probabilities:

$$\psi, \phi \in \mathcal{B} \to \rho(\psi), \; \rho(\phi) \in \mathcal{R} \xrightarrow{\Omega} \rho(\psi') = \Omega(\rho(\psi)), \; \rho(\phi') = \Omega(\rho(\phi)),$$

$$\text{Tr}(\rho(\phi')\rho(\psi')) = |(\phi', \psi')|^2 = \text{Tr}(\rho(\phi)\rho(\psi)) = |(\phi, \psi)|^2. \quad (9.18)$$

Here for given ψ, ϕ each of ψ', ϕ' is determined by Ω up to a phase. The W-S Ω is by definition a ray space mapping $\mathcal{R} \to \mathcal{R}$ which is one-to-one. If we wish, by composition with the projection $\pi : \mathcal{B} \to \mathcal{R}$ we can construct a many-to-one map $\tilde{\Omega} : \mathcal{B} \to \mathcal{R}$:

$$\tilde{\Omega} = \Omega \cdot \pi : \psi \in \mathcal{B} \xrightarrow{\pi} \rho(\psi) \in \mathcal{R} \xrightarrow{\Omega} \Omega(\rho(\psi)) = \rho(\psi'), \; \psi' \in \mathcal{B}, \quad (9.19)$$

with, of course, ψ' determined up to a phase. But neither Ω nor $\tilde{\Omega}$ is a map $\mathcal{B} \to \mathcal{B}$, much less a map $\mathcal{H} \to \mathcal{H}$ at the level of vectors.

Now the *Wigner theorem* says: if Eq. (9.18) is obeyed, i.e., quantum mechanical probabilities are preserved, then it is possible to 'lift' Ω acting on unit rays to a map $\omega : \mathcal{B} \to \mathcal{B}$, further extended to act $\mathcal{H} \to \mathcal{H}$, such that

$$\Omega(\rho(\psi)) = \rho(\omega(\psi)) = \omega|\psi\rangle\langle\psi|\omega^{-1} = \omega\rho(\psi)\omega^{-1}, \tag{9.20}$$

with ω *either linear unitary or antilinear antiunitary*. This lift ω is not unique, but for given Ω its linear or antilinear character is completely determined. We have the composition rule to be compared with (9.19):

$$\tilde{\Omega} = \Omega \cdot \pi = \pi \cdot \omega : \mathcal{B} \to \mathcal{R}. \tag{9.21}$$

We say the following diagram 'commutes': Ω obeying (9.18) means ω can be constructed:

$$
\begin{array}{ccc}
\mathcal{B}(\text{or } \mathcal{H}) & \xrightarrow{\ \omega\ } & \mathcal{B}(\text{or } \mathcal{H}) \\
\pi \downarrow & & \pi \downarrow \\
\mathcal{R} & \xrightarrow{\ \Omega\ } & \mathcal{R}
\end{array}
$$

Here, $\tilde{\Omega} = \Omega \circ \pi$ is the transition NW→SE via SW, while $\pi \circ \omega$ is NW→SE via NE, and they are equal.

Any unitary transformation U on $\mathcal{H}$ leads to a W-S $\mathcal{U}$ in this way:

$$\rho \in \mathcal{R} \to \mathcal{U}(\rho) = U\rho U^{-1}:$$
$$\text{Tr}(\mathcal{U}(\rho_1)\mathcal{U}(\rho_2)) = \text{Tr}(U\rho_1 U^{-1} U\rho_2 U^{-1}) = \text{Tr}(\rho_1\rho_2). \tag{9.22}$$

One can also see easily that the inverse Ω^{-1} of a W-S Ω, and the composition $\Omega' \circ \Omega$ of two W-S's, are also W-S's.

The only physically interesting case where the antiunitary possibility shows up is time reversal. A consequence is that in quantum mechanics parity is an observable whose eigenvalues can be measured, but time reversal is not!

9.3 Proofs of Wigner's Theorem

Wigner's original 1931 proof (though not complete in all details but essentially so) constructs ω using an orthonormal set of vectors in $\mathcal{H}$ defined once for all. Over the years many proofs have been presented. Bargmann's 1964 proof is very elegant but quite intricate. It is based on repeated use of Eq. (9.18), and draws successively more detailed conclusions. Its economy lies in using limited orthonormal sets of at most two or three vectors at a time, adjusted to the vectors under consideration. We give below two recent proofs which, in Bargmann's words, do not 'obscure the quite elementary nature of Wigner's theorem', but have an attractive global quality.

9.3.1 Proof 1

We allow the dimension N of $\mathcal{H}$ to be finite or infinite. Let a W-S Ω and its associated map $\tilde{\Omega} : \mathcal{B} \to \mathcal{R}$ (Eq. (9.19)) be given. We have

$$\psi \in \mathcal{B}: \ \tilde{\Omega}(\psi) = \pi(\psi'), \tag{9.23}$$

where $\psi' \in \mathcal{B}$ is determined up to a phase. We use this fact repeatedly. The basic condition (9.18), hereafter called the symmetry condition or SC, reads:

$$\psi, \phi \in \mathcal{B}: \ \mathrm{Tr}(\tilde{\Omega}(\phi)\tilde{\Omega}(\psi)) = |(\phi, \psi)|^2. \tag{9.24}$$

Our strategy in analysing Ω is to take it to a *simpler or canonical form* by composition with a unitary symmetry *naturally suggested* by Ω itself. This simpler W-S is then examined step by step until we see that the canonical form of Ω is either the identity map or complex conjugation (in some orthonormal basis).

Let $\{|n\rangle, n = 1, 2, \cdots, N\}$ be (any) orthonormal basis (ONB) for $\mathcal{H}$. Relative to it, for each pair (j, k) with $1 \le j < k \le N$ we define a two dimensional linear subspace $\mathcal{H}_{jk} \subset \mathcal{H}$:

$$\mathcal{H}_{jk} = \{\alpha|j\rangle + \beta|k\rangle | \alpha, \beta \in \mathbb{C}\} \subset \mathcal{H}. \tag{9.25}$$

The set of unit vectors in $\mathcal{H}_{jk}$ is its intersection with $\mathcal{B}$:

$$\mathcal{B}_{jk} = \mathcal{B} \cap \mathcal{H}_{jk} = \{\alpha|j\rangle + \beta|k\rangle |, \ |\alpha|^2 + |\beta|^2 = 1\} \subset \mathcal{B}. \tag{9.26}$$

Upon projection π this maps onto a subset $\mathcal{R}_{jk} \subset \mathcal{R}$ which can be conveniently parametrised by spherical polar angles on $\mathbb{S}^2$:

$$\mathcal{R}_{jk} = \pi(\mathcal{B}_{jk}) = \{\rho(|\theta; \phi\rangle_{jk})\} \subset \mathcal{R},$$

$$|\theta; \phi\rangle_{jk} = \cos\frac{\theta}{2}|j\rangle + e^{i\phi} \sin\frac{\theta}{2}|k\rangle \in \mathcal{B}_{jk}, \ 0 \le \theta \le \pi, \ 0 \le \phi < 2\pi. \tag{9.27}$$

Thus each $\mathcal{R}_{jk}$ for $j < k$ has the form of a Poincaré sphere.

The proof now involves six steps, each quite elementary.

Step 1: Choose any ONB $\{|n\rangle\}$ for $\mathcal{H}$. From the SC (9.24) for pairs of basis vectors, and Eq. (9.23), we find that

$$\tilde{\Omega}(|n\rangle) = \pi(|n;\Omega\rangle), \quad n = 1, 2, \cdots, N, \qquad (9.28)$$

where each $|n;\Omega\rangle$ is determined up to a phase and the collection $\{|n;\Omega\rangle\}$ is also an ONB for $\mathcal{H}$. Make any choices for the vectors $|n;\Omega\rangle$ and define a unitary transformation U on $\mathcal{H}$ by

$$U|n;\Omega\rangle = |n\rangle, \quad n = 1, 2, \cdots, N. \qquad (9.29)$$

Thus U is designed to 'partly undo' the effect of $\tilde{\Omega}$ on the vectors $|n\rangle$. We now define a W-S Ω' by

$$\Omega' = \mathcal{U} \circ \Omega, \quad \tilde{\Omega}' = \mathcal{U} \circ \tilde{\Omega}, \qquad (9.30)$$

and reduce the study of Ω to that of Ω'. This W-S has a simple action on the vectors $\{|n\rangle\}$, namely

$$\tilde{\Omega}'(|n\rangle) = \pi(|n\rangle), \quad n = 1, 2, \cdots, N, \qquad (9.31)$$

and using the SC (9.24) for $|n\rangle$ and any $|\psi\rangle \in \mathcal{B}$ we have

$$\tilde{\Omega}'\left(\sum_{n=1}^{N} c_n |n\rangle\right) - \pi\left(\sum_{n=1}^{N} c_n' |n\rangle\right), \quad |c_n'| = |c_n|. \qquad (9.32)$$

At this point, we see that $\tilde{\Omega}'$ leaves the standard set of orthonormal rays $\{\pi(|n\rangle)\}$ unchanged, and this can be achieved for any W-S Ω. We still have the freedom of a phase factor in each vector $|n;\Omega\rangle$, which means that U can be post multiplied by *any* unitary operator diagonal in the basis $\{|n\rangle\}$. Such a change in U will preserve (9.31). While Ω' in (9.30) is not yet the desired canonical form for Ω, this form will be determined in Step 3 after using (and exhausting) the phase freedom just mentioned.

Step 2: Next we choose some pair (j, k) with $j < k$ and study the action of $\tilde{\Omega}'$ on vectors in $\mathcal{B}_{jk}$. It suffices to look at the action on a (latitude) circle of vectors $|\theta_0; \phi\rangle_{jk}$, Eq. (9.27), for fixed $\theta_0 \in (0, \pi)$ and varying $\phi \in [0, 2\pi)$. From

Eq. (9.32) we see that

$$\tilde{\Omega}'(|\theta_0;\phi\rangle_{jk}) = \pi(|\theta_0;\phi'\rangle_{jk}), \tag{9.33}$$

with θ_0, j, k unchanged and ϕ' dependent on ϕ (and possibly also on θ_0, j, k) in an invertible manner. Since

$$\left|_{jk}\langle\theta_0;\phi_1|\theta_0;\phi_2\rangle_{jk}\right|^2 = 2\cos^2\frac{\theta_0}{2}\sin^2\frac{\theta_0}{2}\cos(\phi_1-\phi_2) + \cos^4\frac{\theta_0}{2} + \sin^4\frac{\theta_0}{2}, \tag{9.34}$$

use of the SC (9.24) for this pair of vectors shows that ϕ_1, ϕ_2 are carried by Eq. (9.33) into ϕ_1', ϕ_2' such that

$$\cos(\phi_1' - \phi_2') = \cos(\phi_1 - \phi_2). \tag{9.35}$$

It follows that the change $\phi \to \phi'$ in Eq. (9.33) caused by $\tilde{\Omega}'$ action is of the form

$$\phi' = \phi_{jk} + \epsilon_{jk}\phi, \quad \phi_{jk} \in [0, 2\pi), \quad \epsilon_{jk} = \pm 1. \tag{9.36}$$

(This and similar later equations are understood to be valid mod 2π). Thus the action (9.33) reads

$$\tilde{\Omega}'(|\theta_0;\phi\rangle_{jk}) = \pi(|\theta_0;\phi_{jk} + \epsilon_{jk}\phi\rangle_{jk}). \tag{9.37}$$

For fixed θ_0, j, k this means that $\tilde{\Omega}'$ acts on ϕ via some element of the group $O(2)$: ϕ_{jk} determines an $SO(2)$ element, while ϵ_{jk} determines whether we have a proper or an improper rotation.

Step 3: We now use the freedom noted at the end of Step 1, to multiply each vector $|n\rangle$ in the ONB $\{|n\rangle\}$ by an independent phase factor, preserving the structure of the results obtained up to this point. Thus we define a (diagonal) unitary transformation U' on $\mathcal{H}$ by

$$U'|n\rangle = e^{-i\phi_n}|n\rangle, \quad n = 1, 2, \cdots, N, \tag{9.38}$$

and pass from the W-S Ω' to the (final) W-S Ω'' by

$$\Omega'' = \mathcal{U}' \circ \Omega' \ . \ \tilde{\Omega}'' = \mathcal{U}' \circ \tilde{\Omega}'. \tag{9.39}$$

This helps us simplify the $SO(2)$ angles ϕ_{jk} in Eq. (9.37) to some extent. We have

$$\tilde{\Omega}''(|\theta_0; \phi\rangle_{jk}) = \tilde{U}'\left(|\theta_0; \phi_{jk} + \epsilon_{jk}\phi\rangle_{jk}\right)$$
$$= \pi\left(|\theta_0; \phi_{jk} + \phi_j - \phi_k + \epsilon_{jk}\phi\rangle_{jk}\right). \tag{9.40}$$

Remembering that $k \geq 2$, we choose

$$\phi_1 = 0, \quad \phi_n = \phi_{1n} \text{ for } n \geq 2. \tag{9.41}$$

Then we have

$$\tilde{\Omega}''(|\theta_0; \phi\rangle_{jk}) = \pi(|\theta_0; \phi'_{jk} + \epsilon_{jk}\phi\rangle_{jk}),$$
$$\phi'_{jk} = \phi_{jk} + \phi_j - \phi_k, \phi'_{1k} = 0 \text{ for } k \geq 2. \tag{9.42}$$

Compared to Eq. (9.37), the sign factors ϵ_{jk} are unchanged, while the $SO(2)$ angles ϕ_{jk} have been simplified to ϕ'_{jk} with $\phi'_{1k} = 0$.

Step 4: Since Ω'' is related to Ω' by the diagonal (in the ONB $\{|n\rangle\}$) unitary transformation U', Eq. (9.38), it is clear that the structure of Eq. (9.32) is retained for $\tilde{\Omega}''$:

$$\tilde{\Omega}''\left(\sum_{n=1}^{N} c_n |n\rangle\right) = \pi\left(\sum_{n=1}^{N} c''_n |n\rangle\right), \quad |c''_n| = |c_n|. \tag{9.43}$$

Now we show that in the action (9.42) by $\tilde{\Omega}''$, not only is $\phi'_{1k} = 0$ but in fact $\phi'_{jk} = 0$ for all $j < k$. Here we see the power of the SC (9.24). We choose a single special 'real' normalised vector $\boldsymbol{r}^{(0)} = (r_1^{(0)}, r_2^{(0)}, \cdots r_N^{(0)})^T$ and define $|\boldsymbol{r}^{(0)}\rangle \in \mathcal{B}$ as

$$|\boldsymbol{r}^{(0)}\rangle = \sum_{n=1}^{N} r_n^{(0)} |n\rangle, \quad r_n^{(0)} \text{ real nonzero for all } n. \tag{9.44}$$

Under $\tilde{\Omega}''$ action we have, from Eq. (9.43).

$$\tilde{\Omega}''(|\boldsymbol{r}^{(0)}\rangle) = \pi\left(\sum_{n=1}^{N} r_n^{(0)} e^{i\eta_n} |n\rangle\right), \tag{9.45}$$

for some phases η_n. We now invoke the SC (9.24) for the pair of vectors $|\theta_0; \phi\rangle_{jk}$, $|\mathbf{r}^{(0)}\rangle$, any $j < k$, under $\tilde{\Omega}''$ action to get:

$$\left| r_j^{(0)} e^{i\eta_j} \cos \frac{\theta_0}{2} + r_k^{(0)} e^{i\eta_k} \sin \frac{\theta_0}{2} e^{-i(\phi_{jk}' + \epsilon_{jk}\phi)} \right|^2$$

$$= \left| r_j^{(0)} \cos \frac{\theta_0}{2} + r_k^{(0)} \sin \frac{\theta_0}{2} e^{-i\phi} \right|^2,$$

i.e., $\cos\left(\epsilon_{jk}\phi + \phi_{jk}' + \eta_j - \eta_k \right) = \cos\phi, \quad 0 \le \phi < 2\pi.$ $\qquad$ (9.46)

This implies

$$\phi_{jk}' = \eta_k - \eta_j, \quad j < k \qquad (9.47)$$

For $j = 1$, from Eq. (9.42) we get $\eta_k = \eta_1$ independent of k. Putting this back into Eq. (9.47) gives $\phi_{jk}' = 0$ for all $j < k$. Thus the actions (9.42, 9.45) by $\tilde{\Omega}''$ simplify to

$$\tilde{\Omega}''(|\theta_0; \phi\rangle_{jk}) = \pi(|\theta_0; \epsilon_{jk}\phi\rangle_{jk}),$$

$$\tilde{\Omega}''(|\mathbf{r}^{(0)}\rangle) = \pi(|\mathbf{r}^{(0)}\rangle). \qquad (9.48)$$

For each pair $j < k$, only a sign factor ϵ_{jk} remains. We will soon find that this cannot depend on j, k.

Step 5: Let us next apply the SC (9.24) under Ω'' for the pair of vectors $|\mathbf{c}\rangle$, $|\theta_0; \phi\rangle_{jk}$ where $|\mathbf{c}\rangle$ is the general normalised linear combination on the left in Eqs. (9.32, 9.43). Using Eqs. (9.43, 9.48) we have:

$$\left| c_j'' \cos \frac{\theta_0}{2} + c_k'' \sin \frac{\theta_0}{2} e^{-i\epsilon_{jk}\phi} \right|^2 = \left| c_j \cos \frac{\theta_0}{2} + c_k \sin \frac{\theta_0}{2} e^{-i\phi} \right|^2,$$

i.e., $c_j'' c_k''^* e^{i\epsilon_{jk}\phi} + c_j''^* c_k'' e^{-i\epsilon_{jk}\phi} = c_j c_k^* e^{i\phi} + c_j^* c_k e^{-i\phi}, \quad 0 \le \phi < 2\pi.$ $\quad$ (9.49)

Therefore the change $\{c_n\} \to \{c_n''\}$ caused by $\tilde{\Omega}''$ in Eq. (9.43) must obey

$$c_j'' c_k''^* = c_j c_k^* \quad \text{if } \epsilon_{jk} = +1,$$

$$c_j'' c_k''^* = c_j^* c_k \quad \text{if } \epsilon_{jk} = -1, \qquad (9.50)$$

Step 6: This is the last step. We show that consistency demands that ϵ_{jk} cannot depend on the pair j, k; it must be $+1$ or -1 uniformly for all pairs. Choose any j, k, l

with $j < k < l$. For any $|c\rangle \in \mathcal{B}$ we trivially have

$$c_j c_k^* \cdot c_k c_l^* \cdot (c_j c_l^*)^* = |c_j c_k c_l|^2 = \text{real} \geq 0. \qquad (9.51)$$

Therefore when $\tilde{\Omega}''$ 'takes' $|c\rangle$ to $|c''\rangle$ as in Eq. (9.43) we must necessarily have

$$c_j'' c_k''^* \cdot c_k'' c_l''^* \cdot (c_j'' c_l''^*)^* = \text{real} \geq 0. \qquad (9.52)$$

Depending on the values of $\epsilon_{jk}, \epsilon_{kl}, \epsilon_{jl}$, by Eq. (9.50) this requires for any c:

$$(c_j c_k^* \text{ or } c_j^* c_k)(c_k c_l^* \text{ or } c_k^* c_l)(c_j c_l^* \text{ or } c_j^* c_l)^* = \text{real} \geq 0. \qquad (9.53)$$

In each factor we have the first entry if $\epsilon = +1$, the second if $\epsilon = -1$. We see easily that if $\epsilon_{jk} = \epsilon_{kl} = \epsilon_{jl} = +1$ or if $\epsilon_{jk} = \epsilon_{kl} = \epsilon_{jl} = -1$, this condition is obeyed. But in every other case the left hand side of (9.53) is in general complex. Therefore we conclude that the only consistent possibilities for $\tilde{\Omega}''$ action are:

$$\tilde{\Omega}'' \left(\sum_{n=1}^{N} c_n |n\rangle \right) = \pi \left(\sum_{n=1}^{N} c_n |n\rangle \right) \text{ if all } \epsilon_{jk} = +1,$$

$$= \pi \left(\sum_{n=1}^{N} c_n^* |n\rangle \right) \text{ if all } \epsilon_{jk} = -1. \qquad (9.54)$$

To lift Ω'' to an ω'' acting on vectors in $\mathcal{B}$ (and by extension to vectors in $\mathcal{H}$) is trivially easy: ω'' is either the identity operator or K, the operator of complex conjugation of the expansion coefficients of a general vector in the ONB $\{|n\rangle\}$. Now the relation between Ω'' and the originally given W-S Ω is, from Eqs. (9.30, 9.39)

$$\Omega'' = \mathcal{U}' \circ \mathcal{U} \circ \Omega. \qquad (9.55)$$

Therefore the two possible consistent lifts ω of Ω are either the unitary transformation $U^{-1} U'^{-1}$ or the antiunitary transformation $U^{-1} U'^{-1} K$, both defined on $\mathcal{H}$. Thus Wigner's theorem is established.

We make a note at this point on the origin of the BI's defined in Eqs. (9.14, 9.15). These were first introduced by Bargmann in 1964 in the course of his proof of Wigner's theorem, as a way to determine, once the action of a W-S Ω on $\mathcal{R}$ is given, whether after lifting to an ω on $\mathcal{H}$ it would turn out to be of the unitary or the antiunitary type. As noted after Eq. (9.14), in general $\Delta_3(\psi_1, \psi_2, \psi_3)$ is a complex quantity and one can see that

$$\Delta_3(\psi_1, \psi_2, \psi_3) = Tr(\rho_1\rho_2\rho_3) \xrightarrow{\Omega} Tr(\Omega(\rho_1)\Omega(\rho_2)\Omega(\rho_3))$$

$$= \Delta_3(\psi_1, \psi_2, \psi_3) \text{ in unitary case,}$$

$$\Delta_3(\psi_1, \psi_2, \psi_3)^* \text{ in antiunitary case.} \qquad (9.56)$$

It is interesting that in the present treatment these invariants have appeared already in the analysis of the Superposition Principle, in advance of the discussion and proof of Wigner's theorem.

9.3.2 Proof 2

The crucial step in the previous proof is the first one – the passage from Ω to Ω' in Eq. (9.30) resulting in Eqs. (9.31, 9.32). The truth of this statement is clearly brought out by the second proof based on induction on the dimension of $\mathcal{H}$.

For a two dimensional quantum system, as mentioned after Eq. (9.8), the space $\mathcal{R}$ is the Poincaré sphere $\mathbb{S}^2$. As in Eq. (9.27), each $\rho \in \mathcal{R}$ corresponds one-to-one to a point on this sphere:

$$\hat{n} \in \mathbb{S}^2 \longleftrightarrow \rho(\hat{n}) = \frac{1}{2}(\mathbb{1} + \hat{n} \cdot \sigma),$$

$$Tr(\rho(\hat{n}')\rho(\hat{n})) = \frac{1}{2}(1 + \hat{n}' \cdot \hat{n}). \qquad (9.57)$$

(Therefore orthogonal states correspond to antipodal points.) Thus a W-S in this case is a one-to-one onto map $\Omega : \mathbb{S}^2 \to \mathbb{S}^2$ preserving angles:

$$\Omega(\hat{n}') \cdot \Omega(\hat{n}) = \hat{n}' \cdot \hat{n}. \qquad (9.58)$$

As is well known, such maps are either proper rotations belonging to $SO(3)$, induced by $SU(2)$ transformations on $\mathcal{H}$, or improper rotations in $O(3)$, involving in addition complex conjugation on $\mathcal{H}$ (mirror reflection on $\mathbb{S}^2$). This is Wigner's theorem in this case.

For general finite dimension we use induction. We assume the theorem is true in N dimensions, and prove it for $N+1$ dimensions. Let $\mathcal{H}^{(N+1)}$ be an $(N+1)$-dimensional Hilbert space, and Ω a W-S for the corresponding quantum system. Choose any ONB $\{|n\rangle, n = 1, 2, \cdots, N+1\}$ for $\mathcal{H}^{(N+1)}$. As in the initial steps of Proof 1, pass from

Ω to another W-S Ω' by composition with a suitable unitary symmetry. Then as in Eqs. (9.31, 9.32) we have:

$$\tilde{\Omega}'(|n\rangle) = \pi(|n\rangle), \quad n = 1, 2, \cdots, N + 1;$$

$$\tilde{\Omega}'\left(\sum_{n=1}^{N+1} c_n|n\rangle\right) = \pi\left(\sum_{n=1}^{N+1} c'_n|n\rangle\right), \quad |c_n| = |c'_n|. \tag{9.59}$$

Let us identify $\mathcal{H}^{(N)} \subset \mathcal{H}^{(N+1)}$ as the subspace spanned by the first N basis vectors:

$$\mathcal{H}^{(N)} = \mathrm{Sp}\{|n\rangle, n = 1, 2, \cdots, N\} \subset \mathcal{H}^{(N+1)}. \tag{9.60}$$

By Eq. (9.59), Ω' is a W-S for $\mathcal{H}^{(N)}$. By assumption, possibly after composition with a diagonal unitary symmetry which is suppressed for simplicity, we have one of two possibilities:

$$\tilde{\Omega}'\left(\sum_{n=1}^{N} c_n|n\rangle\right) = \pi\left(\sum_{n=1}^{N} c_n|n\rangle\right), \quad \text{all } \boldsymbol{c} = (c_1, c_2, \cdots, c_N)^T,$$

$$\text{or } \tilde{\Omega}'\left(\sum_{n=1}^{N} c_n|n\rangle\right) = \pi\left(\sum_{n=1}^{N} c_n^*|n\rangle\right), \quad \text{all } \boldsymbol{c} = (c_1, c_2, \cdots, c_N)^T. \tag{9.61}$$

Suppose the first choice holds. To deal with $\mathcal{H}^{(N+1)}$, a general vector here is

$$|\boldsymbol{c}\rangle + c_{N+1}|N + 1\rangle \in \mathcal{H}^{(N+1)}, \quad |\boldsymbol{c}\rangle = \sum_{n=1}^{N} c_n|n\rangle \in \mathcal{H}^{(N)}. \tag{9.62}$$

Let

$$\tilde{\Omega}'(|\boldsymbol{c}\rangle + c_{N+1}|N + 1\rangle) = \pi(|\boldsymbol{c}'\rangle + c'_{N+1}|N + 1\rangle),$$

$$|c'_j| = |c_j|, \quad j = 1, 2, \cdots, N + 1. \tag{9.63}$$

The SC (9.24) for a general $|\boldsymbol{c}^{(1)}\rangle \in \mathcal{H}^{(N)}$ and the vector (9.62) in $\mathcal{H}^{(N+1)}$ yields

$$\left|\sum_{n=1}^{N} c_n^{(1)*} c'_n\right|^2 = \left|\sum_{n=1}^{N} c_n^{(1)*} c_n\right|^2, \quad \text{all } \boldsymbol{c}^{(1)}$$

$$\Rightarrow \boldsymbol{c}^{(1)\dagger}\boldsymbol{c}'\boldsymbol{c}'^{\dagger}\boldsymbol{c}^{(1)} = \boldsymbol{c}^{(1)\dagger}\boldsymbol{c}\boldsymbol{c}^{\dagger}\boldsymbol{c}^{(1)}, \quad \text{all } \boldsymbol{c}^{(1)}$$

$$\Rightarrow \boldsymbol{c}' = (\text{phase}) \, \boldsymbol{c}. \tag{9.64}$$

Then Eq. (9.63) simplifies to

$$\tilde{\Omega}'(|c\rangle + c_{N+1}|N+1\rangle) = \pi(|c\rangle + c''_{N+1}|N+1\rangle), \; |c''_{N+1}| = |c_{N+1}|. \tag{9.65}$$

Next, the SC for two vectors in $\mathcal{H}^{(N+1)}$ gives:

$$|c^{(1)}\rangle + c^{(1)}_{N+1}|N+1\rangle, \; |c^{(2)}\rangle + c^{(2)}_{N+1}|N+1\rangle \in \mathcal{H}^{(N+1)},$$

$$|c^{(1)\dagger}c^{(2)} + c^{(1)''*}_{N+1}c^{(2)''}_{N+1}|^2 = |c^{(1)\dagger}c^{(2)} + c^{(1)*}_{N+1}c^{(2)}_{N+1}|^2, \text{ all } c^{(1)}, \; c^{(2)};$$

$$\text{i.e., } c^{(1)''*}_{N+1}c^{(2)''}_{N+1} = c^{(1)*}_{N+1}c^{(2)}_{N+1}. \tag{9.66}$$

So at the level of $\mathcal{H}^{(N+1)}$ we have

$$c''_{N+1} = e^{i\phi}c_{N+1},$$

$$\tilde{\Omega}'(|c\rangle + c_{N+1}|N+1\rangle) = \pi(|c\rangle + e^{i\phi}c_{N+1}|N+1\rangle), \tag{9.67}$$

the phase ϕ being independent of $|c\rangle$. Then by a diagonal unitary phase transformation we can pass to a W-S Ω'' which acts trivially 'on' $\mathcal{H}^{(N+1)}$. This proves, in the case of the first option in Eq. (9.61), that validity of Wigner's theorem in N dimensions leads to its validity in $N+1$ dimensions.

For the second option in Eq. (9.61), a similar argument holds. Now validity of Wigner's theorem for $N=2$ is evident as noted after Eq. (9.58). Hence by induction the theorem holds for all finite N.

For a W-S Ω on a Hilbert space $\mathcal{H}$ of infinite dimension, choose an ONB $\{|n\rangle, n = 1, 2, \cdots\}$, denote by ψ a column vector $(\psi_1, \psi_2, \cdots)^T$, and pass to Ω' such that

$$\tilde{\Omega}'(|n\rangle) = \pi(|n\rangle), \; n = 1, 2, \cdots;$$

$$\tilde{\Omega}'\left(\sum_{n=1}^{\infty} \psi_n|n\rangle\right) = \pi\left(\sum_{n=1}^{\infty} \psi'_n|n\rangle\right), \; |\psi'_n| = |\psi_n|. \tag{9.68}$$

This ONB gives a sequence of subspaces $\mathcal{H}^{(2)} \subset \mathcal{H}^{(3)} \cdots \subset \mathcal{H}^{(N)} \subset \mathcal{H}^{(N+1)} \cdots$. So as in the above we can pass from Ω' to an Ω'' whose action on $\mathcal{H}^{(N)}$ is either trivial for all finite N or is complex conjugation for all finite N. Now in the trivial case the SC for $|c\rangle \in \mathcal{H}^{(N)}$, $|\psi\rangle \in \mathcal{H}$ is

$$\tilde{\Omega}''(|c\rangle \text{ or } |\psi\rangle) = \pi(|c\rangle \text{ or } |\psi'\rangle), \; |\psi'_n| = |\psi_n|, \text{all } n;$$

$$c^{\dagger}\psi'\psi'^{\dagger}c = c^{\dagger}\psi\psi^{\dagger}c. \tag{9.69}$$

Therefore $c^{\dagger}\psi' = (\text{phase})c^{\dagger}\psi$, and this is so for *all* c and *all* finite N. Further, the condition (9.69) implies $\psi'_n = e^{i\phi_n}\psi_n$. To see that the phase ϕ_n cannot really depend on n, assume the contrary and suppose $\phi_m \neq \phi_n$ for some m, n. One can then easily choose a vector c whose only nonvanishing components are c_m, c_n such that $c^\dagger \psi'$ and $c^\dagger \psi$ do not have the same magnitude, leading to a contradiction. That is, the SC (9.69) forces ϕ_n to be independent of n and

$$\psi' = (\text{phase})\, \psi, \quad \text{all } \psi \in \mathcal{H}. \tag{9.70}$$

It may appear at this stage that the phase here could depend on ψ. That this cannot happen is seen by applying the SC to two different $|\psi\rangle$ vectors and their superpositions. This establishes the triviality of $\tilde{\Omega}''$ action on $\mathcal{H}$. A similar argument holds when complex conjugation on all $\mathcal{H}^{(N)}$ is involved, so the proof is complete.

9.4 Applications to Quantum Mechanics – Ray Representations and Neutral Elements

Now if a Lie group G is the symmetry group of some quantum system, we ask: When can $g \in G$ be represented unitarily, when by an antiunitary transformation? In general we have the results that the product of two transformations – both unitary or both antiunitary – is unitary; while the product of a unitary and an antiunitary transformation is antiunitary. In the component G_0 of G containing elements connected continuously to the identity, certain important results hold:

(i) There is a neighbourhood $\mathcal{N} \subset G_0$ containing the identity such that every element $g \in \mathcal{N}$ lies on some OPS and so is the square of a unique 'square root' element $g' : g = (g')^2$.

(ii) Each $g \in G_0$ can be expressed as the product $g = g_1 g_2 \cdots g_s$ of a finite number of elements $g_j \in \mathcal{N}, j = 1, \cdots, s$, where s may depend on g.

As a result of all these known facts, we can see every element $g \in G_0$ can only be represented by a unitary transformation. So the antiunitary option can only arise for elements in G outside of G_0. In actual fact, *only* time reversal symmetry is represented by an antilinear antiunitary transformation in quantum mechanics. We now exclude this and consider only the unitary option.

So let a Lie group G be represented by unitary transformations on the Hilbert space $\mathcal{H}$ of some quantum system. For each $g \in G$, starting from its action on unit pure states or pure rays in the Wigner sense and then working one's way up, one arrives at a unitary operator $\bar{U}(g)$ acting on $\mathcal{H}$, with a certain amount of phase freedom in the construction of $\bar{U}(g)$: this is the freedom present in the 'lifting' of Ω on $\mathcal{R}$ to ω on $\mathcal{B}$ and then on to $\mathcal{H}$, depicted in the diagram accompanying Eq. (9.21). Next, in considering the successive actions of two elements g' and g we have to keep in mind the fact that the state vector to physical state map is unavoidably many to one, ψ and $e^{i\alpha}\psi$ being physically indistinguishable. This means that we must allow for a phase in the composition law

$$g',g \in G: \ \bar{U}(g')\bar{U}(g) = e^{i\omega(g';g)}\bar{U}(g'g),$$

$$\omega(g';g) = \text{some phase}. \tag{9.71}$$

Thus we can only ask for *ray representations* or *representations up to phases* of symmetry groups in quantum mechanics. The classic study of such representations in the case of Lie groups is the one carried out by Bargmann [Bargmann, 1954].

We can agree to arrange that $\bar{U}(e) = \mathbb{1}$ and $\bar{U}(g^{-1}) = \bar{U}(g)^{-1} = \bar{U}(g)^{\dagger}$ for every $g \in G$. These entail:

$$\omega(e;g) = \omega(g;e) = \omega(g;g^{-1}) = 0. \tag{9.72}$$

Now the associativity of group composition leads to a condition on $\omega(g';g)$ for any three elements $g,g',g'' \in G$:

$$\bar{U}(g'')\bar{U}(g')\bar{U}(g) = e^{i\omega(g'';g')}\bar{U}(g''g')\bar{U}(g)$$

$$= e^{i\omega(g'';g')+i\omega(g''g';g)}\bar{U}(g''g'g)$$

$$\text{also} \ = e^{i\omega(g';g)}\bar{U}(g'')\bar{U}(g'g)$$

$$= e^{i\omega(g';g)+i\omega(g'';g'g)}\bar{U}(g''g'g),$$

$$\text{i.e., } \omega(g'';g') + \omega(g''g';g) = \omega(g'';g'g) + \omega(g';g), \text{ mod } 2\pi. \tag{9.73}$$

This is a consistency condition in the form of a functional relation on possible phases $\omega(g';g)$. On the other hand, one such ω is completely equivalent to another ω' if we obtain it by merely attaching phase factors to individual $\bar{U}(g)$'s, which can always be done:

$$\bar{U}'(g) = e^{i\alpha(g)}\bar{U}(g) \Rightarrow \omega'(g';g) = \omega(g';g) + \alpha(g') + \alpha(g) - \alpha(g'g). \tag{9.74}$$

The phases $\alpha(g)$ can be chosen arbitrarily except for continuity and differentiability conditions, and the following consequences of (9.72):

$$\alpha(e) = 0, \ \alpha(g^{-1}) = -\alpha(g). \tag{9.75}$$

Then if $\omega(g';g)$ obeys (9.73), so will $\omega'(g';g)$, and we say they are *equivalent*.

We now ask: if a ray representation of G with phase $\omega(g';g)$ is given, is it possible to use the freedom (9.74) to reduce $\omega'(g';g)$ to zero? If the answer is 'yes', then we would have reduced the ray representation to a true (or vector) representation. If the answer is 'no', we can at best simplify $\omega(g';g)$ as much as possible by suitable choice of $\alpha(g)$. *This* is how and why $SO(3)$ rotational symmetry in quantum mechanics is able to 'make use of' UIR's and UR's of $SU(2)$, which when true are double-valued representations of $SO(3)$. In the case $G = SO(3)$ (because of its being doubly connected), the freedom (9.74) allows us to achieve only $e^{i\omega(g';g)} = \pm 1$.

What happens at the level of the Lie algebra $\mathbf{G}$ of G is this. Let $\{\mathbf{e}_j\}$ be a basis for $\mathbf{G}$, and C_{jk}^l the corresponding structure constants, as in Eq. (4.98). Then for each $\mathbf{e}_j$ we ask for a hermitian generator X_j to represent it, however, these only have to obey the necessary commutation relations up to additive c-numbers;

$$\mathbf{e}_j \in \mathbf{G} \longrightarrow X_j \text{ on } \mathcal{H}, \ X_j^\dagger = X_j:$$

$$[\mathbf{e}_j, \mathbf{e}_k] = C_{jk}^l \mathbf{e}_l \longrightarrow [X_j, X_k] = i(C_{jk}^l X_l + d_{jk}),$$

$$d_{jk} = \text{real numbers}. \tag{9.76}$$

These d_{jk} are called 'neutral elements', and they must obey the infinitesimal form of Eq. (9.73) which turns out to be

$$d_{jk} = -d_{kj},$$

$$\sum_{(jkl)} C_{jk}^m d_{ml} = 0. \tag{9.77}$$

The d_{jk} are the Lie algebra analogues of $\omega(g';g)$ in Eq. (9.71). The analogue of the equivalence transformation (9.74) is the fact that the X_j are needed only up to additive neutral elements, again c-numbers. If we shift the X_j by

$$X_j' = X_j + d_j \Rightarrow d_{jk}' = d_{jk} - C_{jk}^l d_l, \tag{9.78}$$

we get a physically equivalent representation of $\mathbf{G}$. We naturally regard d_{jk}' and d_{jk} as equivalent for any choice of d_j. The Jacobi identities for the structure constants ensure that d_{jk}' will obey Eq. (9.77) if d_{jk} do so. The freedom (9.78) can then be used to make d_{jk}' as simple as possible by suitable choice of d_j.

This whole situation has been studied beautifully by Wigner and Bargmann, and here are some important conclusions:

(i) If G is compact and simply connected, we can always transform $\omega(g';g)$ to zero, and need to consider only true or vector UR's. This is the case for $SU(n)$ and $USp(2n)$, and also for the two-fold covers $\overline{SO(2l)}$, $\overline{SO(2l+1)}$.

(ii) For $G = SO(3)$ or $SO(3,1)$: after maximum simplification we are left with a possible double-valuedness, $e^{i\omega(g';g)} = \pm 1$. This is also true for the Euclidean group $E(3)$, and the Poincaré group $\mathcal{P}$, which we take up in Chapter 10.

(iii) If G, and so $\boldsymbol{G}$ as well, is abelian, any neutral elements present in the commutation relations (9.76) cannot be transformed away since $C_{jk}^l = 0$, and so are definitely nontrivial. Thus, for example, the Heisenberg CCR's which we studied in Chapter 8 are the results of nontrivial neutral elements in the commutation relations for the abelian group of phase space translations.

(iv) As we will see in Chapter 10, the Galilei group $\mathcal{G}$ is a very important case where the commutation relations can contain nontrivial neutral elements.

(v) For semisimple $\boldsymbol{G}$, neutral elements can be transformed to zero.

9.5 Neutral Elements in Classical Mechanics

The concept of ray representations of Lie groups in quantum mechanics has led to the appearance of neutral elements in the commutation relations (9.76). One may be tempted to conclude that this is a specifically quantum mechanical feature since it has arisen from the nonobservability of overall phases in state vectors in Hilbert space. However, such neutral elements appear also in the classical case and we indicate briefly how this happens.

For purposes of classical canonical mechanics we are interested in realisations of Lie groups in which each group element acts as a canonical transformation on the phase space of the concerned system. These canonical transformations must (for consistency) obey the same composition law as in the group. Now we know that every infinitesimal canonical transformation possesses an infinitesimal generator, a phase space function; and the changes in the canonical variables are given in terms of Poisson brackets in the form

$$\delta q_r = \epsilon\{q_r, G(q,p)\}, \ \delta p_r = \epsilon\{p_r, G(q,p)\}. \tag{9.79}$$

Here ϵ is a small parameter and $G(q,p)$ the generator. We see immediately that $G(q,p)$ is determined only up to an additive constant by the transformation.

It now turns out after a detailed analysis that for a realisation of a Lie group G by canonical transformations, at the level of the Lie algebra $\mathbf{G}$ we only need a Poisson bracket realisation up to neutral elements. Thus, the classical replacement for Eq. (9.76) is:

$$e_j \in \mathbf{G} \longrightarrow G_j(q, p) = \text{real phase space function:}$$

$$[e_j, e_k] = C^l_{jk} e_l \longrightarrow \{G_j(q, p), G_k(q, p)\} = C^l_{jk} G_l(q, p) + d_{jk}. \tag{9.80}$$

We see that the sources of these neutral elements may be different in the two forms of mechanics; but finally their appearance and the conditions on them and freedom in them are similar, indeed identical, in the two cases. Thus there is nothing specifically quantum mechanical in them, and allowed nontrivial systems of neutral elements are actually associated with the group G and its Lie algebra $\mathbf{G}$ and can play a role in both classical and quantum mechanics.

Problems

P9.1 Show that in the generic case the Bargmann invariant

$$\Delta_m(\phi_1, \phi_2, \cdots, \phi_m)$$

for $m \geq 4$, Eq. (9.15), can be expressed in terms of lower order invariants as $\Delta_m = \Delta_{m-1} \Delta_3 / \Delta_2$ for suitably chosen arguments.

P9.2 A UR of $Sp(2, \mathbb{R})$ involves three hermitian generators J_0, K_1, K_2 obeying Eq. (8.13):

$$[J_0, K_1] = iK_2, \ [J_0, K_2] = -iK_1, \ [K_1, K_2] = -iJ_0.$$

Show that any real neutral elements which might be included on the right hand sides can be transformed away by simple redefinitions of the generators.

Bibliography

Bargmann, V. (1954). On Unitary Ray Representations of Continuous Groups. *Annals of Mathematics.* 59: 1

Bargmann, V. (1964). Note on Wigner's Theorem on Symmetry Operations. *J. Math. Phys.* 5: 862.

Simon, R., Mukunda, N., Chaturvedi, S. and Srinivasan, V. (2008). Two Elementary Proofs of the Wigner Theorem on Symmetry in Quantum Mechanics. *Phys. Lett. A.* 372: 6847

Simon, R., Mukunda, N., Chaturvedi, S. et al. (2014). Comment on: 'Two Elementary Proofs of the Wigner Theorem on Symmetry in Quantum Mechanics'. *Phys. Lett. A.* 378: 2332.

Wigner, E. P. (1931). *Gruppentheorie und ihre Anwendung auf die Quanten Mechanik der Atomspektren.* Braunschweig: Vieweg. pp. 251–254; English translation: (1959). Group Theory and Its Applications to the Quantum Mechanics of Atomic Spectra. New York: Academic Press. pp. 233–236.

10

Groups Related to Spacetime

The aim of this chapter is to look at the structures of Lie groups related to space and spacetime. We look very briefly at $SO(3)$ which we have studied earlier quite extensively, then the Euclidean group $E(3)$ which acts on physical space; the Galilei group on spacetime underlying nonrelativistic Galilean–Newtonian mechanics; the homogeneous Lorentz group $SO(3,1)$ and its double cover $SL(2,\mathbb{C})$; and finally the Poincaré group $\mathcal{P}$ basic to special relativity. In each case we look at the defining representation resulting from action on spacetime; useful descriptions of the group, the composition law and inverses; the structure of the Lie algebra and possible neutral elements which are permitted; and then a study of the UIR's of the concerned group. This involves constructing Casimir invariants, physical interpretation, etc. In the $SO(3,1)$ and $SL(2,\mathbb{C})$ cases, we also look at all their finite dimensional nonunitary representations. We will see similarities to the discussion of induced group representations in Chapter 7, in connection with representations of the group $E(3)$ and the Poincaré group.

10.1 $SO(3)$ **and** $SU(2)$

We studied these two groups and their UIR's in some detail in Chapter 3. We saw that $SU(2)$ is a two-fold covering of $SO(3)$. Here we first deal with the possible presence of neutral elements in a hermitian representation of the Lie algebra generators J_j. As we have seen in Chapter 9, Eq. (9.76), to handle ray representations of $SO(3)$ and $SU(2)$ in quantum mechanics we must allow for the presence of neutral elements d_{jk} in the basic angular momentum commutation relations:

$$[J_j, J_k] = i(\epsilon_{jkl}J_l + d_{jk}). \tag{10.1}$$

But these are immediately and easily eliminated: antisymmetry $d_{jk} = -d_{kj}$ implies $d_{jk} = \epsilon_{jkl}d_l$ for some real d_l; then if we redefine $J'_j = J_j + d_j$ we get the standard commutation relations without any neutral elements:

$$d_{jk} = -d_{kj} \Rightarrow d_{jk} = \epsilon_{jkl}d_l \Rightarrow J'_j = J_j + d_j : [J'_j, J'_k] = \epsilon_{jkl}J'_l. \tag{10.2}$$

This happens since physical space is three dimensional, there being no need to invoke the Jacobi identity explicitly. At the same time, no further shifts in J'_j are permitted.

We can therefore always assume, whenever $SO(3)$ occurs as a subgroup of a larger group, that there are no neutral elements in its commutation relations. So we find, as $SU(2)$ is simply connected, every ray representation of $SU(2)$ can be converted to a true or vector representation. And for $SO(3)$, only the double-valuedness has to be allowed for when $j = \frac{1}{2}, \frac{3}{2}, \cdots$.

There is one other item of unfinished business – the concept of spherical tensor operators. We have seen in Eqs. (3.43, 3.93) that under conjugation the generators J_j transform in a particular way, namely by the defining UIR of $SO(3)$ which is the adjoint representation for both $SO(3)$ and $SU(2)$:

$$A \in SO(3) : \mathcal{D}(A)J_j\,\mathcal{D}(A)^{-1} = A_{kj}J_k;$$

$$u \in SU(2) : \mathcal{D}(u)J_j\,\mathcal{D}(u)^{-1} = A_{kj}(u)J_k. \tag{10.3}$$

We say that J_j are the components of an (irreducible) tensor operator of rank 1. More generally, an irreducible tensor operator of rank j is a set of $(2j+1)$ operators T^j_m, say, behaving as follows under conjugation in any UR or UIR of $SO(3)$ or $SU(2)$:

$$\mathcal{D}(A)T^j_m\mathcal{D}(A)^{-1} = \sum_{m'} \mathcal{D}^{(j)}_{m'm}(A)T^j_{m'}. \tag{10.4}$$

Here it is understood that each T^j_m is a linear operator on the space $\mathcal{V}$ on which the $\mathcal{D}(A)$ act. Such a collection $\{T^j_m\}$ constitutes a rank j irreducible tensor. This can also be expressed in terms of the generators of $\mathcal{D}(A)$:

$$[J_3, T^j_m] = mT^j_m, \quad [J_1 \pm iJ_2, T^j_m] = \sqrt{(j \mp m)(j \pm m + 1)}\, T^j_{m\pm 1}. \tag{10.5}$$

Indeed we find:

$$e^{-i\boldsymbol{\alpha}\cdot\boldsymbol{J}}\,T_m^j\,e^{i\boldsymbol{\alpha}\cdot\boldsymbol{J}} = T_m^j - i[\boldsymbol{\alpha}\cdot\boldsymbol{J},\ T_m^j] - \frac{1}{2!}[\boldsymbol{\alpha}\cdot\boldsymbol{J},[\boldsymbol{\alpha}\cdot\boldsymbol{J},T_m^j]] + \cdots$$

$$= \sum_{m'} \mathcal{D}^{(j)}_{m'm}(A(\boldsymbol{\alpha}))\,T_{m'}^j. \tag{10.6}$$

Unlike a numerical tensor with components defined to transform in a prescribed manner, here the various components T_m^j are all nonvanishing linear operators. In the former case it can well happen that in a specially chosen 'frame' or coordinate system one (or more) of the components of a tensor may vanish. However, for a nontrivial tensor operator T_m^j, each component has to be nonzero, as is especially clear from Eq. (10.5).

A possible 'reality' condition one might impose is, for instance,

$$T_m^{j\dagger} = (-1)^{j-m}\,T_{-m}^j, \tag{10.7}$$

as the two sides transform in the same way. Such a condition is consistent with Eq. (10.5).

10.2 The Euclidean Group $E(3)$

This is the group of proper orthogonal rotations and rigid translations acting on physical three dimensional space. An element $g \in E(3)$ is a pair made up of some $A \in SO(3)$ and some translation $\boldsymbol{a} \in \mathbb{R}^3$, acting on a point $\boldsymbol{x}$ in Euclidean space by:

$$g = (A, \boldsymbol{a}) \in E(3) : \boldsymbol{x} \to \boldsymbol{x}' = g\boldsymbol{x} = A\boldsymbol{x} + \boldsymbol{a},$$

$$x_j' = A_{jk}x_k + a_j. \tag{10.8}$$

(Here, of course, the choice of an origin and right handed Cartesian coordinate axes are both understood.) Clearly this is a six parameter Lie group. The composition law follows from (10.8):

$$(A', \boldsymbol{a}')(A, \boldsymbol{a}) = (A'A, \boldsymbol{a}' + A'\boldsymbol{a}). \tag{10.9}$$

Associativity is easy to check, while the identity and inverses are given by

$$e \to (\mathbb{1}, \boldsymbol{0}); \ (A, \boldsymbol{a})^{-1} = (A^T, -A^T\boldsymbol{a}). \tag{10.10}$$

$SO(3)$ is a subgroup, and translations T_3 an invariant subgroup, of $E(3)$:

$$SO(3) = \{(A, \mathbf{0}) \mid A \in SO(3)\} \subset E(3),$$

$$T_3 = \{(\mathbb{1}, \mathbf{a}) \mid \mathbf{a} \in \mathbb{R}^3\} \subset E(3),$$

$$SO(3) = E(3)/T_3; \tag{10.11}$$

and $E(3)$ is their semidirect product:

$$E(3) = SO(3) \rtimes T_3. \tag{10.12}$$

In a UR or UIR of $E(3)$, we have unitary operators $\overline{U}(A, \mathbf{a})$ representing elements $(A, \mathbf{a}) \in E(3)$. As any such element is uniquely expressible as a product of elements drawn from the two subgroups,

$$(A, \mathbf{a}) = (A, \mathbf{0})(\mathbb{1}, A^T \mathbf{a}) = (\mathbb{1}, \mathbf{a})(A, \mathbf{0}), \tag{10.13}$$

we have unitary operators $\overline{U}(A), \overline{U}(\mathbf{a})$, respectively:

$$(\mathbb{1}, \mathbf{a}) \in T_3 \subset E(3) \to \overline{U}(\mathbf{a}),$$

$$(A, \mathbf{0}) \in SO(3) \subset E(3) \to \overline{U}(A). \tag{10.14}$$

Their rules of composition are easily read off from Eq. (10.9):

$$\overline{U}(A, \mathbf{a}) = \overline{U}(\mathbf{a})\overline{U}(A):$$

$$\overline{U}(A', \mathbf{a}')\overline{U}(A, \mathbf{a}) = \overline{U}(A'A, \mathbf{a}' + A'\mathbf{a}) \Rightarrow$$

$$\overline{U}(A')\overline{U}(A) = \overline{U}(A'A),$$

$$\overline{U}(\mathbf{a}')\overline{U}(\mathbf{a}) = \overline{U}(\mathbf{a}' + \mathbf{a}),$$

$$\overline{U}(A)\overline{U}(\mathbf{a})\overline{U}(A^{-1}) = \overline{U}(A\mathbf{a}). \tag{10.15}$$

Thus, $\overline{U}(A)$ give a UR of $SO(3)$, $\overline{U}(\mathbf{a})$ of T_3, and the last relation expresses the semidirect product structure. Now to the generators and their commutation relations corresponding to the Lie algebra $\mathbf{E(3)}$.

Let us write

$$\overline{U}(A(\boldsymbol{\alpha})) = \exp(-i\boldsymbol{\alpha} \cdot \mathbf{J}), \quad \overline{U}(\mathbf{a}) = \exp(-i\mathbf{a} \cdot \mathbf{P}). \tag{10.16}$$

So J and P generate rotations and translations, respectively. Then from the first and second of the three relations in (10.15) we get the commutation relations

$$[J_j, J_k] = i\epsilon_{jkl}J_l,$$
$$[P_j, P_k] = 0. \tag{10.17}$$

The remaining J-P commutators follow from the third relation in (10.15):

$$\overline{U}(A)P_j\overline{U}(A)^{-1} = A_{kj}P_k. \tag{10.18}$$

Taking $A = A(\boldsymbol{\alpha})$ for infinitesimal $\boldsymbol{\alpha}$ we immediately get:

$$[J_j, P_k] = i\epsilon_{jkl}P_l. \tag{10.19}$$

Thus the $E(3)$ relations are the combined set $(10.17, 10.19)$ with hermitian J_j and P_j.

Now we ask if nontrivial neutral elements can be present. We already know that by suitable choices of J_j there are none in the $SO(3)$ relations. Next we consider the possibility

$$[J_j, P_k] = i(\epsilon_{jkl}P_l + d_{jk}). \tag{10.20}$$

The Jacobi identities

$$[J_m, [J_j, P_k]] + [J_j, [P_k, J_m]] + [P_k, [J_m, J_j]] = 0 \tag{10.21}$$

lead to restrictions on the neutral elements d_{jk} in (10.20):

$$\epsilon_{jkl}d_{ml} - \epsilon_{mkl}d_{jl} + \epsilon_{jml}d_{lk} = 0. \tag{10.22}$$

This is explicitly antisymmetric in jm, so we contract it with $\frac{1}{2}\epsilon_{jmp}$ to get:

$$\frac{1}{2}(\delta_{km}\delta_{lp} - \delta_{kp}\delta_{lm})d_{ml} + \frac{1}{2}(-\delta_{kp}\delta_{lj} + \delta_{kj}\delta_{lp})d_{jl} + d_{pk} = 0,$$

$$\text{i.e.,} \quad d_{kp} + d_{pk} = \delta_{kp}d_{ll}. \tag{10.23}$$

Setting $k = p$ and summing shows that $d_{ll} = 0$, hence $d_{jk} = -d_{kj} = \epsilon_{jkl}d_l$, say, for some neutrals d_l. Use this in (10.20), absorb d_l in P_l by redefining $P_l \to P_l + d_l$, so all neutrals are eliminated from the $[J, P]$ brackets. At the same time, we see that there is no residual freedom to shift either J or P. Now turn to the last set, the second line in Eq. (10.17), and for simplicity use d_{jk} again to write:

$$[P_j, P_k] = i d_{jk} = -i d_{kj}.$$ (10.24)

The Jacobi identities to be used now involve one J and two P's:

$$[P_m, [J_j, P_k]] + [J_j, [P_k, P_m]] + [P_k, [P_m, J_j]] = 0,$$

i.e., $[P_m, [J_j, P_k]] = [P_k, [J_j, P_m]],$

i.e., $\epsilon_{jkl} d_{ml} = \epsilon_{jml} d_{kl},$

i.e., $\epsilon_{jrs} \epsilon_{jkl} d_{ml} = \epsilon_{jrs} \epsilon_{jml} d_{kl},$

i.e., $(\delta_{kr}\delta_{ls} - \delta_{ks}\delta_{lr}) d_{ml} = (\delta_{mr}\delta_{ls} - \delta_{ms}\delta_{lr}) d_{kl},$

i.e., $\delta_{kr} d_{ms} - \delta_{ks} d_{mr} = \delta_{mr} d_{ks} - \delta_{ms} d_{kr}.$ (10.25)

Here the indices *mkrs* are all free. If we set $k = r$ and sum, we get:

$$3 d_{ms} - d_{ms} = d_{ms} \Rightarrow d_{jk} = 0.$$ (10.26)

This is just the statement that there is no nonzero numerical second rank *antisymmetric* tensor with respect to $SO(3)$, unlike δ_{jk} in the symmetric case.

Thus in all the $E(3)$ Lie algebra relations the neutrals can be transformed to zero by suitably shifting the J_j and P_j, and one has the simple relations (10.17, 10.19). Globally $E(3)$ has the structure of $SO(3) \times \mathbb{R}^3$, so the nontriviality resides only in the $SO(3)$ factor. For quantum mechanical purposes one must permit double-valued UR's of $SO(3)$, which means any faithful or nonfaithful UR's of $SU(2)$. At the Lie algebra level we have Eqs. (10.17, 10.19). In later situations if $E(3)$ is present as a subalgebra we can assume right away that there are no neutrals in its commutation relations, and also no freedom to shift its generators by neutral elements.

Now we turn to the problem of constructing all quantum mechanically acceptable UIR's of $E(3)$, using a combination of generator commutation relations and actions by finite group elements. To begin with, from the commutation relations (10.17, 10.19) we find we can construct two hermitian quadratic Casimir invariants:

$$\mathcal{C}_1 = \mathcal{C}_1^\dagger = \boldsymbol{P} \cdot \boldsymbol{P} = P_j P_j,$$

$$\mathcal{C}_2 = \mathcal{C}_2^\dagger = \boldsymbol{J} \cdot \boldsymbol{P} = \boldsymbol{P} \cdot \boldsymbol{J} = J_j P_j :$$

$$[\mathcal{C}_1 \text{ or } \mathcal{C}_2, J_j \text{ or } P_j] = 0.$$ (10.27)

It turns out that the numerical values of $\mathcal{C}_1$ and $\mathcal{C}_2$ can be used to label the UIR's. In case $\mathcal{C}_1 = 0$, $P_j = 0$ identically, the translations T_3 are represented trivially, so the $E(3)$ UIR reduces to some UIR of $SU(2)$. Thus every UIR of $SU(2)$ is, as it stands, a nonfaithful finite dimensional UIR of $E(3)$, with $\mathcal{C}_1 = \mathcal{C}_2 = 0$.

Let us next look at UIR's of $E(3)$ with $\mathcal{C}_1 > 0$. It is now convenient to work with a combination of generator commutation relations and actions by finite group elements. We need an irreducible representation of the system of operator relations

$$A', A \in SO(3): \overline{U}(A')\overline{U}(A) = \pm\overline{U}(A'A),$$

$$\overline{U}(A)P_j\overline{U}(A)^{-1} = A_{kj}P_k,$$

$$[P_j, P_k] = 0. \tag{10.28}$$

with unitary $\overline{U}(A)$ and hermitian P_j. We will soon see that the values of $\mathcal{C}_1, \mathcal{C}_2$ have to be of the forms

$$\mathcal{C}_1 = p_0^2, \; p_0 \text{ real } > 0;$$

$$\mathcal{C}_2 = \lambda p_0, \; \lambda = 0, \pm\frac{1}{2}, \; \pm 1, \cdots; \tag{10.29}$$

and then the pair $\{p_0, \lambda\}$ denotes one infinite dimensional nontrivial UIR of $E(3)$. The physical meanings are that p_0 is the magnitude of the momentum, and λ is the helicity (i.e., component of angular momentum along momentum). To construct this UIR $\{p_0, \lambda\}$ we use an ideal basis in which the commuting momenta P_j are all diagonal:

$$P_j|\boldsymbol{p}, \cdots\rangle = p_j|\boldsymbol{p}, \cdots\rangle, \; \boldsymbol{p} \cdot \boldsymbol{p} = p_0^2,$$

$$\langle\boldsymbol{p}', \cdots|\boldsymbol{p}, \cdots\rangle \simeq \delta^{(2)}(\hat{\boldsymbol{p}}', \hat{\boldsymbol{p}}) \cdots. \tag{10.30}$$

Here the vector $\boldsymbol{p}$ ranges over a sphere of radius p_0, the dots in the vectors denote any needed additional quantum numbers or state labels, and $\delta^{(2)}(\cdots)$ is the 'surface delta function'. Then the second relation in (10.28) is obeyed by schematically setting

$$\overline{U}(A)|\boldsymbol{p}, \cdots\rangle \sim |A\boldsymbol{p}, \cdots\rangle. \tag{10.31}$$

and in fact implies this. Now choose $\boldsymbol{p}^{(0)} = (0, 0, p_0)$ along the positive z-axis, to which 'position' or configuration any $\boldsymbol{p}$ can be rotated. The subgroup of $SO(3)$ which preserves $\boldsymbol{p}^{(0)}$, its stability group or 'little group', consists of $SO(2)$ rotations about the

z-axis. In the notation of Eq. (3.20),

$$A(e_3, \alpha) \in SO(2, e_3): A(e_3, \alpha)p^{(0)} = p^{(0)}, \tag{10.32}$$

so the basis vectors $|p^{(0)}, \cdots\rangle$ must be just such as to provide a UIR of this subgroup. Since $SO(2)$ is abelian, its UIR's are all one dimensional, so we can omit the dots in $|p, \cdots\rangle$ and define the helicity λ by the statement

$$\overline{U}(A(e_3, \alpha))|p^{(0)}\rangle = e^{-i\alpha\lambda}|p^{(0)}\rangle,$$
$$J_3|p^{(0)}\rangle = \lambda|p^{(0)}\rangle, \tag{10.33}$$

where $\lambda = 0$ or $\pm\frac{1}{2}$ or $\pm 1 \cdots$.

Next, for each numerical momentum p with magnitude p_0, we choose some element $A_0(p) \in SO(3)$ carrying $p^{(0)}$ to p:

$$p = A_0(p)p^{(0)} \tag{10.34}$$

For instance, using spherical polar coordinates θ, ϕ for p we are led to a simple choice for $A_0(p)$:

$$p = p_0(\sin\theta\cos\phi, \sin\theta\sin\phi, \cos\theta), \ (0 \le \theta \le \pi, \ 0 \le \phi \le 2\pi) \rightarrow$$
$$A_0(p) = A(e_3, \phi)A(e_2, \theta). \tag{10.35}$$

This is shown in the figure below:

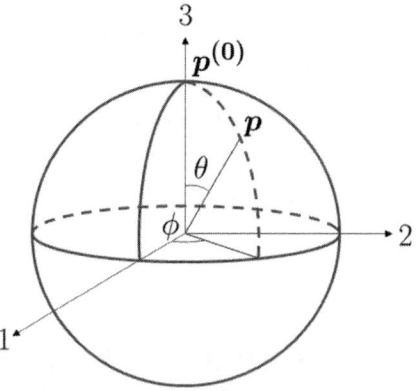

Figure 10.1   A choice for the rotation $A_0(p)$

Then we set by definition:

$$|\boldsymbol{p}\rangle = \overline{U}(A_0(\boldsymbol{p}))|\boldsymbol{p}^{(0)}\rangle,$$

$$\langle \boldsymbol{p}'|\boldsymbol{p}\rangle = \delta^{(2)}(\hat{\boldsymbol{p}}',\hat{\boldsymbol{p}}). \tag{10.36}$$

With this, all the relations (10.28) are obeyed, and we have the UIR $\{p_0,\lambda\}$ of $E(3)$. To 'see' this explicitly, for a general A and $\boldsymbol{p}$ we develop, in addition to Eq. (10.30):

$$\overline{U}(A)|\boldsymbol{p}\rangle = \overline{U}(A)\overline{U}(A_0(\boldsymbol{p}))|\boldsymbol{p}^{(0)}\rangle$$

$$= \text{by (10.31), expected to be a phase times } |A\boldsymbol{p}\rangle$$

$$= \overline{U}(A_0(A\boldsymbol{p}))\overline{U}(A_0(A\boldsymbol{p}))^{-1}\overline{U}(A)\overline{U}(A_0(\boldsymbol{p}))|\boldsymbol{p}^{(0)}\rangle$$

$$= \overline{U}(A_0(A\boldsymbol{p}))\overline{U}(A(\boldsymbol{e}_3,\alpha(A,\boldsymbol{p})))|\boldsymbol{p}^{(0)}\rangle, \tag{10.37}$$

where necessarily

$$A_0(A\boldsymbol{p})^{-1}AA_0(\boldsymbol{p}) = A(\boldsymbol{e}_3,\alpha(A,\boldsymbol{p})), \tag{10.38}$$

resulting in

$$\overline{U}(A)|\boldsymbol{p}\rangle = e^{-i\lambda\alpha(A,\boldsymbol{p})}|A\boldsymbol{p}\rangle. \tag{10.39}$$

This is an instance of the 'Wigner rotation' which we will encounter again later.

As we expect, $\overline{U}(A)$ has the effect of rotating the momentum eigenvalue by A, and in addition there is a phase due to the helicity λ. From Eqs. (10.30, 10.37) we know the effects of both $\overline{U}(\mathbb{I},\boldsymbol{a})$ and $\overline{U}(A,\boldsymbol{0})$ on the momentum eigenvectors $|\boldsymbol{p}\rangle$, so all the $\overline{U}(A,\boldsymbol{a})$ are in hand.

We can give a description of the Hilbert space $\mathcal{H}$ carrying this UIR $\{p_0,\lambda\}$ as a space of suitable square integrable functions in two equivalent and convenient ways. First we describe $\mathcal{H}$ via square integrable functions on $\mathbb{S}^2$. Each $\boldsymbol{p}$ which occurs is of the form $\boldsymbol{p} = p_0\hat{\boldsymbol{p}}(\theta,\phi)$, so we can write:

$$|f\rangle \in \mathcal{H}: f(\boldsymbol{p}) = \langle p_0\hat{\boldsymbol{p}}|f\rangle,$$

$$\|f\|^2 = \int_0^{2\pi} d\phi \int_0^{\pi} \sin\theta \, d\theta |f(\boldsymbol{p})|^2,$$

$$(P_j f)(\boldsymbol{p}) = p_j f(\boldsymbol{p}),$$

$$(\overline{U}(A)f)(\boldsymbol{p}) = \langle \boldsymbol{p}|\overline{U}(A)|f\rangle$$

$$= e^{i\lambda\alpha(A^{-1},\boldsymbol{p})}f(A^{-1}\boldsymbol{p}). \tag{10.40}$$

where the adjoint of Eq. (10.37) was used after the change $A \to A^{-1}$. Notice again the presence of the helicity dependent phase factor.

Now another elegant way to describe $\mathcal{H}$ is to 'elevate' $f(\boldsymbol{p})$ to a function $\tilde{f}(A)$ on $SO(3)$, or better to $\tilde{f}(u)$ on $SU(2)$, with a special simplifying feature. In Chapter 3 we introduced the Euler angles parametrisations of $SO(3)$ and $SU(2)$. Let us change the 'names' of the variables so that the 'usual' spherical angles on $\mathbb{S}^2$ turn out to be θ and ϕ, as in Eqs. (10.35, 10.40) above. So, in place of Eq. (3.24) let us write:

$$A(\phi,\theta,\psi) = A(\boldsymbol{e}_3,\phi)A(\boldsymbol{e}_2,\theta)A(\boldsymbol{e}_3,\psi) \in SO(3),$$

$$0 \le \phi, \ \psi \le 2\pi, \ \ 0 \le \theta \le \pi. \tag{10.41}$$

Then Eqs. (10.33, 10.36) become.

$$\overline{U}(A(\phi,\theta,\psi))|\boldsymbol{p}^{(0)}\rangle = e^{-i\lambda\psi}|\boldsymbol{p}(\theta,\phi)\rangle. \tag{10.42}$$

Now represent each $|f\rangle \in \mathcal{H}$ by a 'wave function' $\tilde{f}(A)$ over $SO(3)$ (or over $SU(2)$ if necessary):

$$\tilde{f}(A) = \langle \boldsymbol{p}^{(0)}|\overline{U}(A)^{-1}|f\rangle = \tilde{f}(\phi,\theta,\psi) = e^{i\lambda\psi}f(\boldsymbol{p}(\theta,\phi)),$$

i.e., $\quad \tilde{f}(\phi,\theta,\psi) = e^{i\lambda\psi}\tilde{f}(\phi,\theta,0). \tag{10.43}$

This is called a 'covariance condition' with respect to $SO(2,\boldsymbol{e}_3)$ on the right:

$$\tilde{f}(A\,A(\boldsymbol{e}_3,\alpha)) = e^{i\lambda\alpha}\tilde{f}(A). \tag{10.44}$$

Thus these are wave functions on $SO(3)$ with a simple prescribed ψ-dependence. Their (left) $SO(3)$ behaviour is quite simple:

$$B \in SO(3): (\overline{U}(B)\tilde{f})(A) = \langle \boldsymbol{p}^{(0)}|\overline{U}(A^{-1})\overline{U}(B)|f\rangle$$

$$= \tilde{f}(B^{-1}A). \tag{10.45}$$

Clearly this $SO(3)$ action does not 'interfere' with the covariance condition controlling the ψ dependence, since B^{-1} is on the left of A in Eq. (10.45). It also immediately follows that

$$\|f\|^2 = \int_0^{2\pi} d\phi \int_0^\pi \sin\theta\,d\theta|f(\boldsymbol{p}(\theta,\phi))|^2$$

$$= 4\pi \int_{SO(3)} dA|\tilde{f}(A)|^2. \tag{10.46}$$

The 4π factor occurs because $dA = \sin\theta\,d\theta\,d\phi\,d\psi/8\pi^2$.

Now $\tilde{f}(A)$ possesses an expansion (3.54) or (3.98a–3.98c). Adopting the latter:

$$\tilde{f}(A(\phi,\theta,\psi)) = \sum_{jmm'} (2j+1)^{1/2} f^{(j)}_{mm'} \, \mathcal{D}^{(j)}_{mm'}(\phi,\theta,\psi). \tag{10.47}$$

But the prescribed ψ-dependence in Eq. (10.43) means that only terms with $m' = -\lambda$ survive, so for wave functions in the UIR $\{p_0, \lambda\}$ of $E(3)$ we have the expressions

$$\tilde{f}(\phi,\theta,\psi) = \sum_{j=|\lambda|,|\lambda|+1,\cdots}^{\infty} \sum_{m=-j}^{j} (2j+1)^{1/2} f^{(j)}_m \mathcal{D}^{(j)}_{m,-\lambda}(\phi,\theta,\psi),$$

$$f(\boldsymbol{p}(\theta,\phi)) = \sum_{j=|\lambda|,|\lambda|+1,\cdots}^{\infty} \sum_{m=-j}^{j} (2j+1)^{1/2} f^{(j)}_m \mathcal{D}^{(j)}_{m,-\lambda}(\phi,\theta,0). \tag{10.48}$$

(Recall that $\mathcal{D}^{(j)}_{mm'}(\phi,\theta,\psi) = e^{-im\phi-im'\psi} d^{(j)}_{mm'}(\theta)$. Formulae for the expansion coefficients, Parseval's formula, etc., are easy to obtain.)

Thus for each fixed helicity $\lambda = 0, \pm\frac{1}{2}, \pm 1, \cdots$, the functions $\{(2j+1)^{1/2}\mathcal{D}^{(j)}_{m,-\lambda}(\phi,\theta,0)\}$ with $j \geq |\lambda|$, $m = j, \cdots, -j$ form an orthonormal set over $\mathbb{S}^2$, a helicity generalisation of the familiar spherical harmonics $\{Y_{lm}(\theta,\phi)\}$ which correspond to $\lambda = 0$. Of course, $\lambda = $ integer/half odd integer implies $j = $ integer/half odd integer, but in both cases we get an orthonormal complete set of functions on $\mathbb{S}^2$.

10.3 The Galilei Group $\mathcal{G}$

This   is
the group of spacetime transformations basic to nonrelativistic Galilean–Newtonian mechanics. To the elements of the Euclidean group $E(3)$ we add Galilean velocity transformations and time translations. The pattern of the analysis is similar to the $E(3)$ case, but now we find there is room for nontrivial neutral elements.

A general Galilean transformation acts on the Cartesian coordinates of points in three dimensional Euclidean space and on (absolute) time in this way:

$$A \in SO(3), \ \boldsymbol{v} \in \mathbb{R}^3, \ \boldsymbol{a} \in \mathbb{R}^3, \ b \in \mathbb{R}:$$

$$\boldsymbol{x}' = A\boldsymbol{x} - \boldsymbol{v}t + \boldsymbol{a}, \ t' = t - b. \tag{10.49}$$

So we denote elements $g \in \mathcal{G}$ by quadruples:

$$g = (A, \boldsymbol{v}, b, \boldsymbol{a}). \tag{10.50}$$

This is clearly a ten parameter continuous group, three each in $A, \boldsymbol{v}, \boldsymbol{a}$ and one for b. The structure of the group manifold is

$$\mathcal{G} \sim SO(3) \times \mathbb{R}^3 \times \mathbb{R} \times \mathbb{R}^3. \qquad (10.51)$$

The composition law, identity and inverses are easily found:

$$\begin{aligned}
g'g &= (A', \boldsymbol{v}', b', \boldsymbol{a}')(A, \boldsymbol{v}, b, \boldsymbol{a}) \\
&= (A'A, \boldsymbol{v}' + A'\boldsymbol{v}, b' + b, \boldsymbol{a}' + A'\boldsymbol{a} + b\boldsymbol{v}'); \\
e &= (\mathbb{1}, \boldsymbol{0}, 0, \boldsymbol{0}); \\
(A, \boldsymbol{v}, b, \boldsymbol{a})^{-1} &= (A^T, -A^T\boldsymbol{v}, -b, -A^T\boldsymbol{a} + bA^T\boldsymbol{v}).
\end{aligned} \qquad (10.52)$$

Associativity is easily checked.

There are several significant subgroups:

$$\begin{aligned}
SO(3) &= \{(A, \boldsymbol{0}, 0, \boldsymbol{0})\}; \\
E(3) &= \{(A, \boldsymbol{0}, 0, \boldsymbol{a})\}; \\
E^{(v)}(3) &= \{(A, \boldsymbol{v}, 0, \boldsymbol{0})\}; \\
T_4 &= \{(\mathbb{1}, \boldsymbol{0}, b, \boldsymbol{a})\}.
\end{aligned} \qquad (10.53)$$

We are already familiar with $SO(3)$ and $E(3)$ acting on space alone; $E^{(v)}(3)$ consists of rotations and pure Galilean transformations and has the same structure as $E(3)$, namely as in Eq. (10.9),

$$(A', \boldsymbol{v}', 0, \boldsymbol{0})(A, \boldsymbol{v}, 0, \boldsymbol{0}) = (A'A, \boldsymbol{v}' + A'\boldsymbol{v}, 0, \boldsymbol{0}). \qquad (10.54)$$

T_4 is the abelian subgroup of spacetime translations:

$$(\mathbb{1}, \boldsymbol{0}, b', \boldsymbol{a}')(\mathbb{1}, \boldsymbol{0}, b, \boldsymbol{a}) = (\mathbb{1}, \boldsymbol{0}, b' + b, \boldsymbol{a}' + \boldsymbol{a}). \qquad (10.55)$$

We have another interesting seven parameter subgroup made up of all elements with $A = \mathbb{1}$, their composition law given by

$$(\mathbb{1}, \boldsymbol{v}', b', \boldsymbol{a}')(\mathbb{1}, \boldsymbol{v}, b, \boldsymbol{a}) = (\mathbb{1}, \boldsymbol{v}' + \boldsymbol{v}, b' + b, \boldsymbol{a}' + \boldsymbol{a} + b\boldsymbol{v}'). \qquad (10.56)$$

This is an invariant nonabelian subgroup, and $\mathcal{G}$ is the semidirect product of $SO(3)$ with it.

A general $g \in \mathcal{G}$ is the product of four factors in a particular order: we can check from Eq. (10.52) that

$$g = (A, \boldsymbol{v}, b, \boldsymbol{a}) = (\mathbb{1}, \boldsymbol{0}, b, \boldsymbol{0})(\mathbb{1}, \boldsymbol{v}, 0, \boldsymbol{0})(\mathbb{1}, \boldsymbol{0}, 0, \boldsymbol{a})(A, \boldsymbol{0}, 0, \boldsymbol{0}). \qquad (10.57)$$

Therefore in a UR or UIR of $\mathcal{G}$ we have unitary operators $\overline{U}(g)$ expressed as products:

$$\overline{U}(g) = \overline{U}(\mathbb{1}, \boldsymbol{0}, b, \boldsymbol{0})\overline{U}(\mathbb{1}, \boldsymbol{v}, 0, \boldsymbol{0})\overline{U}(\mathbb{1}, \boldsymbol{0}, 0, \boldsymbol{a})\overline{U}(A, \boldsymbol{0}, 0, \boldsymbol{0})$$
$$= e^{-ibH} e^{-i\boldsymbol{v}\cdot\boldsymbol{G}} e^{-i\boldsymbol{a}\cdot\boldsymbol{P}} e^{-i\boldsymbol{\alpha}\cdot\boldsymbol{J}}. \qquad (10.58)$$

In the second step we have introduced ten hermitian generators $H, \boldsymbol{G}, \boldsymbol{P}, \boldsymbol{J}$ generating time translations, pure velocity transformations, space translations, and space rotations, respectively. The subsets generating each of the subgroups listed in Eq. (10.53) are easily identified. We can write down relations analogous to Eq. (10.15) in the $E(3)$ case, but will consider selected ones as we need them in the sequel.

Now we examine the commutation relations among the ten generators and the possible presence of neutral elements in them. From the $SO(3)$ and $E(3)$ analyses we know that using the freedom to add neutral elements to $\boldsymbol{J}, \boldsymbol{G}, \boldsymbol{P}$ we can secure the following commutation relations, with no remaining freedom to shift these generators:

$$[J_j, J_k \text{ or } P_k \text{ or } G_k] = i\epsilon_{jkl} (J_l \text{ or } P_l \text{ or } G_l),$$
$$[P_j, P_k] = [G_j, G_k] = 0. \qquad (10.59)$$

The first set leads to the expected conjugation relations

$$\overline{U}(A)(J_j \text{ or } P_j \text{ or } G_j)\overline{U}(A)^{-1} = A_{kj} (J_k \text{ or } P_k \text{ or } G_k). \qquad (10.60)$$

The remaining commutators are $[J_j, H], [P_j, G_k], [P_j, H]$ and $[G_j, H]$. For each of these, we should see what the Lie algebra $\mathcal{G}$ says about them, then whether there could be added neutral elements.

Since space rotations and time translations commute, the commutator $[J_j, H]$ can at best be a neutral element:

$$[J_j, H] = id_j. \qquad (10.61)$$

The Jacobi identity with one more generator J_k gives:

$$[J_k,[J_j,H]]+[J_j,[H,J_k]] + [H,[J_k,J_j]] = 0 \Rightarrow$$

$$\epsilon_{kjl}d_l = 0 \Rightarrow$$

$$d_j = 0 \Rightarrow$$

$$[J_j,H] = 0. \tag{10.62}$$

This amounts to the statement that there is no nonzero invariant numerical three dimensional vector.

Next, since T_4 is abelian, the commutator $[P_j, H]$ can at best be a neutral element but for the same reason as presented above, it must vanish:

$$[P_j,H] = id_j : [J_k,[P_j,H]] + [P_j,[HJ_k]] + [H,[J_k,P_j]] = 0 \Rightarrow$$

$$\epsilon_{kjl}d_l = 0 \Rightarrow$$

$$d_j = 0 \Rightarrow$$

$$[P_j,H] = 0. \tag{10.63}$$

The remaining commutators are $[H, G_j]$ and $[G_j, P_k]$. For the former we start with the composition law involving time translations and Galilean transformations:

$$(\mathbb{1},\mathbf{0},b,\mathbf{0})(\mathbb{1},\boldsymbol{v},0,\mathbf{0}) = (\mathbb{1},\boldsymbol{v},b,\mathbf{0}),$$

$$(\mathbb{1},\boldsymbol{v},0,\mathbf{0})(\mathbb{1},\mathbf{0},b,\mathbf{0}) = (\mathbb{1},\boldsymbol{v},b,b\boldsymbol{v});$$

$$(\mathbb{1},\mathbf{0},b,\mathbf{0})(\mathbb{1},\boldsymbol{v},0,\mathbf{0})(\mathbb{1},\mathbf{0},b,\mathbf{0})^{-1}(\mathbb{1},\boldsymbol{v},0,\mathbf{0})^{-1}$$

$$= (\mathbb{1},\boldsymbol{v},b,\mathbf{0})(\mathbb{1},\mathbf{0},-b,\mathbf{0})(\mathbb{1},-\boldsymbol{v},0,\mathbf{0})$$

$$= (\mathbb{1},\boldsymbol{v},b,\mathbf{0})(\mathbb{1},-\boldsymbol{v},-b,\mathbf{0})$$

$$= (\mathbb{1},\mathbf{0},0,-b\boldsymbol{v}). \tag{10.64}$$

Therefore apart from possible neutrals we should have

$$e^{-ibH}e^{-i\boldsymbol{v}\cdot\boldsymbol{G}}e^{ibH}e^{i\boldsymbol{v}\cdot\boldsymbol{G}} = e^{ib\boldsymbol{v}\cdot\boldsymbol{P}},$$

$$\text{i.e.,} \quad [G_j,H] = iP_j. \tag{10.65}$$

Now let us allow for a neutral element, take the commutator with J_k and use the Jacobi identity:

$$[G_j, H] = i(P_j + d_j) : \ [J_k, [G_j, H]] + [G_j, [H, J_k]] + [H, [J_k, G_j]] = 0 \Rightarrow$$

$$\epsilon_{kjl} P_l - \epsilon_{kjl} P_l - \epsilon_{kjl} d_l = 0 \Rightarrow$$

$$d_j = 0. \tag{10.66}$$

The reason for this result in the same as with Eqs. (10.62, 10.63), and (10.65) is obeyed.

Now we are left with $[G_j, P_k]$. But space translations and Galilean transformations commute,

$$(\mathbb{1}, \mathbf{0}, 0, \boldsymbol{a})(\mathbb{1}, \boldsymbol{v}, 0, \mathbf{0}) = (\mathbb{1}, \boldsymbol{v}, 0, \mathbf{0})(\mathbb{1}, \mathbf{0}, 0, \boldsymbol{a}) = (\mathbb{1}, \boldsymbol{v}, 0, \boldsymbol{a}), \tag{10.67}$$

(more simply, set $b' = b = 0$ in (10.56)), so at best $[G_j, P_k]$ can be neutral. Behavior under $SO(3)$ (expressed in Eq. (10.60)) fixes it to be a numerical multiple of δ_{jk} and this cannot be transformed away:

$$[G_j, P_k] = iM\delta_{jk}, \quad M \text{ neutral.} \tag{10.68}$$

From Eqs. (10.59, 10.62, 10.63, 10.65, 10.68) we get the complete list of commutation relations:

$$[J_j, J_k \text{ or } P_k \text{ or } G_k] = i\epsilon_{jkl}(J_l \text{ or } P_l \text{ or } G_l);$$

$$[J_j, H] = [P_j, H] = [P_j, P_k] = [G_j, G_k] = 0;$$

$$[G_j, H] = iP_j;$$

$$[G_j, P_k] = i\delta_{jk} M. \tag{10.69}$$

At the end we see that $\boldsymbol{J}, \boldsymbol{P}$ and $\boldsymbol{G}$ cannot be shifted by neutrals; but as H does not appear on the right hand side in any of these commutation relations, it can be shifted by any constant (neutral) amount.

Thus $\mathcal{G}$ has nontrivial ray representations, and the composition law (10.52) is obeyed by the unitary operators (10.58) up to phases. As in Eq. (9.71),

$$\overline{U}(g')\overline{U}(g) = e^{i\omega(g', g)}\overline{U}(g'g),$$

$$\omega(g', g) = \frac{b}{2}M\boldsymbol{v}'^2 + M\boldsymbol{v} \cdot (\boldsymbol{a}' + b\boldsymbol{v}'), \text{ mod } 2\pi. \tag{10.70}$$

This expression for $\omega(g',g)$ is the result of the following sequence of steps. From Eqs. (10.57, 10.58) we need to find $\omega(g',g)$ using

$$e^{-ib'H}e^{-iv'\cdot G}\, e^{-ia'\cdot P}\, e^{-i\alpha'\cdot J}\, e^{-ibH}e^{-iv\cdot G}\, e^{-ia\cdot P}\, e^{-i\alpha\cdot J}$$

$$= e^{i\omega}\, e^{-i(b'+b)H}\, e^{-i(v'+A'v)\cdot G}\, e^{-i(a'+A'a+bv')\cdot P}\, e^{-i\alpha'\cdot J}\, e^{-i\alpha\cdot J}. \qquad (10.71)$$

Here: cancel the $b'H$ and $\alpha\cdot J$ factors; on the left, move the $\alpha'\cdot J$ factor to the extreme right, replacing $v\to A'v$, $a\to A'a$ in the process; cancel the $\alpha'\cdot J$ factors on both sides, set $A'=\mathbb{1}$ to get

$$e^{-iv'\cdot G}\, e^{-ia'\cdot P}\, e^{-ibH}\, e^{-iv\cdot G}\, e^{-ia\cdot P} = e^{i\omega}\, e^{-ibH}\, e^{-i(v'+v)\cdot G}\, e^{-i(a'+a+bv')\cdot P}. \qquad (10.72)$$

Now cancel the $a\cdot P$ factors; shift the $v'\cdot G$ and $v\cdot G$ factors from left to right, use

$$e^{iv'\cdot G}\, H e^{-iv'\cdot G} = H - v'\cdot P + \frac{1}{2}Mv'^2,$$

$$e^{-iv\cdot G}P\, e^{iv\cdot G} = P + Mv, \qquad (10.73)$$

to get:

$$e^{-ia'\cdot P}\, e^{-ibH} = e^{i\omega}\, e^{-ib(H-v'\cdot P+\frac{1}{2}Mv'^2)}\, e^{-i(a'+bv')\cdot(P+Mv)}, \qquad (10.74)$$

which then leads to the expression for $\omega(g',g)$ in Eq. (10.70).

In case $M=0$, we have a true or vector UR of $\mathcal{G}$, but these are rather unphysical. The parameter M is the 'mass', which could be positive or negative. For physical reasons we assume $M>0$. Then we have to find irreducible hermitian solutions to the system (10.69), M being a preassigned real positive parameter. We analyse the situation as follows. First we see that we have a Casimir invariant

$$\mathcal{C}_1 = H - \frac{P_j P_j}{2M}, \qquad (10.75)$$

so this can be set equal to a constant ϵ_0, the 'internal energy':

$$H = \frac{P_j P_j}{2M} + \epsilon_0. \qquad (10.76)$$

This is the free particle Hamiltonian with a 'zero point energy'. Next we set

$$Q_j = G_j/M \qquad (10.77)$$

and find that the Galilei Lie algebra contains the Heisenberg canonical commutation relations

$$[Q_j, P_k] = i\delta_{jk}, \quad [Q_j, Q_k] = [P_j, P_k] = 0. \qquad (10.78)$$

With $\mathbf{Q}$ interpreted as the position three-vector operator, we construct the orbital angular momentum as

$$L = \mathbf{Q} \wedge \mathbf{P}, \tag{10.79}$$

and subtracting this from the total angular momentum $\mathbf{J}$ we have a spin angular momentum:

$$S = \mathbf{J} - \mathbf{L} = \mathbf{J} - \mathbf{Q} \wedge \mathbf{P} = \mathbf{J} - \frac{1}{M} \mathbf{G} \wedge \mathbf{P}. \tag{10.80}$$

We easily find the commutation relations

$$[J_j \text{ or } S_j, \ S_k] = i\epsilon_{jkl} S_l,$$
$$[S_j, H \text{ or } P_k \text{ or } G_k] = 0. \tag{10.81}$$

These show that $\mathbf{S}$ is a vector operator, an angular momentum, invariant under spacetime translations and pure Galilean transformations.

To build up a UIR of $\mathcal{G}$ with the nontrivial phase factor (10.70), all we require are an irreducible representation of the Heisenberg commutation relations (10.78) and the spin commutation relations in (10.81). The 'square of the spin' is the second Casimir invariant:

$$\mathcal{C}_2 = S_j S_j. \tag{10.82}$$

Reversing these arguments, we see that we can build the above UIR of $\mathcal{G}$ (up to phases!) starting with primitive irreducible hermitian operators $\mathbf{q}, \mathbf{p}, \mathbf{S}$ obeying

$$[q_j, p_k] = i\delta_{jk}, \quad [S_j, S_k] = i\epsilon_{jkl} S_l,$$
$$[q_j, q_k] = [p_j, p_k] = [q_j \text{ or } p_j, S_k] = 0. \tag{10.83}$$

Then we take the generators of $\mathcal{G}$ to be:

$$J = S + \mathbf{q} \wedge \mathbf{p} \ , \quad P = \mathbf{p}, \quad G = M\mathbf{q}, \quad H = \frac{\mathbf{p}^2}{2M} + \epsilon_0, \tag{10.84}$$

and all the relations (10.69) will be obeyed. If the S_j are the spin s solution of the $SU(2)$ Lie algebra, for some $s = 0, \frac{1}{2}, 1, \cdots$, then this UIR of $\mathcal{G}$ can be labelled as $\{M, s, \epsilon_0\}$ – mass, magnitude of spin, internal energy. The Hilbert space consists of

$(2s + 1)$-component complex Schrödinger wave functions:

$$\mathcal{H} = \{\psi(\boldsymbol{q}') = \text{complex } (2s + 1)\text{-component column vector } |$$

$$\|\psi\|^2 = \int d^3 q' \; \psi(\boldsymbol{q}')^\dagger \psi(\boldsymbol{q}') < \infty\}, \tag{10.85}$$

and the actions of $\boldsymbol{q}, \boldsymbol{p}, \boldsymbol{S}$ are as expected:

$$(q_j \psi)(\boldsymbol{q}') = q'_j \psi(\boldsymbol{q}'), \; (p_j \psi)(\boldsymbol{q}') = -i\frac{\partial}{\partial q'_j} \psi(\boldsymbol{q}'),$$

$$(S_j \psi)(\boldsymbol{q}') = (j^{\text{th}} \text{ matrix generator of spin } s \text{ UIR of } SU(2))\psi(\boldsymbol{q}'). \tag{10.86}$$

We see that irreducible ray representations of $\mathcal{G}$ (with $M > 0$) correspond to nonrelativistic massive point particles *with spin*, with the nonrelativistic energy apart from the constant ϵ_0, describing free particles. The operators $\boldsymbol{J}, \boldsymbol{P}, H, \boldsymbol{S}$ are all conserved. We can of course give a description of the Hilbert space $\mathcal{H}$ using momentum space wave functions. The generators $\boldsymbol{G}$ are not constants of motion, they describe uniform straight line motion according to Newton's First Law.

We can construct a ray representation of $\mathcal{G}$ by forming the direct product of several single particle UIR's, with various masses $M_1, M_2, \cdots$ and spins $s_1, s_2, \cdots$. These would be noninteracting systems, and in the commutation relations (10.69) we have $M = M_1 + M_2 + \cdots$. We can then see whether interaction terms could be included in the Hamiltonian H, while preserving the forms of the other generators and still obeying (10.69). This is possible, and leads to general Galilean invariant interactions among nonrelativistic particles with spin.

The fact that the total mass M of a Galilean invariant system appears as a neutral element in a commutation relation leads to the Mass Conservation Law, also called the Bargmann superselection rule. This is why total energy and total mass are separately conserved in Galilean relativistic theory.

10.4 Homogeneous Lorentz Group $SO(3, 1)$, and $SL(2, \mathbb{C})$

The groups $SO(3)$ and $SU(2)$ are both compact. They are the 'homogeneous parts' of the groups $E(3)$ and $\mathcal{G}$ (and their covering groups) which are both noncompact 'because of' their abelian invariant subgroups (in the case of $\mathcal{G}$, the seven parameter invariant subgroup in (10.56) is nonabelian). Now we study the groups $SO(3, 1)$ and $SL(2, \mathbb{C})$ which are the 'homogeneous parts' of the Poincaré group $\mathcal{P}$ and its covering